Mirror Neurons and the Evolution of Brain and Language

# Advances in Consciousness Research

Advances in Consciousness Research provides a forum for scholars from different scientific disciplines and fields of knowledge who study consciousness in its multifaceted aspects. Thus the Series will include (but not be limited to) the various areas of cognitive science, including cognitive psychology, linguistics, brain science and philosophy. The orientation of the Series is toward developing new interdisciplinary and integrative approaches for the investigation, description and theory of consciousness, as well as the practical consequences of this research for the individual and society.

Series B: Research in progress. Experimental, descriptive and clinical research in consciousness.

Volume 42

Mirror Neurons and the Evolution of Brain and Language
Edited by Maxim I. Stamenov and Vittorio Gallese

# Mirror Neurons and the Evolution of Brain and Language

*Edited by*

Maxim I. Stamenov

Georg-August-Universität Göttingen/Bulgarian Academy of Sciences

Vittorio Gallese

Università di Parma

John Benjamins Publishing Company
Amsterdam/Philadelphia

The paper used in this publication meets the minimum requirements of American
National Standard for Information Sciences – Permanence of Paper for Printed
Library Materials, ANSI z39.48-1984.

Library of Congress Cataloging-in-Publication Data

Mirror neurons and the evolution of brain and language / edited by Maxim I. Stamenov,
Vittorio Gallese.
    p.  cm. (Advances in Consciousness Research, ISSN 1381–589X ; v. 42)
    Selected contributions to the symposium on "Mirror neurons and the evolution of brain
    and language" held on July 5–8, 2000 in Delmenhorst, Germany.
    Includes bibliographical references and index.
    1. Neural circuitry. 2. Brain--Evolution. 3. Language and languages. I. Stamenov,
Maxim. II. Gallese, Vittorio. III. Series.

QP363.3.M57   2002
153-dc21                                                    2002074572
ISBN 90 272 5166 5 (Eur.) / 1 58811 242 X (US) (Hb; alk. paper)
ISBN 90 272 5162 2 (Eur.) / 1 58811 215 2 (US) (Pb; alk. paper)

John Benjamins Publishing Co. · P.O. Box 36224 · 1020 ME Amsterdam · The Netherlands
John Benjamins North America · P.O. Box 27519 · Philadelphia PA 19118-0519 · USA

# Table of contents

## IV.   Applications

# Introduction

This collective volume brings together selected contributions to the symposium on "Mirror Neurons and the Evolution of Brain and Language". The meeting took place July 5–8, 2000, at the premises and with the financial support of the Hanse Institute for Advanced Study, Delmenhorst, Germany. The aim of the symposium was to discuss the status of the recent scientific discovery of the so called 'mirror neurons' and its potential consequences, more specifically from the point of view of our understanding of the evolution of brain, aspects of social intelligence (like imitation, behavioral and communicative role identification and theory of mind) and language from monkeys to primates to humans.

It is hard to overestimate the importance of this discovery. First of all it concerns the way and level of implementation of mental functions in the brain. Currently it is widely believed that such specifically human capacities like language, social intelligence, and invention and use of tools are dependent on the wide scale developments and re-organization of neural functioning involving cascades of networks of neural circuits at the highest level of unification, effectively engaging 'the brain as a whole'. This point was made, e.g., by Dennett (1991) in discussing the relationship of consciousness to the way of brain performance in implementing it, but the logic of the argument could be applied equally well to the other specifically human cognitive, social and behavioral capacities. As Roth points out (this volume), however, the differences in the brains of humans and other biological species turn out to be quite elusive. As a rule, they are not due to the development of new and easy to identify anatomical structures in the brain. They seem, rather, to involve much more subtle re-organizations of the way of functioning of the already existing neural networks. What is really challenging in considering Mirror Neurons System (MNS) in this respect is that the re-organization and development of new functions supported by corresponding brain circuits seems to influence the way of performance of these structures not just at the macroscale, but also at the microscale of single neurons performance. Here the latest report of the study of mirror neurons (cf. Fogassi & Gallese, this volume) speaks for itself in showing the scale of functional specialization in the way of performance of single neurons in different areas of monkey's brain. Although it is impossible to study humans with such direct-invasion methods, one could infer that MNS functions *both* the same way (as is the case with monkeys) as well as in a more flexible way (as in the case of humans). If this is the case, the uniqueness of human MNS consists in its capacity to function as a component structure of a cognitive module and a central cognitive system. The challenge, then, would be further to develop new technological means for a non-invasive *in vivo* study of single neuron functioning in the brain. In investigating this aspect of brain performance we will add more flesh from the neurophysiological point of view to the point made by Weigand (this volume) that humans are complex adaptive systems on a scale unprecedented in the biological kingdom. The comparative study of the behavior of mirror neurons in humans and monkeys promises to become, in the near future, an even more challenging and controversial enterprise for the way we envisage how mental functions are implemented by neurophysiological processes.

Turning to the mental aspect of MNS discovery, we find no less surprising consequences of it for our understanding of conscious and unconscious functions of the mind. Consciousness is sometimes figuratively envisaged as a specifically human faculty to function as 'the mirror of nature', i.e., to have a capacity symmetrically to re-present the world, building up a mental replica of it in this way. Recent investigations have shown, however, that this 'mirroring capacity' of the brain originates at a much deeper level than the level of phenomenal consciousness. The 'mirroring' can be enacted not only completely unconsciously, but is also coded at quite a low level of brain functioning – at the microscale level of its neural performance. The mirror neurons become activated independently of the agent of the action – the self or a third person whose action is observed. The peculiar (first-to-third-person) 'intersubjective character' of the performance of mirror neurons and their surprising complementarity to the functioning of the strategic (intentional, conscious) communicative face-to-face (first-to-second person) interaction may help shed light from a different perspective on the functional architecture of the conscious vs. unconscious mental processes and the relationship between behavioral and communicative action in monkeys and humans. And they may help to re-arrange, at least to a certain degree, some aspects of the big puzzle of the emergence of language faculty, the relation of the latter to other specifically human capacities like social intelligence and tool use and their neural implementation.

***

The present volume discusses the nature of MNS, as presented by the members of the research team of Prof. Giacomo Rizzolatti, some consequences of this discovery for how we should understand the evolution of brain and mind in monkeys, apes, hominids and man, as well as some possibilities to simulate the behavior of mirror neurons by the means of contemporary technology.

The contributions to the symposium went in four main directions that formed the corresponding parts of the present book.

In the first part the discoverers of mirror neurons are given the floor to present a state-of-the-art report about the status of MNS in the brain of monkeys (Fogassi & Gallese) and humans (Rizzolatti, Craighero, & Fadiga). Fogassi and Gallese illustrate the functional properties of mirror neurons, discuss a possible cortical circuit for action understanding in the monkey brain, and propose a hypothetical developmental scenario for the emergence of such a matching system. They speculate on the bearings of these results on theoretical issues such as action representation and mindreading. Rizzolatti, Craighero and Fadiga review a series of brain imaging, TMS, and psychological studies showing the existence of a mirror-matching system in the human brain. The authors propose that such a matching system could be at the basis of fundamental cognitive capacities such as action understanding and imitation.

In the second part of the book contributions are grouped that offer further developments to the study of MNS and interpretations of its possible functions. In the first article Roth makes the point how controversial is the question of what makes the human brain indeed unique compared to the brain of the other biological species on earth. Upon closer scrutiny, it turns out that all popular beliefs about the sources of the uniqueness of human brain are either misconceived or require further systematic study and significant in-depth elaboration. The most plausible changes during the hominid evolution, most probably, targeted a re-organization of the frontal–prefrontal cortex networks connecting the facial and oral motor cortices and the related subcortical speech centers with those controlling the temporal sequence of events including sequence of action.

One example of a possible wide-scale 're-wiring' in the brain is offered and discussed in the contribution of Gruber. He shows with experimental data and modelling that the evolution of the human language faculty must have involved a massive re-organization in the way of performance of the hominid-cum-human working memory (WM). In other words, the first level of traces we could possibly identify are on the level of 're-wiring' of the widely distributed neural circuits, but not of the emergence of new anatomical structures and/or 'encephalization' (positive change in the ratio of brain/body weight) in its gross physical form. Gruber introduces an alternative model of human WM that emphasizes the special evolutionary role of the rehearsal mechanism in it. He points out that the results of the reported experimental studies strongly suggest that Broca's area and other premotor cortices constitute not only a support for aspects of a sophisticated language system, but also a very efficient WM mechanism. The well-known effect of articulatory suppression on memory performance can be taken, in this respect, as an indication for the clearly higher capacity of this memory mechanism compared to the phylogenetically older WM mechanism, which human subjects have to rely on when verbal rehearsal is prevented. In this sense, Gruber points out in conclusion, that a co-evolution of language and WM capacity has taken place in the human brain.

Senkfor further develops the point about the 'wide-scale re-wiring of the brain circuits in humans'. She reports results of experiments showing that different brain circuits are engaged depending on the nature of the prior experience in performing an action, in watching the experimenter perform an action, in imagining an action, or in a cognitive control task of cost estimation. In non-invasive scalp ERP recordings, clear binary (oppositive) distinctions were observed between retrieval of episodes with and without action over premotor cortex, between episodes with and without visual motion over posterior cortex, and between episodes with and without motor imagery over prefrontal cortex. The results suggest the high degree of specificity of action memory traces that should be subserved by a coordinated action of multiple systems in memory (coding different aspects of it online and offline). The functioning of this 'rich' central system in humans may have involved a sort of 're-interpretation' of the way of functioning of the MNS that was already in place with the monkeys and primates.

Wohlschläger and Bekkering present a theory of goal-directed imitation that states that perceived acts are decomposed into a hierarchical order of goal aspects in monkeys. Primary goals become the goal of the imitative act, whereas secondary goals are only included in the imitative act, if there is sufficient processing capacity. The results from the experiments and their discussion provide clues to a new view on imitation: Imitation is the copying of goals and intentions of others rather than the copying of movements. This new view implies that action understanding is a prerequisite and a precursor of imitation in a double sense: (1) during a single act of imitation, action understanding is a necessary but not sufficient condition and it precedes action execution and (2) a neural system for action understanding is necessary but not sufficient for imitation behavior to occur in a certain species. The population of mirror-neurons in macaque monkeys can be considered such a system for action understanding.

Knoblich and Jordan continue the discussion and elaboration of this trend in the study and interpretation of the potential functions of MNS in the evolution of human cognitive abilities. They hypothesize that the notion of joint action (proposed by the psycholinguist Herbert Clark) might prove useful in tracing the origin of the sophisticated language faculty in an earlier system for action understanding, offered originally by Rizzolati and Arbib (1998). The notion of joint action suggests that the successful coordination of self- and other-generated actions might have provided an evolutionary advantage because coordinated action allows achieving effects in the environment that cannot be achieved by individual action alone. They point out also that although MNS may turn out not sufficient for the full establishment of successful social action coordination, it is still possible to envisage scenarios where an additional system that codes joint action effects might modulate and be modulated by mirror-like activations in order to coordinate self- and other-generated actions.

Using functional magnetic resonance imaging (fMRI), McGlone, Howard and Roberts investigated the neural basis of a proposed mirror (i.e. observation-action) system in humans, in which participants observed an actor using her right hand in grasping actions related to two classes of objects. Action-specific activation was observed in a network of visual and motor areas including sectors of the inferior frontal gyrus, a putative 'mirror' area. These results are interpreted as evidence in favor of the point that there is a MNS area in the human brain. The activity of MNS in humans may be covert, as well as overt, supporting a whole set of behavioral and cognitive capacities, e.g. imitation and empathy. The authors also argue that there may be developmental links of MNS to the immature brains of infants not having developed the inhibitory circuitry so they can learn by mimicry in echolalia.

Vogeley and Newen address a problem that further extends the domain of the studies associated with the discovery of MNS. They point out that up to now the concept of mirror neurons was not used in addressing the question whether there is a specific difference between the other individual observed and 'myself', between third-person and first-person perspective. They present a functional magnetic resonance imaging study that varies a person's perspective systematically. The experimental data suggest that these different perspectives are implemented at least in part in distinct brain regions. With respect to the debate on simulation theory, the results reject the exclusive validity of simulation theory (based on 'direct' mirror-like activation).

Rotondo and Boker observe that individuals initiate actions toward as, well as adaptively respond to, conversational partners to maintain and further communications. As part of the nonverbal process, individuals often display mirror or matched positions during conversation. Whereas previous research has examined the personal and situational characteristics that influence such symmetry, few studies have investigated the specific sequencing and timing details of these processes with respect to dyad composition. The present study of Rotondo and Boker examines how gender and dominance orientation, two previously suggested effects for differences in nonverbal communication, influence the formation and breaking of symmetry in conversational partners.

In a continuation and extension of the previous study, Boker and Rotondo analyse the phenomenon of mirror symmetry in human conversation from a different perspective: individuals tend to mimic each other's postures and gestures as a part of a shared dialog. The present article studies the process of symmetry building and symmetry breaking in the movements of pairs of individuals while imitating each others' movements in dance. Spatial and temporal symmetries are found in the overall velocities from the results of full body-motion tracking.

The third part of this volume includes articles dealing with the evolution of brain, language and communication. In their article, Li and Hombert offer an overview of the last 6 million years of hominid evolution and a sketch of a diverse array of information from different disciplines that are relevant to the evolutionary origin of language. They first distinguish the problem of the origin of language from the problems associated with the study of evolution of language. The former is an enterprise concerned with the evolution of the communicative behavior of our hominid ancestors, not the evolution of language. This latter concerns linguistic change and is the subject matter of diachronic linguistics. Thus the study of the evolutionary change of communication is not a study of linguistic change (within certain human language that already has the critical features qualifying it as such *in toto*). Li and Hombert discuss a set of fundamental problems related to the emergence of language capacity, e.g., the emergence of language and the emergence of anatomically modern humans, the four evolutionary processes leading to the emergence of language (the reduction of the gastrointestinal tract, the enlargement of the vertebral canal, the descent of the larynx, and the increase of encephalization), as well as the three evolutionary mechanisms underlying the emergence of language (the duplication of Hometic genes, the change of the developmental clock, and the causal role of behavior in evolution). On the basis of this broad biological background, they proceed with the consideration of the foundational aspects of symbolic communication as such. Here they introduce a core explanatory concept – that of cognitive reserve, by which they mean the cognitive capability that is not fully utilized or manifested in the normal repertoire of behavior of a mammal. Li and Hombert also discuss some important steps toward the 'crystallization' of language during the hominid evolution.

Studdert-Kennedy makes the point that the unbounded semantic scope of human language rests on its dual hierarchical structure of phonology and syntax grounded in discrete, particulate phonetic units. According to the theory of articulatory phonology discussed in his contribution, the basic particulate units of language are not consonants and vowels, but the so - called 'dynamic gestures' that compose them. A gesture is defined as the formation and release of a constriction, at a discrete locus in the vocal tract, by one of five articulators (lips, tongue tip, tongue body, velum, larynx), so as to form a characteristic configuration of the tract. Evidence for the gesture as an independent, somatotopically represented unit of phonetic function, consistent with its possible representation in a system of mirror neurons, comes from several sources, including children's early imitations of spoken words. How might a somatotopically organized mirror system supporting the capacity for vocal imitation, a capacity unique among primates to humans, have arisen evolutionarily? The paper proposes a path from brachio-manual imitation, grounded in the mirror neuron system of non-human primates, through analogue facial imitation (also unique among primates to humans) to differentiation of the vocal tract and digital (particulate) vocal imitation.

At the beginning of her contribution Weigand makes the important claim that mirror neurons are not 'simple components' but themselves implement complex units integrating several different dimensions of neural and cognitive processing. It is from a complex integrated whole, the MNS, that the evolution of the language faculty has started. The consequences of positing this hypothesis are explicitly pointed out and discussed. The basic point to be made is that the basis for language emergence should have been a grammar of a rather different heritage – a dialogue grammar. Weigand offers a description of the criterial features of such a dialogue grammar, including those of intentionality, social motivation and meaning indeterminacy. She also offers a prolegomena for a theory of dialogic action games that break down the simple imitative symmetry of mirror imitation and serve as the most probable evolutionary rationale for the emergence of human language proper as a distinct human faculty.

Stamenov presents the challenging idea that the discovery of the MNS does not *per se* help explain any of the higher cognitive capacities of humans, as the MNS in monkeys functions very well without giving them access to the theory of mind, language, imitation, communicative gesturing, etc. The paradox involved in the implementation of the human language faculty remains in the form of the claim that one and the same class of neurons performs mutually contradictory functions in different biological species – functions associated with the way of performance of cognitive modules vs. cognitive central systems (in the sense of Fodor 1983).

Bråten proposes a model that implies that conversation partners simulate one another's complementary processes in terms of the virtual participation in the partner's executed act. A preverbal parallel to this (possibly highest) level of imitation is established in the infants' re-enactment from altercentric perception of their caregivers as if they had been guided from the caregivers' stance. Bråten also offers a speculation as to why such an adaptation in support of learning by altercentric participation may have afforded a selective advantage in hominid and human evolution.

In his contribution, Bichakjian discusses the proposal that the evolution of speech and language may have followed distinct evolutionary routes of development. He points out that Broca's area is the control center of articulatory movements as well as at least partially of grammatical organization. If this is the case, the action-gesture-articulation scenario according to which MNS plays its part in the evolution of the language capacity can be held responsible only for the speech articulation aspect of communicative interaction. And what about the 'content' aspect of this capacity? Bichakjian cites recent results of brain imaging studies in order to support the point that the near congruence of spatial and verbal intelligence areas in the left hemisphere and the fact that they fall largely outside Broca's area suggest that verbal intelligence and speech articulation are two different things, and the rise of the latter does not explain the development of the former.

Anderson, Koulomzin, Beebe and Jaffe report that self-grooming manual activity among four-month-olds correlates with increased duration of attentive gaze fixation on the mother's face. Eight four-month-old infants were selected from a larger study and were coded second-by-second for infant gaze, head orientation and self-touch/mouthing behavior during face-to-face play with the mother. Among these infants it was found that attentive head/gaze coordination is contingent upon self-touch/mouthing behavior. Episodes of mutual gaze are especially prominent up to the age of 4 months, before which infants exhibit an obligatory, automatic tendency to remain totally gaze-locked on the maternal face. Regular repetition of these mutual gaze episodes offer ample opportunity for the infant to coordinate the image of the face with her tactile signals of concern and comfort. If mirror neurons are involved, it is predicted that functional motor structures controlling the mother's manual activities may also be prefigured among neurons in the infant's ventral premotor cortex. Indirect evidence for this would be the appearance of maternal grooming patterns – mouthing, stroking, rubbing, etc. – executed by the infants themselves, and later evolving into autonomous resonant motor patterns.

Vihman discusses the hypothesis that children's own vocal patterns play a key role in the development of segmental representations of adult words. She reports and discusses studies relating the discovery of MNS with the requirement of 'articulatory filter' for learning to produce and comprehend speech patterns. The article outlines a developmental shift in perception from prosodic to segmental processing over the first year of life and relates that shift to the first major maturational landmark in vocal production – the emergence of canonical syllables. A challenging point is the suggestion that the activation of a mirror-neuron(-like) mechanism makes possible this uniquely human shift to segmentally based responses to speech and then to first word production.

McCune points out that in the course of the evolution of language, pre-human hominids and early humans needed to achieve the capacity of representational consciousness, such that meanings might be symbolized and expressed by some external medium. Human infants face these same challenges with a difference: the prior existence of an ambient language. In humans the capacity for conscious mental states undergoes a transition from limitation to the here-and-now to possible consideration of past, future, and counterfactual states. The MNS discovery and up-to-the-date findings point to a neurological basis for a fundamental mapping of meaningful spatial categories through motion during human infants' second year of life. The development of their representational consciousness allows the energence of meaning in relation to a simple vocal signal, the grunt, which accompanies attended actions, and subsequently in relation to words and utterances in the ambient language. One of the critical steps, in this respect, is learning to represent spatial categories (related among other to self- and other-enacted behavioral actions). McCune has in mind here, among others, the learning and representing of the cognitive structure associated with the relational words like verbs and prepositions in natural language. To the degree MNS supports and 'represents' the control structure of a behavioral action (like grasping a small object), it may have served phylogenetically and ontogenetically, and/or also serve during online processing here-and-now as a support for fixing the syntactic–semantic skeleton of the cognitive representation.

Morrison offers an account of a potential main venue for cultural transmission involving routes of unconscious imitation of some cultural patterns named 'catchy memes'. She conjectures that if MNS exists in humans, it could not serve just one rather fixed (encapsulated) function or a list of several discrete functions. It rather may help support a gamut of social cognitive phenomena. One of its roles would be in understanding conspecific behavior with respect to nonconcrete, as well as object-oriented, goals. Cultural transmission certainly requires representing (with or without understanding and interpretation) the intentions and actions of others.

The final, fourth, part of the book includes three applications. Billard and Arbib point out that in order to better understand the leap between the levels of imitation in different animals and humans, there is a need better to describe and simulate the neural mechanisms underlying imitation. They approach this problem from the point of view of computational neuroscience. They focus, in particular, on the capacity for representation and symbolization which underlies that of imitative action and investigate via computer simulation and robotic experiments the possible neural mechanisms by which it could be generated. The authors take as a point of departure the way MNS functions in order to build a computational model of the human ability to learn by imitation. The possibility to extrapolate the way of MNS functioning to complex movements and 'true' imitation is also considered.

In their article, Womble and Wermter develop a model of syntax acquisition using an abstracted MNS. They apply a simple language acquisition task, and show that for the basic system using only correct examples drawn from the language even a carefully constructed connectionist network has great difficulty in fully learning the grammar. However when the full mirror system is used and feedback can be incorporated into the learning mechanism through comparison of correct examples and examples tentatively produced by the learning agent, the performance of the system can efficiently be brought into a state of parity with an agent possessing perfect knowledge of the language. Womble and Wermter also demonstrate that a necessary part of this system is a filtering mechanism, which, when disabled, generates a system which produces examples with errors that are characteristic of Broca's aphasics.

Sugita and Tani present their novel connectionist model developed for the linguistic communication between robots and humans and demonstrate its experimental results. They maintain that the meaning of sentences can be embedded in the coupled dynamics of the behavioral system, which acquires the grounded forward model of the environment, and the linguistic system, which acquires the forward model of the structure of language. It is important to note that both systems interact during learning, thus their structures are inevitably generated in a co-dependent manner. Thus it becomes possible to investigate and verify three essential claims. First, the grounded semantic structure of language is self-organized through the experience of behavioral interaction. Second, at the same time, some imaginary sensory-motor sequences which robots never experienced in the experiments could be generated in an arbitrary way depending on the acquired linguistic structure. And, third, the role of mirror neurons could be explained by the context units activation which unifies the behavioral and the linguistic processes. The above claims are examined through the experimental studies of communication between humans and a real mobile robot.

***

This volume offers a selection of contributions discussing aspects of the function and way of implementation of MNS. The editors hope that this overview of the state of the art in the study of MNS will serve as a paradigmatic example of how a discovery in a certain context and discipline in the cognitive sciences can reach for an interdisciplinary impact and fruitful development of ideas in empirical experimental and clinical research, as well as in computer simulation and industrial application.

## References

Dennett, D. (1991). *Consciousness Explained*. Boston: Little, Brown.

Fodor, J. (1983). *The Modularity of Mind*. Cambridge, MA: MIT Press.

Rizzolatti, G., & Arbib, M. A. (1998). Language within our grasp. *Trends in Neuroscience, 21*, 188–194.

Part I

# Mirror neurons system
# Past, present, and future of a discovery

# The neural correlates of action understanding in non-human primates

Leonardo Fogassi and Vittorio Gallese
Istituto di Fisiologia Umana, Università di Parma, Italy

## 1. Introduction

In everyday life we are commonly exposed to actions performed by other individuals. Depending on the context or the circumstances, we may be witnessing different types of actions. For example, if we are walking on a street it is common to observe other persons walking, stopping or going into or out of buildings, shops, etc. When we are at our working place, we may observe individuals who use their arms and hands to accomplish a variety of tasks such as writing, typing, modeling, repairing, etc. Any time we interact with other people we commonly observe them using different facial expressions and limb gestures to convey meaningful messages to us.

Although we normally understand the features and meaning of other individuals' actions, the neural mechanisms that underlie this ability are, however, by no means obvious. At first glance one could assume that the visual analysis of observed biological movements made by our nervous system should be sufficient to assign, at the final stage of the cortical visual processing, a semantic value to those same movements. The mnemonic storage of many different types of biological movements would then allow us to recognize them each time we observe them again. This assumption, however, does not explain why we are able to understand that the observed biological movements are indeed *goal-related* movements made by conspecifics and not simply objects moving towards other objects.

A possible way to address this issue is to consider the relationship between acting and perceiving action. In spite of a certain degree of variability in action execution among different individuals, manifest in various motor parameters such as acceleration, velocity, movement smoothness, etc., we certainly share with others the neural circuits responsible for programming, controlling and executing similar actions. Moreover, part of these common neural circuits could be active also when the action is not overtly executed, but simply imagined. In other words, these circuits could contain the *representation* of those actions that, when the internal drive and the context are suitable, are overtly executed. If such motor "representations" are present in the motor system, they could be automatically retrieved not only when we execute or mentally rehearse a specific action, but also when we observe the same action performed by other individuals. This mechanism may constitute the basis for action understanding.

A neural mechanism for understanding the actions made by others is a necessary prerequisite also for non-human primates such as monkeys. Particularly so for those living in large social groups, in which individuals need to recognize gestures related to hierarchy, food retrieval, defense from predators, etc. Understanding actions made by others would enable the observer to be faster in competition, to learn new motor abilities or, possibly, to begin an inter-individual gestural communication.

If these premises are accepted, the following empiric questions arise: (1) Does a mechanism for action understanding exist and where is it localized in the monkey brain? (2) How this mechanism could develop in ontogenesis? (3) Which are the possible implications of such a mechanism for social cognition?

In this article we will summarize the properties of the neural centers possibly involved in action understanding, their role in coding the action goal and the anatomical circuit that could possibly ground action understanding. We will conclude by proposing some hypothesis on how such a mechanism might have evolved and its bearing on a new definition of the concept of representation.

## 2.   Action-related neurons in area F5

About fifteen years ago Rizzolatti and colleagues demonstrated that in the rostral part of the monkey ventral premotor cortex there is an area the neurons of which discharge during hand and mouth actions (Rizzolatti et al. 1988). This area, which is shown in Figure 1, identified by means of the cytochrome oxidase staining technique, was called F5 (Matelli et al. 1985). The functional properties of F5 are different from those of the caudally located premotor area F4, where proximal, axial and facial movements are represented (Gentilucci et al. 1988).

The most important feature of F5 motor neurons is that they do not code elementary movements, as neurons of the primary motor cortex (area F1) do: F5 neurons fire during goal related *actions* such as grasping, manipulating, holding, tearing objects. Most of them discharge during grasping actions. Some of them discharge, for example, when the monkey grasps food with the hand or with the mouth, thus coding the action "grasp" in an abstract way, independently from the

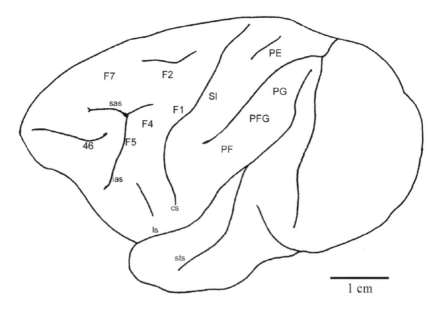

**Figure 1.** Lateral view of the left hemisphere of a standard macaque monkey brain. The agranular frontal cortex is parcellated according to Matelli et al. (1985, 1991). The posterior parietal cortex is parcellated according to Pandya and Seltzer (1982). Abbreviations: cs = central sulcus; ias = inferior arcuate sulcus; ls = lateral sulcus; sas = superior arcuate sulcus; sts = superior temporal sulcus.

effector used for executing that action. Other F5 motor neurons code actions in a more specific way such as, for example, grasping a small object using a precision grip. A neuron of this latter type does not discharge when the monkey grasps food using a whole hand prehension.

Beyond purely motor neurons, which constitute the overall majority of all F5 neurons, area F5 contains also two categories of visuomotor neurons. Neurons of both categories have motor properties that are indistinguishable from those of the above-described purely motor neurons, while they have peculiar visual properties. The first category is made by neurons responding to the presentation of objects of particular size and shape. Very often the size or the shape of the object effective in triggering the neurons discharge is congruent with the specific type of action they code (Rizzolatti et al. 1988; Murata et al. 1997). These neurons were named "canonical" neurons (Rizzolatti & Fadiga 1998; Rizzolatti et al. 2000).

The second category is made by neurons that discharge when the monkey *observes* an action made by another individual and when it *executes* the same or a similar action. These visuomotor neurons were called "mirror" neurons (Gallese et al. 1996; Rizzolatti et al. 1996a).

In the following sections we will summarize the visual and motor properties of F5 mirror neurons.

## 3.  Visual properties of F5 mirror neurons

Mirror neurons discharge when the monkey observes another individual (a human being or another monkey) performing a hand action in front of it (see Figure 2). Differently from canonical neurons, they do not discharge to the simple presentation of food or of other interesting objects. They also do not discharge, or discharge much less, when the observed hand mimics the action without the target object. The response is generally weaker or absent also when the effective action is executed by using a tool instead of the hand. Summing up, the only effective visual stimulus is a hand-object interaction (Gallese et al. 1996).

The response of mirror neurons is largely independent from the distance and the spatial location at which the observed action is performed, although in a minority of neurons the response is modulated by the direction of the observed action or by the hand used by the observed individual (Gallese et al. 1996).

Mirror neurons were subdivided on the basis of the observed action they code (see Table 1) (Gallese et al. 1996). Using this classification criterion, it appears that the coded actions in general coincide with or are very similar to those "motorically" coded in F5 motor neurons (see above): grasping, manipulating, tearing, holding objects.

This classification reveals also that more than half of F5 mirror neurons responds to the observation of only one action, while the remaining ones respond

**Figure 2.**  Example of the visual and motor responses of a F5 mirror neuron. The behavioral situation during which the neural activity was recorded is illustrated schematically in the upper part of each panel. In the lower part rasters and the relative peristimulus response histograms are shown. **A:** A tray with a piece of food placed on it was presented to the monkey; the experimenter grasped the food and then moved the tray with the food toward the monkey, which grasped it. A strong activation was present during observation of the experimenter's grasping movements and while the same action was performed by the monkey. Note that the neural discharge was absent when the food was presented and moved toward the monkey. **B:** As **A**, except that the experimenter grasped the food with pliers. Note that only a weak discharge was elicited when the observed action was performed with a tool. Rasters and histograms are aligned (vertical bar) with the moment in which the experimenter touched the food. Abscissae: time. Ordinate: spikes/bin. Bin width: 20 ms. (Modified from Gallese et al. 1996).

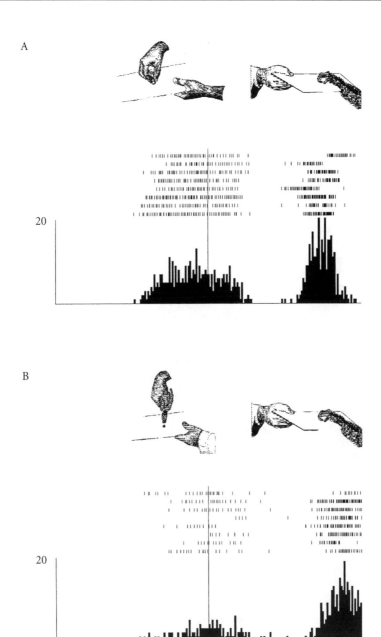

A

B

20

20

1 s

Table 1. Mirror neurons subdivided according to the observed hand actions effective in activating them

| Observed hand actions | No. of neurons |
|---|---|
| Grasping | 30 |
| Placing | 7 |
| Manipulating | 7 |
| Hands interaction | 5 |
| Holding | 2 |
| Grasping/Placing | 20 |
| Grasping/Manipulating | 3 |
| Grasping/Hands interaction | 3 |
| Grasping/Holding | 5 |
| Grasping/Grasping with the mouth | 3 |
| Placing/Holding | 1 |
| Hands interaction/Holding | 1 |
| Grasping/Placing/Manipulating | 1 |
| Grasping/Placing/Holding | 4 |
| Total | 92 |

to the observation of two or more actions. Observation of grasping action, alone or associated to other actions, is by far the most effective in driving the neurons' discharge. Among neurons responding to the observation of grasping action there are some very specific, since they code also the *type* of observed grip. Thus, mirror neurons can present different types of visual selectivity: selectivity for the observed action, and selectivity for the way in which the observed action is accomplished.

## 4. Motor properties of F5 mirror neurons

Although visual responses of F5 mirror neurons are quite surprising, especially so if one considers that they are found in a premotor area, their most important property is that these "visual" responses are matched, at the single neuron level, with motor responses which, as emphasized above, are virtually indistinguishable from that of F5 purely motor or canonical neurons.

An analysis of the congruence between the observed and the executed action effective in triggering the neuron response was carried out (Gallese et al. 1996). The comparison revealed that most of mirror neurons show a good congruence between visual and motor responses, thus allowing to divide them in the categories of "strictly congruent" and "broadly congruent" neurons. "Strictly congruent" neurons are those neurons in which observed and executed actions coincide. For example, a neuron discharged both when the experimenter (observed by the monkey) or the monkey itself executed a precision grip to grasp a small piece of food. Strictly congruent neurons represent about 30% of all F5 mirror neurons. As "broadly congruent" we defined those neurons in which the coded observed action and the coded executed action are similar but not identical. For example a neuron could discharge when the monkey executed a grasping action and when it observed an experimenter grasping and taking away a piece of food. In some cases there is congruence according to a logical or "causal" sense: for example, a neuron responded when the monkey observed an experimenter placing a piece of food on a tray and when the monkey grasped the same piece of food. The two actions can be considered to be part of a logical sequence. Broadly congruent neurons represent about 60% of all F5 mirror neurons. Finally, in about 10% of F5 mirror neurons there is no clear-cut relationship between the effective observed and executed action.

The congruence found between the visual and motor responses of mirror neurons suggests that every time an action is observed, there is an activation of the motor circuits of the observer coding a similar action. According to this interpretation, strictly congruent mirror neurons are probably crucial for a detailed analysis of the observed action. In contrast, broadly congruent neurons appear to generalize across different ways of achieving the same goal, thus probably enabling a more abstract type of action coding. Moreover, these neurons could be very important for other two functions: (a) to appropriately react within a social environment, where normally understanding the actions made by conspecifics is crucial for survival; (b) to communicate, responding with gestures to other individuals gestures. In both cases what is crucial for any individual belonging to a social group is to understand and discriminate the different types of actions made by another conspecific in order to react appropriately. When a monkey observes another monkey throwing an object away, the former can react by grasping the same object. When a monkey of higher hierarchical rank performs a threatening gesture when facing another monkey of lower rank, this latter will not respond with the same gesture but, for example, with a gesture of submission. All these different types of social behaviors could benefit of a mechanism such as that instantiated by broadly congruent mirror neurons. In fact, these neurons "recognize" one or more observed actions, and produce an output that can be ethologically related to them.

If congruence is explained in terms of these different, ethologically meaningful, functions, mirror neurons may constitute a "tool" for understanding the actions made by others, for choosing the appropriate behavior in response to these latter actions, and, in principle, to imitate them. The first two functions apply to monkeys, apes and humans, while, as far as imitation is concerned, experiments made on monkeys show that they apparently lack this ability (Visalberghi & Fragaszy 1990; Whiten & Ham 1992; Tomasello & Call 1997; Whiten 1998).

## 5.    Mirror neurons and action understanding

The triggering feature that evokes the mirror neurons' discharge is the sight of a hand-object interaction. For most mirror neurons the response is independent from the hand used by the observed agent to perform the action and also from the orientation of the observed hand. The discharge is present both when the agent's hand executing the action is seen frontally or from a side view. What matters is that a target is grasped, tore apart, manipulated, or held by the agent. In this respect, it is very important to note that when the agent mimics the action in absence of the

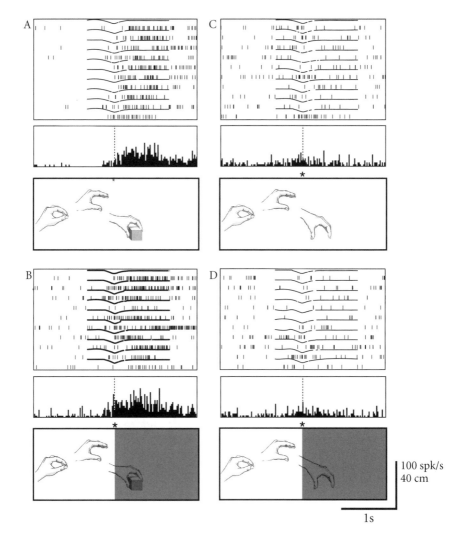

target, the response of mirror neurons is much weaker or absent. We can suppose that as monkeys do not act in absence of a target at which to direct their movements, they do not interpret observed mimicking as a goal-directed action. These observations suggest that mirror neurons may have a crucial role in *goal detection*, and therefore in *action understanding*. According to this hypothesis, the goal of an action made by another individual is recognized by the observer by means of the activation of his/her motor representation of the goal.

**Figure 3.** Example of a F5 mirror neuron responding to action observation in Full vision and in Hidden condition.

The lower part of each panel illustrates schematically the experimenter's action as observed from the monkey's vantage point: the experimenter's hand starting from a fixed position, moving toward an object and grasping it (panels A and B), or mimicking grasping (panels C and D). The behavioral paradigm consisted of two basic conditions: Full vision condition (A) and Hidden condition (B). Two control conditions were also performed: Mimicking in full vision (C), and Mimicking hidden (D). In these last two conditions the monkey observed the same movements as in A and B, but without the target object.

The black frame depicts the metallic frame interposed between the experimenter and the monkey in all conditions. In panels B and D the gray square inside the black frame represents the opaque sliding screen that prevented the monkey from seeing the experimenter's action performed behind it. The asterisk indicates the location of a marker on the frame. In hidden conditions the experimenter's hand started to disappear from the monkey's vision when crossing the marker position.

The upper part of each panel shows rasters display and histograms of ten consecutive trials recorded during the corresponding experimenter's hand movement illustrated in the lower part. Above each raster kinematics recordings (black traces) of the experimenter's hand are shown. The black trace indicates the experimenter's hand movements recorded using a motion analysis system. This system recognized also the position of a fixed marker and referred to it the experimenter's hand trajectory. This marker indicated when, in the Hidden condition, the experimenter's hand began to disappear from monkey's vision. Rasters and histograms are aligned (interrupted vertical line) with the moment at which the experimenter's hand was closest to the fixed marker.

The illustrated neuron responded to the observation of grasping and holding in Full vision (A) and in the Hidden condition (B), in which the interaction between the experimenter's hand and the object occurred behind the opaque screen. The neuron response was virtually absent in the two conditions in which the observed action was mimed (C and D). Histograms bin width = 20 ms. Ordinates: spikes/s; abscissae: time (Modified from Umiltà et al. 2001).

Goal detection can be achieved also when visual information about the observed action is incomplete. In everyday life objects move into and out of sight because of interposition of other objects. However, even when an object, target of the action, is not visible, an individual is still able to understand which action another individual is doing. For example, if one observes a person making a reaching movement toward a bookshelf, he/she will have little doubt that the person in question is going to pick up a book, even if the book is not visible. Full visual information about an action is not necessary to understand its goal.

If mirror neurons are indeed the neural substrate for action understanding, they (or a subset of them) should become active also during the observation of partially hidden actions. To empirically address this hypothesis, we recently carried out a series of experiments (see Umiltà et al. 2001). The experiments consisted of two basic experimental conditions (see Figure 3). In one, the monkey was shown a fully visible action directed toward an object ("Full vision" condition). In the other, the same action was presented, but with its final critical part (hand-object interaction) hidden behind an occluding screen ("Hidden" condition). In two control conditions ("Mimicking in Full vision", and "Hidden mimicking") the same action was mimed without object, both in full vision and behind the occluding screen, respectively.

The main finding was that the majority of tested F5 mirror neurons responded to the observation of hand actions even when the final part of the action, i.e. the part triggering the response in full vision, was hidden from the monkey's vision. However, when the hidden action was mimed, with no object present behind the occluding screen, there was no response. An example of one of these neurons is shown in Figure 3. Two requirements were to be met in order to activate the neurons in hidden condition. The monkey had to "know" that there was an object behind the occluder, and the monkey should see the experimenter's hand disappearing behind the occluder. Once these requirements were met, most mirror neurons discharged even if the monkey did not see the late part of the action. Furthermore and most importantly, in Hidden condition neurons maintained the functional specificity they had in Full vision.

It appears therefore that the mirror neurons responsive in Hidden condition are able to generate a motor representation of an observed action, not only when the monkey sees that action, but also when it "knows" its outcome without seeing its most crucial part (i.e. hand-object interaction). For these neurons therefore out of sight does not mean "out of mind". These results further corroborate the hypothesis, previously suggested, that the mirror neurons mechanism could underpin action understanding (Gallese et al. 1996; Rizzolatti et al. 1996a; Rizzolatti et al. 2000).

## 6.  A cortical circuit for action understanding

Mirror neurons are endowed with both visual and motor properties. What is the origin of their visual input? Data from Perrett and coworkers (Perrett et al. 1989, 1990) show that in the anterior part of the superior temporal sulcus (STSa) there are neurons responding to the sight of hand-object interactions. These neurons apparently do not discharge during monkey's actions, although it must be stressed that they were never systematically tested for the presence of motor properties. These neurons could constitute an important part of a cortical circuit involved in matching action observation with action execution. STSa has no direct connections with the ventral premotor cortex, where area F5 is located. Thus, a functional connection between STSa and F5 could be possibly established only indirectly by means of two pathways: one throughout the prefrontal cortex, the other through the inferior parietal lobule, since STSa is connected with both these cortical regions (Cavada & Goldman-Rakic 1989; Seltzer & Pandya 1994). Of these two pathways the first one seems the most unlike, since the connections between area F5 and prefrontal cortex are present but very weak (Matelli et al. 1986). In contrast, much stronger are the connections between the inferior parietal lobule, and in particular area PF (7b), and the ventral premotor cortex (Matsumura & Kubota 1979; Muakkassa & Strick 1979; Petrides & Pandya 1984; Matelli et al. 1986; Cavada & Goldman-Rakic 1989; see also Rizzolatti et al. 1998).

On the basis of this anatomical evidence, we decided to look for mirror properties in area PF. In this area, as previously demonstrated by Hyvärinen and coworkers (Leinonen & Nyman 1979; Leinonen et al. 1979; Hyvärinen 1981), and subsequently shown by others (Graziano & Gross 1995), there are neurons with bimodal, visual and somatosensory, properties. Most of them respond to tactile stimuli applied to the face and to visual stimuli introduced in the space around the tactile receptive field. Many neurons respond also during mouth and hand movements. We were able to confirm these findings and, in addition, we found also neurons responding to the sight of hand-object interactions (Fogassi et al. 1998; Gallese et al. 2001). Among all visually responsive neurons, about 40% discharged during action observation. Of them, 70% had also motor properties, being activated when the monkey performed mouth or hand actions or both. These neurons were therefore designated as "PF mirror neurons".

PF mirror neurons, similarly to F5 mirror neurons, respond to the observation of several types of single or combined actions. Grasping action, alone or in combination with other actions, is the most represented one. Figure 4 shows an example of a PF mirror neuron responding to the observation of two actions. This neuron responded during the observation of the experimenter's hand grasping and releasing an object. As for F5 mirror neurons, the observation of a mimed action was not effective.

U345

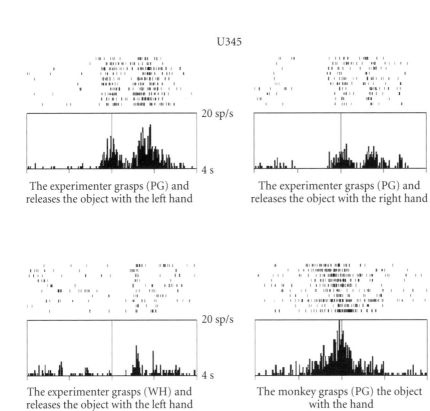

The experimenter grasps (PG) and
releases the object with the left hand

The experimenter grasps (PG) and
releases the object with the right hand

The experimenter grasps (WH) and
releases the object with the left hand

The monkey grasps (PG) the object
with the hand

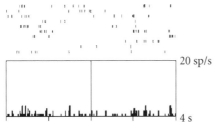

The experimenter presents the object
with the left hand

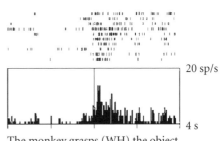

The monkey grasps (WH) the object
with the hand

PF mirror neurons responded during the execution of hand, mouth, or hand and mouth actions. In order to establish the degree of congruence between observed and executed effective actions, the same criterion used to classify F5 mirror neurons was applied. On the basis of this criterion the vast majority of PF mirror neurons present either a strict or a broad congruence between observed and executed action. Among broadly congruent PF mirror neurons two groups are worth discussing. Neurons of the first group respond to the observation and execution of hand actions. However, the executed action could be considered as the logical prolongation of the observed one. For example, the effective observed action could be placing a piece of food on a tray, while the effective executed action could be grasping the piece of food. The response of these neurons does not imply that they are preparatory neurons, because they do not discharge at the simple presentation of the target object. Their response could underpin the most appropriate behavioral monkey's response on the basis of the observation of the other individual's action (see also the section on F5 mirror neurons for a similar interpretation).

Neurons of the second group exhibit a discrepancy between the effector used in the effective observed and executed actions. All neurons of this group discharged during mouth grasping actions, while responding to the observation of hand actions. How can this discrepancy be explained? One possible explanation could be similar to that proposed for the neurons of the first group. There could be a "log-

Figure 4. Example of the visual and motor responses of a PF mirror neuron. This neuron started firing about 300 ms before the experimenter's hand touched the object. The discharge continued until the experimenter's hand took possession of the object, ceased during the holding phase, and started again during the releasing action. This neuron displayed a specificity for the observed grip: the observation of grasping achieved by opposing the index finger to the thumb (precision grip, PG), was much more effective than the observation of grasping achieved by flexing all fingers around the object (whole hand prehension, WH). This selectivity was reciprocated by the neuron's motor selectivity: the neuron's discharge was higher when the monkey grasped the object using a precision grip than when using a whole hand prehension. Note that the most effective observed action determined a higher neural response when executed by the experimenter's left hand (top left panel) than when executed with his right hand (top right panel). Note also that simple object presentation (bottom left panel) did not evoke any response. Rasters and histograms of all panels, but the bottom left one, are aligned with the moment in which either the experimenter's or the monkey's hand touched the object. Rasters and histograms of the bottom left panel are aligned with the moment in which the experimenter's hand started moving to present the object to the monkey. Abscissae: time (each division: 1 s). Ordinate: spikes/s. (Modified from Gallese et al. 2001).

ical" relation between observed hand actions and executed mouth actions. This would imply the observed action to be a placing or a releasing action. However, because both the effective observed and executed actions were grasping actions, the most likely explanation seems to be in terms of a more *abstract* action coding, independent from the used effector (mouth or hand). How could this putative abstract level of action coding have evolved?

A first hypothesis is based on the motor properties of neurons of the rostral part of area PF. Almost all PF motor neurons are activated by hand actions, mouth actions, or both hand and mouth actions. This latter class of motor neurons must have anatomical connections with both the circuits controlling hand movements and those controlling mouth movements. Since PF mirror neurons can be seen as motor neurons responding also to the *observation* of hand actions, one can hypothesize that matching between observed and executed action occurred, during development, not only in neurons endowed with hand motor properties, but also in those controlling both hand and mouth actions. Once neurons of this latter type acquired mirror properties, some of them possibly lost the anatomical connections with the circuit controlling the hand (indeed a feature of motor development is the progressive disappearance of hand and mouth synergism). In the adult these neurons would appear as "mouth" grasping neurons endowed also with the property to respond during the observation of hand actions.

A second, not mutually exclusive, hypothesis is that these PF neurons represent a "primitive" matching system based on mouth movements. This hypothesis will become clearer after having introduced a further striking property of this group of PF broadly congruent mirror neurons, never observed in F5 mirror neurons. These PF mirror neurons responded also to tactile stimuli on the lips and in the region around the mouth and to 3D visual stimuli moved in the peripersonal space around the mouth tactile RF.

A visual peripersonal RF located around a mouth tactile RF can be interpreted as a "motor space", by means of which the visual stimuli that cross it are "translated" into suitable motor plans (e.g. a mouth grasping action), enabling the organism endowed with such RF to successfully interact with the same stimuli (see Fogassi et al. 1996; Rizzolatti et al. 1997). The visual stimulus that most frequently crosses the peripersonal visual RFs of these PF mirror neurons is likely the monkey's own hand, while bringing food to the mouth. A hand approaching the mouth can therefore pre-set the motor programs controlling grasping with the mouth. During development, through a process of generalization between the monkey's own moving hand, treated as a signal to grasp with the mouth, and the object-directed moving hands of others, anytime the monkey observes another individual's hand interacting with food, the same mouth action representation will be evoked. According to this ontogenetic hypothesis, the peripersonal visual RF around the mouth would enable a primitive matching to occur between the vision of a hand and the motor program controlling mouth grasping. Once this equivalence is put in place, a mirror system matching hand actions observation on mouth actions execution can be established. Such a "primitive" matching system, however, would be beneficial also in adulthood, when a more sophisticated hand/hand matching system is developed, in order to provide an "abstract" categorization of the observed actions: what is recognized is a particular action goal, regardless of the effector enabling its achievement.

Thirty percent of PF neurons responding to the observation of actions were devoid of motor properties ("action observation neurons"). Their visual response was very similar to that of PF mirror neurons: they were activated by the observation of a single type or of two or three types of hand actions. These neurons are important because their higher percentage with respect to F5 (30% vs. 22%), probably reflects their proximity to STSa neurons sharing the same properties.

In the light of these findings we propose that a possible circuit for action understanding could be represented by three cortical areas of three different lobes: STSa in the superior temporal cortex, PF in the parietal cortex and F5 in the frontal cortex. Both F5 and PF are endowed with mirror properties and they are reciprocally connected. What is still unclear is where the matching between action execution and action observation occurs first. It could occur in F5 where the visual description of the action fed by STSa through PF could be the input for F5 motor neurons and then be transformed into a pragmatic representation. Alternatively, it could occur in PF from the integration between the visual response to action observation and the efference copy of the motor representation of action coming from F5.

Another important related question worth mentioning is whether the visual response of STSa neurons to action observation is simply the final result of the elaboration of the visual input begun upstream in the higher order visual cortices, or it rather depends in some way from the motor output coming from the frontal cortex, possibly through the inferior parietal lobule. In other words, is the response to hand action observation of STSa neurons influenced by the "motor knowledge" about hand movements? The investigation on the possible presence of motor responses in STSa neurons could help to solve this issue and to give support to our proposal that perception, far from being just the final outcome of sensory integration, is the result of sensorimotor coupling (see Rizzolatti et al. 2001; Rizzolatti & Gallese 2001).

## 7.   Further theoretical implications of the mirror matching system

### 7.1   A new concept for action representation

The discovery of mirror neurons provides a strong argument against the commonly held definition of action, namely, the final outcome of a cascade-like process that starts from the analysis of sensory data, incorporates the result of decision processes, and ends up with responses (actions) to externally- or internally-generated stimuli. The properties of mirror neurons seem to suggest instead that the so-called "motor functions" of the nervous system not only provide the means to control and execute action, but also to internally *represent* it. We submit that this internal representation is crucial for the representation and the knowledge of the external world. According to this view action-control and action-representation become two sides of the same coin (see Gallese 2000).

Let us develop this argument. As we have seen at the beginning of this paper, in a particular sector of the premotor cortex – area F5 – there are three distinct classes of neurons that code goal-related hand movements: purely motor neurons, canonical neurons, and mirror neurons. Why are there three distinct populations of grasping-related premotor neurons? By answering this question we can start to develop a new account of *representation*.

These three neuronal populations have in common their activation during the execution of hand actions. By simply looking at their discharge it would be difficult if not impossible to distinguish a purely motor neuron from a canonical neuron or this latter from a mirror neuron. However, although on the one hand all of them could be involved in movement control, on the other their output seems to convey different meanings. In other words, their discharge *represents,* in a pragmatic way, different aspects of the relationship of an agent with the external world. Purely motor neurons, that could be considered the prototype, *represent* the motor schemas necessary for acting. The target of the action however is not directly specified in their discharge. The other two categories of neurons, both classified as visuomotor neurons, extend their *representational* capabilities to the sensory world, but they acquire this property through the intrinsic motor nature of their discharge. That is, at an early developmental stage both categories of visuomotor neurons are likely just endowed with motor properties, being connected with the external input only at a later developmental stage. In adulthood, canonical neurons *represent objects* in terms of hand motor actions: a small object is a "precision grasp" action, a large object becomes a "whole hand" action. Mirror neurons, instead, *represent hands configurations* in terms of hand actions. It is important to stress that the discharge of canonical and mirror neurons is not necessarily linked to the production of an overt action on the environment. Indeed canonical neurons respond to object presentation also in tasks in which the monkey has only to observe and not to grasp the object. Similarly, the visual discharge of mirror neurons is not directly followed by a monkey action. Therefore, also when an action is not directly executed, the internal motor circuit generates a *representation* of it. It is important to stress that this representation is not limited to area F5, but it is a property of the parieto-frontal circuits of which canonical neurons and mirror neurons are part of. For canonical neurons the circuit is formed by two anatomically connected centers: area AIP in the lateral bank of intraparietal sulcus (Sakata et al. 1995) and area F5 in the premotor cortex (Luppino et al. 1999; see also Rizzolatti et al. 1998). This circuit is involved in the visuomotor transformation for visually guided hand actions (Jeannerod et al. 1995; Rizzolatti et al. 2000). The circuit for action understanding linking F5, PF and possibly part of STS was already introduced in the previous section. In both areas constituting the F5 mirror – PF circuit there are purely motor neurons active only during hand movements, visuomotor neurons (mirror neurons), and purely visual neurons responding to action observation. The higher percent of purely visual neurons in PF could be related to its input from the STS. The presence in both areas F5 and PF of mirror neurons often indistinguishable in their "visual" and "motor" responses suggests that the concept "action representation" should be attributed more to the whole circuit rather than to the individual areas forming it. Due to the their strong anatomical connections, a lesion of either the frontal or the parietal area would damage this common representation. At present, however, there are no lesion data in monkeys confirming this hypothesis.

Data on humans can support our hypothesis of action representation at the level of a sensorimotor circuit. First, in fMRI experiments in which human subjects were asked to simply observe goal-related actions made by others there is a strong activation of both premotor and parietal areas (Rizzolatti et al. 1996b; Buccino et al. 2001), very likely the homologue of the monkey areas in which mirror neurons were found. Second, in humans both lesions in Broca's area (the area homologue of F5, see Rizzolatti & Arbib 1998) and in the inferior parietal lobe (in which area 40 is likely the homologue of monkey's area PF) produce deficits in action recognition (Brain 1961; Gainotti & Lemmo 1966; Heilman et al. 1982; Duffy & Watkins 1984; Heilman & Rothi 1993; Bell 1994).

The issue of how the representational system for action could have emerged deserves a final comment. Coming back to the motor activity of mirror neurons, one could interpret it as being the result of an *efference copy signal*. The efference copy signal enables the motor system to predict the motor consequences of a planned action. If this is the case, it is possible to speculate that this system may have originally developed to achieve a better control of action performance. The coupling between the motor signal and the vision of the agent's own hand, and its later generalization to the hands of others, may have allowed this system to be used also for totally different purposes, namely to represent other individuals' actions. Action representation, following our hypothesis, can be envisaged as the emergence of a new skill that developed by exploiting in new ways resources previously selected for motor control.

Summing up, according to our hypothetical scenario, re-presentational faculties did not primarily originate – neither philogenetically, nor onthogenetically – with a specific *semantic* value. This feature was likely the later result of the functional reorganization of processes originally selected for a different purpose. We submit that this purpose was to achieve a better control of the dynamic relation between an open system – the living organism – and the environment.

## 7.2 Role of the mirror matching system in reading mental states

Until recently, the issue of human social cognition has been addressed mainly as the matter of psychological and/or philosophical investigation. We believe that the functional architecture of the mirror matching system enables to tackle this issue from a new perspective. Let us consider a distinctive feature of human social cognition: the capacity to *represent* mental states of others by means of a conceptual system, commonly designated as "Theory of Mind" (TOM, see Premack & Woodruff 1978).

It is out of the scope of this paper to enter into the debate on which process or substrate could explain this ability. What we would like to emphasize is that when "reading the mind" of conspecifics whose actions we are observing, we rely *also*, if not mostly, on a series of explicit behavioral signals, that we can detect from their observed behavior. These signals may be *intrinsically meaningful* to the extent that they enable the activation of equivalent inner representations on the observer/mind-attributer's side. As we have maintained throughout this paper, we can detect an observed behavior as goal-related, by means of the activation of a motor representation which is *shared* between the agent and the observer (see Gallese 2001). It is only through the activation of this shared representation that we are able to *translate* the pictorial description of fingers approaching to and closing around a spherical solid as a hand grasping an apple. Hence mirror neurons seem to play an important role in recognizing intrinsically meaningful behavioral signals.

One of the behavioral signals that can be linked to TOM is gaze. According to Baron-Cohen (1995) the perception of eye gaze is a crucial step to the development of a mindreading system, allowing individuals to understand not only what another individual is attending to but also what he is thinking about. Recent behavioral and neurophysiological findings seem promising in delineating the evolutionary and neuronal background relating gaze-following behavior to the capacity of understanding intentionality.

In a study of Ferrari et al. (2000) the behavioral responses of macaques to movement of head and eyes of the experimenter were recorded. They found that macaques, as chimpanzees, are able to follow the movements of head and eyes and of the eyes alone. These data correlate well with physiological studies showing that in monkeys there are neurons capable to detect eye direction (Perrett et al. 1985, 1992).

It must be underlined that one of the most important cue for understanding action intentionality of an observed agent is the common direction of its gaze and that of the effector used to perform the action (for example a hand moving to reach and grasp an object). If the two behaviors occur in the same direction, there is a prediction of intentionality, if they are performed in different direction the observed action can be considered unintentional or accidental. Recently Jellema et al. (2000) discovered neurons in STS that respond when the agent performs a reaching action and simultaneously his gaze is directed to the intended target of reaching. In contrast, when the agent performs the same reaching action while gazing at a different direction, the same neurons do not respond. Thus, these neurons seem to combine the activity of two population of neurons, one selective for the observation of a arm reaching action, the other selective for the direction of attention of the observed agent, estimated from his gaze orientation. One may speculate that the sensitivity to both attention direction and reaching direction require the activation of two different set of shared representation, possibly constituted by classes of mirror neurons. Thus, the combined activation of these two types of shared representation could constitute the neural basis for the capacity to detect intentional behavior, an essential component of mind-reading.

## 8.  Conclusions

The mirror system, as reviewed in the present article, appears to support the action understanding ability. It most likely constitutes the neural basis for this fundamental social cognitive function required by the complex social environment typical of primates. We think that the mirror system offers also a new heuristic tool for the empirical investigation of cognitive capacities, such as mindreading, considered to be uniquely human, and still poorly understood.

## Acknowledgement

Supported by M.U.R.S.T. and H.F.S.P.O.

## References

Baron-Cohen, S. (1995). *Mindblindness: an essay on autism and theory of mind*. Cambridge, MA: MIT Press.

Bell, B. D. (1994). Pantomime recognition impairment in aphasia: an analysis of error types. *Brain and Language, 47*, 269–278.

Brain, W. R. (1961). *Speech Disorders: aphasia, apraxia and agnosia*. Washington: Butterworth.

Buccino, G., Binkofski, F., Fink, G. R., Fadiga, L., Fogassi, L., Gallese, V., Seitz, R. J., Zilles, K., Rizzolatti, G., & Freund, H.-J. (2001). Action observation activates premotor and parietal areas in a somatotopic manner: an fMRI study. *European Journal of Neuroscience, 13*, 400–404.

Cavada, C., & Goldman-Rakic, P. S. (1989). Posterior parietal cortex in rhesus monkey: II. Evidence for segregated corticocortical networks linking sensory and limbic areas with the frontal lobe. *Journal of Comparative Neurology, 287*, 422–445.

Duffy, J. R., & Watkins, L. B. (1984). The effect of response choice relatedness on pantomime and verbal recognition ability in aphasic patients. *Brain and Language, 21*, 291–306.

Ferrari, P. F., Kohler, E., Fogassi, L., & Gallese, V. (2000). The ability to follow eye gaze and its emergence during development in macaque monkeys. *Proceedings of the National Academy of Science, 97*, 13997–14002.

Fogassi, L., Gallese, V., Fadiga, L., Luppino, G., Matelli, M., & Rizzolatti, G. (1996). Coding of peripersonal space in inferior premotor cortex (area F4). *Journal of Neurophysiology, 76*, 141–157.

Fogassi, L., Gallese, V., Fadiga, L., & Rizzolatti, G. (1998). Neurons responding to the sight of goal directed hand/arm actions in the parietal area PF (7b) of the macaque monkey. *Society of Neuroscience Abstracts, 24*, 257.5.

Gainotti, G., & Lemmo, M. S. (1976). Comprehension of symbolic gestures in aphasia. *Brain and Language, 3*, 451–460.

Gallese, V. (2000). The inner sense of action: agency and motor representations. *Journal of Consciousness Studies, 7*, 23–40.

Gallese, V. (2001). The "Shared Manifold" Hypothesis: from mirror neurons to empathy. *Journal of Consciousness Studies, 8*(5–7), 33–50.

Gallese, V., Fadiga, L., Fogassi, L., & Rizzolatti, G. (1996). Action recognition in the premotor cortex. *Brain, 119*, 593–609.

Gallese, V., Fogassi, L., Fadiga, L., & Rizzolatti G. (2002). Action representation and the inferior parietal lobule. In W. Prinz & B. Hommel (Eds.), *Attention & Performance XIX. Common mechanisms in perception and action* (pp. 334–355). Oxford: Oxford University Press.

Gentilucci, M., Fogassi, L., Luppino, G., Matelli, M., Camarda, R., & Rizzolatti, G. (1988). Functional organization of inferior area 6 in the macaque monkey. I. Somatotopy and the control of proximal movements. *Experimental Brain Research, 71*, 475–490.

Graziano, M. S. A., & Gross, C. G. (1995). The representation of extrapersonal space: a possible role for bimodal visual-tactile neurons. In M. S. Gazzaniga, (Ed.), *The Cognitive Neurosciences* (pp. 1021–1034). Cambridge, MA: MIT Press.

Heilman, K. M., & Rothi, L. J. (1993). Apraxia. In K. M. Heilman & E. Valenstein (Eds.), *Clinical Neuropsychology* (3rd ed., pp. 141–163). New York: Oxford University Press.

Heilman, K. M., Rothi, L. J., & Valenstein, E. (1982). Two forms of ideomotor apraxia. *Neurology, 32*, 342–346.

Hyvarinen, J. (1981). Regional distribution of functions in parietal association area 7 of the monkey. *Brain Research, 206*, 287–303.

Jeannerod, M., Arbib, M. A., Rizzolatti, G., & Sakata, H. (1995). Grasping objects: the cortical mechanisms of visuomotor transformation. *Trends in Neuroscience, 18*, 314–320.

Jellema, T., Baker, C. I., Wicker, B., & Perrett, D. I. (2000). Neural representation for the perception of the intentionality of actions. *Brain and Cognition, 44*(2), 280–302.

Leinonen, L., & Nyman, G. (1979). II. Functional properties of cells in anterolateral part of area 7 associative face area of awake monkeys. *Experimental Brain Research, 34*, 321–333.

Leinonen, L., Hyvarinen, J., Nyman, G., & Linnankoski, I. (1979). I. Function properties of neurons in lateral part of associative area 7 in awake monkeys. *Experimental Brain Research, 34*, 299–320.

Luppino, G., Murata, A., Govoni, P., & Matelli, M. (1999). Largely segregated parieto-frontal connections linking rostral intraparietal cortex (areas AIP and VIP) and the ventral premotor cortex (areas F5 and F4). *Experimental Brain Research, 128*, 181–187.

Matelli, M., Luppino, G., & Rizzolatti, G. (1985). Patterns of cytochrome oxidase activity in the frontal agranular cortex of the macaque monkey. *Behavioural Brain Research, 18*, 125–136.

Matelli, M., Luppino, G., & Rizzolatti, G. (1991). Architecture of superior and mesial area 6 and the adjacent cingulate cortexin the macaque monkey. *Journal of Comparative Neurology, 311*, 445–462.

Matelli, M., Camarda, R., Glickstein, M., & Rizzolatti, G. (1986). Afferent and efferent projections of the inferior area 6 in the Macaque Monkey. *Journal of Comparative Neurology, 251*, 281–298.

Matsumura, M., & Kubota, K. (1979). Cortical projection of hand-arm motor area from post-arcuate area in macaque monkeys: a histological study of retrograde transport of horse-radish peroxidase. *Neuroscience Letters, 11*, 241–246.

Muakkassa, K. F., & Strick, P. L. (1979). Frontal lobe inputs to primate motor cortex: evidence for four somatotopically organized 'premotor' areas. *Brain Research, 177*, 176–182.

Murata, A., Fadiga, L., Fogassi, L., Gallese, V., Raos, V., & Rizzolatti, G. (1997). Object representation in the ventral premotor cortex (area F5) of the monkey. *Journal of Neurophysiology, 78*, 2226–2230.

Pandya, D. N., & Seltzer, B. (1982). Intrinsic connections and architectonics of posterior parietal cortex in the rhesus monkey. *Journal of Comparative Neurology, 204*, 196–210

Perrett, D. I., Hietanen, J. K., Oram, M. W., & Benson, P. J. (1992). Organization and functions of cells responsive to faces in the temporal cortex. *Philosophical Transactions of the Royal Society of London, 335*, 23–30.

Perrett, D. I., Mistlin, A. J., Harries, M. H., & Chitty, A. J. (1990). Understanding the visual appearance and consequence of hand actions. In M. A. Goodale (Ed.), *Vision and Action: The Control of Grasping* (pp. 163–342). Norwood, NJ: Ablex.

Perrett, D. I., Harries, M. H., Bevan, R., Thomas, S., Benson, P. J., Mistlin, A. J., Chitty, A. J., Hietanen, J. K., & Ortega, J. E. (1989). Frameworks of analysis for the neural representation of animate objects and actions. *Journal of Experimental Biology, 146*, 87–113.

Perrett, D. I., Smith, P. A. J., Potter, D. D., Mistlin, A. J., Head, A. S., Milner, A. D., & Jeeves, M. A. (1985). Visual cells in the temporal cortex sensitive to face view and gaze direction. *Proceedings of the Royal Society of London, 223*, 293–317.

Petrides, M., & Pandya, D. N. (1984). Projections to the frontal cortex from the posterior parietal region in the rhesus monkey. *Journal of Comparative Neurology, 228*, 105–116.

Premack, D., & Woodruff, G. (1978). Does the chimpanzee have a theory of mind? *Behavioral and Brain Sciences, 1*, 515–526.

Rizzolatti, G., & Arbib, M. A. (1998). Language within our grasp. *Trends in Neurosciences, 21*, 188–194.

Rizzolatti, G., & Fadiga, L. (1998). Grasping objects and grasping action meanings: the dual role of monkey rostroventral premotor cortex (area F5). In *Sensory Guidance of Movement (Novartis Foundation symposium 218)* (pp. 81–103). Chichester: Wiley.

Rizzolatti, G., & Gallese, V. (2001). Do perception and action result from different brain circuit? The three visual systems hypothesis. In L. van Hemmen & T. Sejnowski (Eds.), *Problems in Systems Neuroscience*. Oxford, U.K.: Oxford University Press (in press).

Rizzolatti, G., Fogassi, L., & Gallese, V. (2000). Cortical mechanisms subserving object grasping and action recognition: a new view on the cortical motor functions. In M. S. Gazzaniga (Ed.), *The Cognitive Neurosciences, Second Edition* (pp. 539–552). Cambridge, MA: MIT Press.

Rizzolatti, G., Fogassi, L., & Gallese, V. (2001). Neurophysiological mechanisms underlying the understanding and imitation of action. *Nature Neuroscience Reviews, 2*, 661–670.

Rizzolatti, G., Luppino, G., & Matelli, M. (1998). The organization of the cortical motor system: new concepts. *Electroencephalography and Clinical Neurophysiology, 106*, 283–296.

Rizzolatti, G., Fadiga, L., Fogassi, L., & Gallese, V. (1996a). Premotor cortex and the recognition of motor actions. *Cognitive Brain Research, 3*, 131–141.

Rizzolatti, G., Fadiga, L., Fogassi, L., & Gallese, V. (1997). The space around us. *Science, 277*, 190–191.

Rizzolatti, G., Camarda, R., Fogassi, L., Gentilucci, M., Luppino, G., & Matelli, M. (1988). Functional organization of inferior area 6 in the macaque monkey: II. Area F5 and the control of distal movements. *Experimental Brain Research, 71*, 491–507.

Rizzolatti, G., Fadiga, L., Matelli, M., Bettinardi, V., Paulesu, E., Perani, D., & Fazio, F. (1996b). Localization of grasp representation in humans by PET: 1. Observation versus execution. *Experimental Brain Research, 111*, 246–252.

Sakata H., Taira, M., Murata, A., & Mine, S. (1995). Neural mechanisms of visual guidance of hand action in the parietal cortex of the monkey. *Cerebral Cortex, 5*, 429–438.

Seltzer, B., & Pandya, D. N. (1994). Parietal, temporal, and occipital projections to cortex of the superior temporal sulcus in the rhesus monkey: a retrograde tracer study. *Journal of Comparative Neurology, 15*, 445–463.

Tomasello, M., & Call, J. (1997). *Primate Cognition*. Oxford: Oxford University Press.

Umiltà, M. A., Kohler, E., Gallese, V., Fogassi, L., Fadiga, L., Keysers, C., & Rizzolatti, G. (2001). "I know what you are doing": A neurophysiological study. *Neuron, 32*, 91–101.

Visalberghi, E., & Fragaszy, D. (1990). Do monkeys ape? In S. T. Parker, & K. R. Gibson (Eds.), *"Language" and Intelligence in Monkeys and Apes* (pp. 247–273). Cambridge, MA: Cambridge University Press.

Whiten, A. (1998). Imitation of the sequential structure of actions by chimpanzees (Pan troglodites). *Journal of Comparative Psychology, 112*, 270–281.

Whiten, A., & Ham, R. (1992). On the nature and evolution of imitation in the animal kingdom: reappraisal of a century of research. *Advances in the Study of Behavior, 21*, 239–283.

# The mirror system in humans

Giacomo Rizzolatti, Laila Craighero and Luciano Fadiga
Istituto di Fisiologia Umana, Università di Parma, Italy / Dipartimento di
Scienze Biomediche e Terapie Avanzate, Università di Ferrara, Italy

## 1. Introduction

Mirror neurons are a particular class of visuomotor neurons originally discovered in a sector (area F5) of monkey's ventral premotor cortex. Their defining functional characteristics is that they became active both when the monkey makes a particular action (like grasping an object or holding it) and when it observes another individual (monkey or human) making a similar action. Typically, mirror neurons do not respond to the sight of a hand mimicking an action. Similarly, they do not respond to the observation of an object alone, even when it is of interest to the monkey (Gallese et al. 1996; Rizzolatti et al. 1996a). An example of mirror neuron is shown in Figure 1.

The vast majority of F5 mirror neurons shows a marked similarity between the action effective when observed and the action effective when executed. This congruence is sometimes extremely strict. In this cases the effective motor action and the effective observed action coincide both in terms of goal (e.g. grasping) and in terms of how the goal is achieved (e.g. precision grip). For most neurons, however, the congruence is broader and is confined to the goal of the action. These broadly congruent neurons are of particular interest, because they generalize the goal of the observed action across many instances of it.

More recently neurons with properties similar to those of F5 mirror neurons were found also in the monkey's parietal area PF (Fogassi et al. 1998; Gallese et al. 2002).This area is reciprocally connected with area F5, on one side, and with the superior temporal sulcus (STS) cortex, on the other (Seltzer & Pandya 1994). STS cortex is functionally an extremely interesting region. As shown by Perrett and his coworkers, in this cortex there are many neurons that discharge during the observation of a variety of biological actions (Perrett et al. 1989, 1990; see also Carey et al. 1997). These actions in some cases are similar to those coded by F5 (Perrett et al.

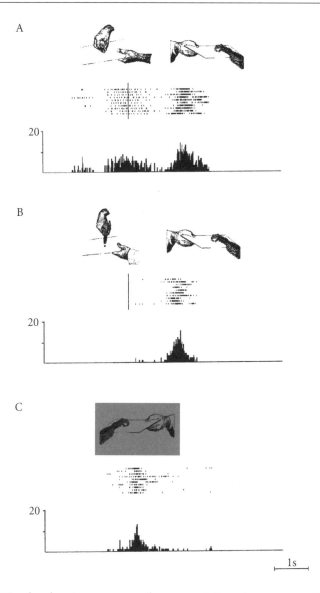

**Figure 1.** Visual and motor responses of a representative mirror neuron. Testing conditions are schematically represented above the rasters. Histograms in each panel represent the sum of eight consecutive trials. A, a tray with a piece of food is presented to the monkey, the experimenter grasps the food, puts the food again on the tray and then moves the tray toward the monkey that grasps the food. B, as above, except that the experimenter grasps the food with pliers. C, active grasping of the monkey in the dark. The presence of a discharge during this last situation demonstrates that the motor discharge of the neuron is not due to monkey's observation of its own hand.

1990). Although this issue was not systematically addressed STS do not appear to discharge during active action, or at least this phenomenon, if present, is not so prominent as in F5. Regardless of this aspect, it is clear that STS, PF and F5 form a system where the biological actions are described in visual terms and then matched on motor neurons coding the same action.

The aim of the present article is to review the available evidence on the existence of a mirror system in humans. It is important to note that when single neuron recording technique is used, information is typically obtained concerning a single brain area or center. Thus, the fact that up to now only one mirror neuron circuit has been defined in the monkey does not exclude the existence of other mirror neuron circuits. This point is important to stress because, as it will be shown below, circuits with mirror properties appear to be more widespread in humans then in monkeys. This difference may be a species difference, but most likely is a consequence of the different technique used in monkeys (single neuron studies) and in humans (brain imaging techniques).

## 2.   Mirror system in humans: Neurophysiological evidence

A first set of evidence, albeit indirect, in favor of a mirror system in humans comes from the study of the reactivity of the cerebral rhythms during movement observation. Traditional EEG studies distinguished two rest rhythms both in the alpha range (8–13 c/s): a posterior alpha rhythm and central mu rhythm. These two rhythms, besides different topography, have different functional significance. The posterior alpha rhythm is present when the sensory systems, the visual one in particular, are not activated, and disappears at the presentation of sensory stimuli. The mu rhythm is present during motor rest and disappears during active movements as well as during somatosensory stimulation (see Chatrian 1976).

Pioneer experiments by Gastaut and Bert (1954) and Cohen-Seat et al. (1954) showed that the observation of actions made by a human being blocks the mu rhythm of the observers. This finding was recently confirmed by Cochin et al. (1998) who showed that during observation of an actor performing leg movements there was a desynchronization of the mu rhythm as well as of beta rhythms (beta 1 = 13–18 Hz and beta 2 = 18–25 Hz) of central-parietal regions. Control

In both A and B, rasters and histograms are aligned with the moment at which the experimenter touches the food either with his hand or with the pliers (vertical line). In C rasters and histograms are aligned with the moment at which the monkey touches the food. Bin width, 20 ms. Ordinates, spikes/bin; abscissas, time.

experiments in which a non-biological motion (e.g. a waterfall) was shown to the recorded subjects did not desynchronize the rhythms. Thus the rhythms that are blocked or desynchronized by movements (for central beta rhythm see Jasper & Penfield 1949) are desynchronized also by movement observation. In a subsequent experiment the same authors compared cortical electrical activity while participants were observing and executing finger movements (Cochin et al. 1999). The results showed that the mu rithm was blocked in correspondence of the central cortex while participants were observing or executing the same movement. Similar data was obtained also by Altschuler and colleagues (Altschuler et al. 1997, 2000).

Further evidence for a matching between action observation and execution comes from magnetoencephalography (MEG) studies. These studies showed that, among the various rhythms recorded from the central region, rhythmic oscillations around 20 Hz originate from the precentral cortex inside the central sulcus (Salmelin & Hari 1994; Hari & Salmelin 1997). The level of the 20-Hz activity enhances bilaterally within 500 ms after median nerve stimulation (Salmelin & Hari 1994; Salenius et al. 1997). This after-stimulation rebound is a highly repeatable and robust phenomenon that can be used as an indicator of the state of the precentral motor cortex. Most interestingly it is abolished when the subject manipulates an object during the median nerve stimulation (Salenius et al. 1997) and is significantly diminished during motor imagery of manipulation movements (Schnitzler et al. 1997).

The post-stimulus rebound method was used to test whether action observation affects the 20 Hz rhythms. Participants were tested in three conditions: (i) rest, (ii) while they were manipulating a small object, (iii) while they were observing another individual performing the same task. The left and right median nerves were stimulated alternatively and the post-stimulus rebound (15–25 Hz activity) was quantified. The results showed that the post-stimulus rebound was strongly suppressed bilaterally during object manipulation and, most interestingly, that it was significantly reduced during action observation. Because the recorded 15–25 Hz activity is known to originate mainly in the precentral motor cortex, these data indicate that human motor cortex is activated both during execution of a motor task and during action observation, a finding strongly supporting the existence of an action observation/execution system in humans.

Another series of evidence in favor of the existence of a mirror system in humans comes from transcranial magnetic stimulation (TMS) studies. Fadiga et al. (1995) stimulated the left motor cortex of normal subjects using TMS while they were observing meaningless intransitive arm movements as well as hand grasping movements performed by an experimenter. Motor evoked potentials (MEPs) were recorded from various arm and hand muscles. As a control, motor cortex was stimulated during the presentation of 3D objects and during an attentionally highly demanding dimming-detection task.

The rationale of the experiment was the following: if the mere observation of the hand and arm movements facilitates the motor system, this facilitation should determine an increase of MEPs recorded from hand and arm muscles. The results confirmed the hypothesis. A selective increase of motor evoked potentials was found in those muscles that the subjects normally use for producing the observed movements.

The MEPs facilitation during movement observation reported by Fadiga et al. (1995) may be explained in two ways. It may result from an enhancement of primary motor cortex excitability due to excitatory cortical connections from a human cortical homologue of monkey of area F5. Alternatively, it may be due not to a cortical facilitation of the primary motor cortex, but to a facilitatory output to the spinal cord originating from the human homologue of F5. Recent data by Strafella and Paus (2000) support the cortico-cortical mechanism. Using a double-pulse TMS technique they showed that the duration of intracortical recurrent inhibition occurring during action observation is similar to that occurring during action execution.

This issue was recently investigated also by Baldissera et al. (2001) from another perspective. These authors examined the modulation of spinal cord excitability during observation of goal directed hand actions by measuring the size of the H-reflex evoked in flexors and extensors muscles in normal human volunteers. They found that, in the absence of any detectable muscle activity, there was a modulation of the reflex amplitude during action observation, specifically related to the different phases of the observed movement. While the H-reflex recorded from flexors rapidly increased in size during hand opening, it was depressed during hand closing and quickly recovered during object lifting. The converse behavior was found in extensors. Thus, while modulation of cortical excitability varies in accordance with the seen movements, the spinal cord excitability changes in the opposite direction.

This apparently paradoxical result obtained by Baldissera et al. (2001) is of great interest because it suggests that, at the spinal cord level, there is a mechanism that prevents execution of the seen actions, leaving, thus, free the cortical motor system to "re-act" the observed actions without the risk of overt movement generation (for the idea of cortex re-acting the observed action see also below).

In conclusion, neurophysiological experiments clearly show that action observation determines in humans an activation of cortical areas involved in motor control. In addition, they indicate that, unlike in monkeys, or at least in monkey F5, where only transitive (i.e. object directed actions) are effective, the observation of intransitive actions (i.e. actions not directed towards an object) may produce an activation of the motor cortex.

### 3.   Mirror system in humans: Brain-imaging studies

The neurophysiological experiments described above, while fundamental in show-ing that action observation elicits a specific, coherent activation of motor system, do not allow the localization of the areas involved in the phenomenon. Data on the localization of human "mirror system" have been obtained, however, using brain-imaging techniques.

The first study in which this issue was addressed was rather disappointing (Decety et al. 1994). In this study, performed using positron emission tomog-raphy (PET), participants were shown grasping movements made by a "hand" generated by a virtual reality system. The generated "hand" was an approximate representation of a human hand. The results, as far as the mirror system is con-cerned, were negative. No motor area was found to be active during action obser-vation that could correspond to monkey ventral premotor cortex or could explain the EEG/MEG desynchronization or TMS excitability increase found in human subjects during action observation.

Subsequent PET experiments, carried out by various groups (including those of the first study), demonstrated that, when the participants observed actions made by human arms or hands, activations were present in the ventral premotor/inferior frontal cortex (Rizzolatti et al. 1996b; Grafton et al. 1996; Decety et al. 1997; Grèzes et al. 1998; Iacoboni et al. 1999). It is likely, therefore, that the initial negative results were due to the fact that human mirror system, as that of the monkey, responds best when the action is made by a biological effector.

As already mentioned in humans both transitive (goal directed) and intransi-tive meaningless gestures activate the mirror system. Grèzes et al. (1998) investi-gated whether the same areas became active in the two conditions. Normal human volunteers were instructed to observe meaningful (pantomimes of bimanual tran-sitive actions) or meaningless (gestures derived from American Sign Language) actions. The results confirmed that the observation of meaningful hand actions ac-tivates the left inferior frontal gyrus (Broca's region), the left inferior parietal lobe plus various occipital and inferotemporal areas. An activation of the left precen-tral gyrus was also found. During meaningless gesture observation there was no Broca's region activation. Furthermore, in comparison with meaningful action ob-servations, an increase was found in activation of the right posterior parietal lobe. It is possible that the activation shift toward the right parietal lobe is related to the fact that the participants, in order to make sense of meaningless movements, took particular care, during observation of these movements, of movement details and matched them with their internal movement proprioceptive templates.

The experiments reviewed up to now tested subjects during action observa-tion. The conclusion that these areas have mirror properties was an indirect con-clusion based on the fact that the activated areas belong to the motor system (see below) and, in the case of Broca's area, by its homology with monkey's area F5. This last inference was recently strongly corroborated by findings showing that Broca's area is an area in which not only speech but also hand movements are represented (see Binkofski et al. 1999).

Direct evidence for an observation/execution system was recently provided by two experiments, one employing fMRI technique (Iacoboni et al. 1999), the other event-related MEG (Nishitani & Hari 2000).

Iacoboni et al. (1999) instructed normal human volunteers to observe and imitate a finger movement and to perform the same movement after a spatial or a symbolic cue (observation/execution tasks). In another series of trials, the same participants were asked to observe the same stimuli presented in the observation/execution tasks, but without giving any response to them (observation tasks). The results showed that activation during imitation was significantly stronger than in the other two observation/execution tasks in three cortical areas: left inferior frontal cortex, right anterior parietal region, and right parietal operculum. The first two areas were active also during observation tasks, while the parietal operculum became active during observation/execution conditions only.

Nishitani and Hari (2000) addressed the same issue using event-related neuromagnetic recordings. In their experiments, normal human participants were requested, in different conditions, to grasp a manipulandum, to observe the same movement performed by an experimenter, and, finally, to observe and simultaneously replicate the observed action. The results showed that during execution, there was an early activation in the left inferior frontal cortex [Brodmann's area 44, BA44] with a response peak appearing approximately 250 ms before the touch of the target. This activation was followed within 100–200 ms by activation of the left precentral motor area and 150–250 ms later by activation of the right one. During observation and during imitation, pattern and sequence of frontal activations were similar to those found during execution, but the frontal activations were preceded by an occipital activation due to visual stimulation occurring in the former conditions.

In all early brain imaging experiments, the participants observed actions made with hands or arms. Recently, experiments were carried out to learn whether mirror system coded actions made by other effectors. Buccino et al. (2001) instructed participants to observe actions made by mouth, foot as well as by hand. The observed actions were biting an apple, reaching and grasping a ball or a small cup, and kicking a ball or pushing a brake. In addition the participants were shown actions not directed toward an object. These actions were again made with the mouth (chewing), the hand and the foot (mimicking reaching to grasp movements, ball kicking and brake pushing). Observation of both object- and non object-related mouth, hand and foot actions (active condition) was contrasted with the

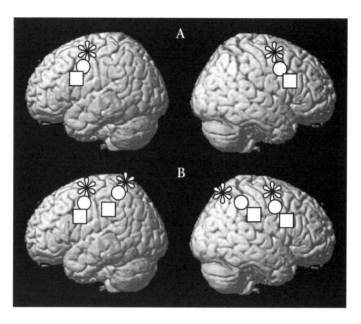

**Figure 2.** Lateral views of left and right hemispheres during observation of actions made with the mouth (square), hand (circle) and foot (asterisk) after subtraction of static mouth, hand and foot observation, respectively. In A intransitive actions are shown, in B object-directed (transitive) actions are represented.

observation of a static face, a static hand and a static foot, respectively, as control conditions. The results are shown in Figure 2.

During non object-related action (chewing) activations were present in areas 6, 44 on both sides and in area 45 in the right hemisphere. Right hemisphere activations were larger and stronger than left hemisphere activations. During object-related action (biting) the pattern of premotor activations was similar, although weaker, to that found during non object-related action. In addition, two activation foci were found in the parietal lobe. These foci were larger in the left than in the right hemisphere. The rostral focus was located in area PF of Von Economo (1929), while the caudal focus was found in area PG.

During the observation of mimicked hand/arm actions (without object) there was a bilateral activation of area 6 that was located dorsal to that found during mouth movement observations. During the observation of object-related arm/hand actions (reaching-to-grasp-movements) there was a bilateral activation of premotor cortex plus an activation site in area 44. As in the case of observation of mouth movements, two activation foci were present in the parietal lobe. The rostral one was located inside the intraparietal sulcus, caudal and dorsal to that found in the mouth movement condition. The caudal focus was again in area PG. This last focus considerably overlapped that of mouth movement condition.

During the observation of mimicked foot actions there was an activation of a dorsal sector of area 6. During the observation of object-related actions, there was as in the condition without object an activation of a dorsal sector of area 6. In addition, there was an activation of the posterior part of the parietal lobe. The parietal activation was in part located in area PE, in part it overlapped the activations seen during mouth and hand actions (area PG).

In conclusion these data indicate that the mirror system is not limited to hand movements. Furthermore, in agreement with previous data by Grèzes et al. (1998) and Iacoboni et al. (1999), they show that the parietal lobe is part of the human mirror systems, and that it is strongly involved when an individual observes object-directed actions.

## 4.   Frontal lobe and parietal lobe mirror areas

The conclusion that a cortical area has mirror properties was frequently reached on the basis of the fact that an area active during action observation is active also, in other experiments, during action execution. In this section we will discuss the cortical motor organization and compare it with the data on the mirror system. Figure 3, 4 and 5 show the lateral view of the monkey and human brain, respectively with indication of the areas of interest.

Broadly speaking, the agranular frontal cortex in the monkey (motor cortex in broad sense) contains three complete movement representations. The first is located in area F1 (area 4), the second in area F3 (SMA proper) and the third on the lateral surface of the frontal cortex. This third representation (which is often referred to as premotor representation) is rather complex and extends over three different cytoarchitectonic areas: F2, where foot and arm movements are represented, F4 where arm and head movements are represented, and area F5 where hand and mouth movements are represented (see for review Rizzolatti et al. 1998).

If one examines the organization of the human agranular frontal cortex with this in mind, a similar representation pattern can be observed. There are two complete motor representations one in area 4 and one in SMA-proper, respectively and a third representation located on the lateral cortical surface. This third representation forms a medio-laterally oriented strip that includes area 6aα (and its subdivisions) and area 44. In this strip leg movements are located medially, while arm, hand and mouth movements are represented progressively more laterally. As in the monkey the representation of different movements shows a considerable overlap (see Rizzolatti et al. 1998).

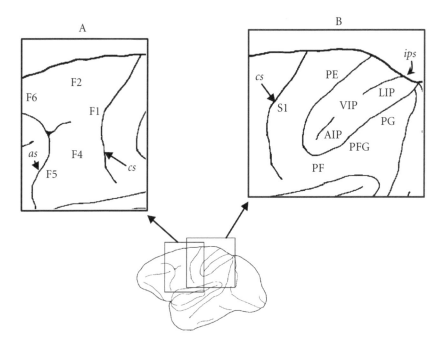

**Figure 3.** Localization of the main motor and posterior parietal areas in the monkey brain. Bottom: the lateral view of the monkey left hemisphere. Top: A; motor areas classify according to Matelli et al. (1985). B; parietal areas, for abbreviations see text. *as*, arcuate sulcus; *cs*, central sulcus; *ips*, intraparietal sulcus. Note that the intraparietal sulcus is opened to show area VIP in the fundus and areas LIP and AIP on the lateral wall.

If one compares the frontal lobe motor organization with the activations during action observation, it is clear that most of the latter activations are located in area 6aα. Furthermore, the activations within area 6aα during action observation are congruent with the different motor fields of this area. Thus, the sectors of area 6aα where foot movements are represented are activated during the observation of foot actions, the sectors where arm/hand movements are represented are activated during the observation of arm/hand actions and the same is true for mouth movements. As far as the arm/hand movements are concerned, the activation of Broca's area was present in the experiments by Buccino et al. (2001) only during the observation of object related actions. This was most likely due to the fact that during intransitive actions, attention was focused on the global movement of the arm, while in transitive actions it was focused on the hand. Thus, in the first case the visuo-motor matching occurred in the motor arm field while in the other in the motor hand field. Broca's area activation during observation of intransitive finger movement was reported by Iacoboni et al. (1999).

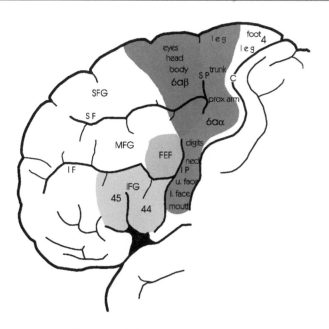

**Figure 4.** Lateral view of human left frontal lobe. The multiple movement somatotopy is indicated by the names of involved body parts. The terminology of Foerster (1936) and Vogt and Vogt (1926) has been adopted to represent the cytoarchitectonical subdivision. Note that, with respect to Brodmann subdivision, the border between area 4 and area 6 is markedly displaced towards the central sulcus (C). Broca's region is represented in pale blue. FEF, frontal eye fields; SFG, superior frontal gyrus; MFG, medium frontal gyrus; IFG, inferior frontal gyrus; SP, superior precentral sulcus; IP, inferior precentral sulcus; SF, superior frontal sulcus; IF, inferior frontal sulcus.

In addition to the agranular frontal cortex, the parietal lobe is also involved in motor organization. This lobe is formed in primates by three main sectors: the postcentral gyrus, the superior parietal lobule and the inferior parietal lobule. The two lobules together are referred to as the posterior parietal lobe. According to Brodmann (1909) the superior parietal lobule of the monkey is formed essentially by one area, area 5, while the inferior parietal lobule is constituted of two areas, area 7a and 7b. Similarly, Von Bonin and Bailey (1947) recognized essentially one area in the superior parietal lobule and two in the inferior parietal lobule. In their study they adopted a nomenclature similar to that used by von Economo (1929) in his study of the cytoarchitectonics of human cortex. They called, therefore, PE the area located in the superior parietal lobule, and PF and PG the two areas located in the inferior parietal lobule.

The different names given to the monkey parietal areas reflect fundamental differences between the two groups in their interpretation of the homologies of

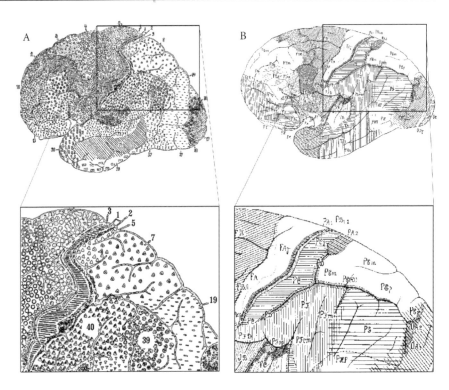

**Figure 5.** Lateral view of human left hemispheres showing enlarged Brodmann's (A) and von Economo's (B) cytoarchitectonical subdivisions of the parietal lobe. See text for symbols.

these areas with the respective human areas. According to Brodmann the two areas (area 5 and 7) that in the monkey form both the superior and inferior parietal lobules, are both located in human superior parietal lobule, while the two areas (39 and 40) that form the human inferior parietal lobule are evolutionary new areas. This view was challenged by von Bonin and Bailey who maintained that there is close homology between the areas forming the superior parietal lobule (area PE) and inferior parietal lobule (PF and PG) in humans and monkey, and, hence, called them in the same way.

There is no doubt that the view of von Bonin and Bailey is much more convincing than that of Brodmann. It is enough to mention here how evolutionary unlikely is the shift of area 7 above the very ancient intraparietal sulcus, that the Brodmann's view implies. In the present chapter we will adopt, therefore, the terminology of von Economo (and von Bonin & Bailey) and will consider that there is a basic homology between the human and monkey inferior and superior parietal lobules. Furthermore, in addition to the basic subdivisions discussed above, the cortex inside the intraparietal sulcus will be parcellated according to recent anatomical (Pandya & Seltzer 1982) and single neuron studies (see Andersen et al. 1997; Colby & Goldberg 1999; Rizzolatti et al. 1998). The areas inside the intraparietal sulcus (areas AIP, LIP, VIP and MIP) discovered in the monkey, appear to be present also in the human parietal lobe (see below).

A fundamental contribution of the classical studies of Hyvärinen (Hyvärinen & Poranen 1974; Leinonen et al. 1979) and Mountcastle (Mountcastle et al. 1975) on monkey posterior parietal lobe was the discovery that the discharge of some parietal lobe neurons is associated with motor activity. It is not known whether this activity is essentially a reflection of motor activity generated in the frontal lobe or represents an endogenous property of the parietal lobe. Regardless of what might be its origin, the fact is that in different parts of the parietal lobe different motor effectors are represented, and, on the basis of monkey data, one can sketch a rough somatotopy of the posterior parietal lobe region, and especially of the inferior parietal lobule that is of major interest here. This organization can be summarized as follows.

Mouth movements are represented in the rostral part of area PF (Leinonen & Nyman 1979; Fogassi et al. 1998). The arm field includes parts of area PF and two areas located in the intraparietal sulcus: area AIP and area MIP. Area AIP is related essentially to distal hand movements (Sakata et al. 1995), while area MIP is involved mostly in the organization of the proximal arm movements (Colby & Duhamel 1991; Snyder et al. 1997). Eye movements are represented in areas LIP (Andersen et al. 1997; Colby & Goldberg 1999) and PG (Mountcastle 1975) although in the last area also responses related to arm movements were described (Mountcastle 1975; MacKay 1992). Finally, head and possibly arm movements are represented in area VIP (Colby et al. 1993; Duhamel et al. 1997, 1998). The representation of leg movements has not been studied. It is likely, however, that this representation should be located more medially in the superior parietal lobe.

The functional organization of the posterior parietal cortex in humans is not known in details. Yet, clinical and brain imaging studies strongly suggest that a segregated pattern of effector organization, as described in the monkey, is present also in humans. Lesions involving the superior parietal lobe and the adjacent areas of the intraparietal sulcus are known to produce reaching deficits (see De Renzi 1982). Although less frequently reported, another impairment in visuomotor behavior following posterior parietal damage is an inadequate hand and finger shaping (Jeannerod 1986; Pause et al. 1989; Seitz et al. 1991). Recent evidence showed that selective deficits in the co-ordination of finger movements for object grasping occur after a lesion located in the anterior part of the lateral bank of the intraparietal sulcus (Binkofski et al. 1998). Patients with these deficits have reaching movement only mildly disturbed. A subsequent fMRI study confirmed this localization (Binkofski et al. 1999). During complex object manipulation an activation was found in the rostral part of the cortex lying in the intraparietal sulcus.

If one now assumes that the organization of human parietal lobe is similar to that of the monkey, the movement organization should be the following. Mouth related action should be localized on the convexity of the inferior parietal lobule, hand movements inside the rostral part of the intraparietal sulcus, and arm movements more medially and posteriorly in correspondence to the medial bank of the intraparietal sulcus. This motor organization corresponds to that found by Buccino et al. (2001) during the observation experiment reported above.

In conclusion, although the evidence of mirror mechanism is only indirect, nevertheless, the overlap between areas with specific motor activity and areas that respond to similar actions performed by others, strongly suggest that in humans a large number of frontal and parietal areas have mirror properties. Furthermore, it appears that, as during action execution (see area AIP), parietal areas also during action observation are more active when the action is directed towards an object.

## 5.   Functional roles of the mirror system in humans

The existence of the mirror neuron system poses two fundamental questions. First, what is its basic functional role? Second, did mirror system acquire new functions in humans?

Mirror neuron system is a system that matches an observed action on a motor representation of the same action. When this matching may be useful? It is likely that a variety of behaviors can be subserved by the matching system. These behaviors may range from the synchronous behavior that some species of animals show when the group leader performs a certain action to some aspects of speech in humans. There are two basic capacities, however, that appear particularly worth noting, especially because they encompass several other functions: action understanding and imitation. We will concentrate in the present article on these two capacities.

### 5.1  Action understanding

As already observed by James (1890), a voluntary movement may occur only if it has been preceded by a series of involuntary movements. These involuntary movements leave a neural trace that later, when the individual acts voluntarily, is reactivated. The trace notion implies that an individual, when performs a voluntary action, must activate, before its execution, a specific neuronal pattern and, most importantly, that he/she is able to predict the impending action outcome.

The assumption at the basis of the hypothesis that mirror system allows action understanding is that the same neuronal pattern that is endogenously activated for action execution is also activated exogenously during action observation. Thus, if one accepts the notion that the acting individual "knows" what will be the results of his/her action, one has to admit also that he/she will be able to know the outcome of the observed action, being the same mechanism involved in both occasions.

The view that the motor system plays a crucial role in action understanding does not deny, of course, that actions are described, coded, and (possibly) recognized in the infero-temporal lobe and, especially, in the areas located in the superior temporal sulcus (STS) (Carey et al. 1997; Fogassi et al. 1998; Gallese et al. 2002). As shown by Perrett and his coworkers (Perrett et al. 1989, 1990) in a very important series of studies, in STS there are many neurons that respond to the observation of biological movements and some, even, to goal-directed actions. The crucial difference between STS and mirror neurons is that STS neurons belong to the visual system, while mirror neurons belong to the motor system, that is to a system that can interact with the external word and create, therefore, a correlation between neural activity and the consequences that this activity produces. This "validation" of the effects of the neural activity creates a basic knowledge of action outcome and therefore of action meaning that subsequently may be used to understand actions of others initially described in the posterior cortical areas. Note that the way in which visual areas, which lack this validation property, might give meaning to the action outcome is highly problematic, admitting that they indeed have this capacity.

A final theoretical point that should be clarified is that, as described above, mirror system includes both frontal and parietal areas. Considering its efferent connections, it is likely that parietal motor activity does not reflect exclusively a corollary discharge coming from the frontal lobe. On the basis of this evidence, the capacity of understanding actions should not be necessarily limited to the frontal lobe, but it should include also the parietal lobe.

## 5.2  Imitation

Another possible functional role of the observation/execution matching system is that of mediating imitation. The term of imitation has different meanings. In every day life it means simply, "to do after the manner of", "to copy" (Oxford English Dictionary. Oxford: Oxford University Press, 1989). There is no specification of what is copied and how. It is obvious that this broad definition includes a large variety of phenomena. Leaving aside phenomena such as stimulus and local enhancements (Spence 1937; Thorpe 1956, 1963) where the apparently imitated behavior is a consequence of a tendency to attend to certain parts of the environment and to emit, unspecifically, the responses appropriate to it, imitation may concern a movement or a sequence of movements. Furthermore, imitation may be preceded or not by the understanding of the action meaning, it may be an approximate or a precise replica of the observed action and, finally, it may concern a sequence of actions never performed before by the observer. In ethology, only this last type of imitation is considered "true" imitation.

According to us a fundamental phenomenon at the basis of many aspects of imitation is what has been referred to as "response facilitation", that is the automatic tendency to reproduce an observed movement (see Byrne 1995). Response facilitation may occur with or without understanding the meaning of the observed action.

*Response facilitation without action meaning understanding.* This type of response facilitation appears to be an ancient function, present in many species of animals. The best-studied example of it is probably the one described in shore birds when a dangerous stimulus appears. As soon as the stimulus is detected one or few birds start flapping their wings, then others repeat the action and, eventually, the whole flock turns in flight (Tinbergen 1953; Thorpe 1963). This "contagious" behavior does not require, necessarily, an "understanding" of the action. What is important here is that the action emitted by the first bird could act as a "release" signal for the behavior of all the other birds (Tinbergen 1953).

Response facilitation without action meaning comprehension is present also in humans. A famous example of this facilitation is the capacity, first described by Meltzoff and Moore (1977), of newborn infants to imitate buccal and manual gestures. Although the response emission is certainly important for the infant because it creates a link between the observing infant and the performing adult, yet the observed buccal or hand action is devoid of any specific meaning.

Phenomena of response facilitation are present also in adult humans. Darwin, for example, mentions in "Expression of emotions" (Darwin 1872) the case of sport fans that, while observing an athlete performing an exercise, tend to "help" him imitating his movements. Similarly many individuals feel a strong urge to copy the "tics" of a person involuntarily making them. Other examples are laughing, yawning, crying, and, as recently shown by Dimberg et al. (2000), involuntarily mimicking of facial expressions.

Our interpretation of the response facilitation without meaning comprehension is based on the notion that some parts of motor system code actions, others code movements. Both these parts may "resonate" in response to visual stimuli congruent with the coded motor activity. When the "resonance" involves neurons coding movements, there is response facilitation without understanding what the observed individual is making, whereas when the "resonance" involves the neurons coding motor acts (see below) the understanding occurs (see Rizzolatti et al.

1999). Neurophysiological evidence in favor of the first type of resonance is scanty in the monkey. Some neurons, however, were found in PF that discharge during meaningless (intransitive) arm movements as well as during observation of similar movements (see Gallese et al. 2002). Stronger evidence in favor of this mechanism comes from human experiments. EEG, MEG, TMS, as well as brain imaging studies (see above), all show that the observation of intransitive meaningless actions activate motor areas.

*Response facilitation with action meaning understanding.* Human observers typically imitate movements made by other individuals, having an understanding what the other individual is doing. At this point, an important theoretical distinction should be considered: Leaving apart the symbolic gestures or "quasi symbolic" gestures, such as the arm movements inviting another individual to approach or to go away, there are two different types of actions with meaning: *motor acts* and *motor actions*. By motor act (see also Rizzolatti et al. 1988) we mean a movement directed towards an object (or the body) which eventually allows an effective interaction between the used effector and the target of the movement. Examples of motor acts are grasping an object, holding it or bringing it to the mouth. By motor action, we mean a sequence of motor acts that at its end determines a reward for the acting individual. An example is the sequence of motor acts (reaching a piece of food, grasping it, holding it, and finally bringing it to the mouth) that allow one to take a piece of food and introduce it into the mouth.

The distinction between motor acts and motor actions is not only logically motivated, but corresponds also to the way in which the motor system is organized. There is evidence from monkey studies that motor acts are coded at single neuron level. For example in area F5 and in the ventro-rostral part of F2 there are neurons that specifically code grasping, holding, tearing etc. In area F4 and in some parts of F2 (Gentilucci et al. 1988; Rizzolatti et al. 1988; Hoshi & Tanji 2000) there are neurons that code arm reaching. We cannot exclude, of course, that motor actions as above defined are also represented at the individual neuron level. Convincing evidence, however, in this sense is at present lacking.

Mirror neurons are the elements that on one side code motor acts, and on the other, when recruited, represent the way through which imitation may take place. From the perspective of mirror system, the mechanism of imitation should be subdivided into three sub-mechanisms: retrieval of a motor act, construction of a sequence of motor acts, refinement of the motor act (or of the motor sequence). All of them are based on the same fundamental mechanism that of response facilitation mediated by the mirror system.

The mere observation of a motor act determines, typically, its retrieval. The difference with action understanding is that the observed act is not only internally copied but also externally manifested. Externally repeating motor acts is in most circumstances of little and in many case even dangerous for the observing individual. Imitation occurs therefore for social reasons or in order to learn from others.

Much more complex is the capacity to imitate a motor action and, even more so, a sequence of motor actions. An interesting hypothesis to explain how this can occurs was recently advanced by Byrne (in press) in his discussion of what he refers to as "action-level-imitation". With this term he indicates the coping of an action not present previously in the behavior repertoire of the observer. According to his suggestion, this behavior can be imitated by dissecting it into a string of components, formed by simpler sequential parts that are already in the observer's repertoire. Specifically, the behavior observed in another individual could be seen as made up of a sequence of simple elements or, using our terminology, of motor acts. The system of mirror neurons would provide the neural basis for recognizing and segmenting actions into strings of discrete elements each one being a motor act in the observer's repertoire. Using Byrne's words, action imitation is "reading the letters of action by mean of response facilitation, action by action".

It is obvious that this proposal leaves open the issue on how the various motor acts are assembled together to construct the new action that an individual become able to perform. Yet, this "mechanistic" theory of imitation opens new empirical possibilities and may clarify why only humans and some species of primates appear to be able to imitate in the proper sense.

A final aspect of imitation is the capacity to modify an action already in the motor repertoire so that the new action become the most similar as possible to that made by the agent of the action. This capacity supports many types of motor learning by imitation: from lifting a finger in a specific way, to playing tennis. In order to do that, individuals should have a copy of their motor actions already coded in visual and proprioceptive terms.

In order to have this type of imitation one should have the capacity to generate a sensory copy of the produced action and to compare it with the action that has to be imitated. The notion that any time we make an action we produce a sensory copy of it is at the basis of the ideomotor theory of action (Greenwald 1970; Prinz 1997; Brass et al. 2000). This theory received recently experimental support from psychological studies. Brass and coworkers (Brass et al. 2001) instructed participants to make a finger movement in response to same or opposite finger movement (lifting or tapping). The results showed that when a subject prepared a given finger movement, the response was faster when the same movement was presented as imperative stimulus. In a further study (Brass et al. 2000) the Munich group addressed the issue of whether the observation of an action facilitates the execution of that action with respect to conditions in which the same action is triggered by other imperative stimuli. The results showed that participants responded faster to the observation of a similar finger movement than to the presentation of a symbolic or spatial imperative stimulus. Further evidence in favor of the notion that during action preparation the individual generates an internal copy of the visual consequences of the action was recently provided by Craighero et al. (2002). Normal human participants were instructed to prepare to grasp a bar, oriented either clockwise or counterclockwise, with their right hand and then to grasp it as fast as possible on presentation of a visual stimulus. The visual stimuli were two pictures of the right hand as seen in a mirror. One of them represented the mirror image of the hand end posture as achieved in grasping the bar oriented clockwise, the other the hand end posture as achieved in grasping the bar oriented counterclockwise. The results showed that when there was a congruence between the prepared hand posture and the picture presented as visual stimulus the reaction times were faster than in incongruent conditions.

The idea that there is an internal sensory copy of the executed action has far reaching consequences for understanding how an observed action can be precisely imitated. If the motor representation of a voluntary action evokes indeed an internal sensory anticipation of its consequences, imitation can be achieved by a mechanism connecting this internal action-related representation with the representation of visually observed movement that has to be imitated, and a subsequent re-activation of the relevant motor representations.

In conclusion, although at this stage all these hypotheses lack of experimental support, a mechanism with the characteristic of the mirror system appears to have the potentiality to give a neurophysiological, mechanistic explanation of imitation.

# References

Altschuler, E. L., Vankov, A., Wang, V., Ramachandran, V. S., & Pineda, J. A. (1997). Person see, person do: Human cortical electrophysiological correlates of monkey see monkey do cell. *Society of Neuroscience Abstracts*, 719.17.

Altschuler, E. L., Vankov, A., Hubbard, E. M., Roberts, E., Ramachandran, V. S., & Pineda, J. A. (2000). Mu wave blocking by observation of movement and its possible use as a tool to study theory of other minds. *Society of Neuroscience Abstracts*, 68.1.

Andersen, R. A., Snyder, A. L., Bradley, D. C., & Xing, J. (1997). Multimodal representation of space in the posterior parietal cortex and its use in planning movements. *Annual Review of Neuroscience, 20*, 303–330.

Baldissera, F., Cavallari, P., Craighero, L., & Fadiga, L. (2001). Modulation of spinal excitability during observation of hand actions in humans. *European Journal of Neuroscience, 13*, 190–194.

Binkofski, F., Dohle, C., Posse, S., Stephan, K. M., Hefter, H., Seitz, R. J., & Freund, H. J. (1998). Human anterior intraparietal area subserves prehension: a combined lesion and functional MRI activation study. *Neurology, 50*, 1253–1259.

Binkofski, F., Buccino, G., Posse, S., Seitz, R. J., Rizzolatti, G., & Freund, H. (1999). A fronto-parietal circuit for object manipulation in man: evidence from an fMRI-study. *European Journal of Neuroscience, 11*, 3276–3286.

Brass, M., Bekkering, H., Wohlschlager, A., & Prinz, W. (2000). Compatibility between observed and executed finger movements: comparing symbolic, spatial and imitative cues. *Brain and Cognition, 44*, 124–143.

Brass, M., Bekkering, H., & Prinz, W. (2001). Movement observation affects movement execution in a simple response task. *Acta Psychologica, 106*, 3–22.

Brodmann, K. (1909). *Vergleichende Lokalisationslehre der Grosshirnrinde in ihren Prinzipien dargerstellt auf Grund des Zellenbaues.* Leipzig: Barth.

Buccino, G., Binkofski, F., Fink, G. R., Fadiga, L., Fogassi, L., Gallese, V., Seitz, R. J., Zilles, K., Rizzolatti, G., & Freund, H.-J. (2001). Action observation activates premotor and parietal areas in a somatotopic manner: an fMRI study. *European Journal of Neuroscience, 13*, 400–404.

Byrne, R. (1995). *The Thinking Ape. Evolutionary origins of intelligence.* Oxford: Oxford University Press.

Byrne, R. W. (in press). *Imitation in Action. Advances in the study of behaviour.*

Carey, D. P., Perrett, D. I., & Oram, M. W. (1997). Recognizing, understanding and reproducing actions. In M. Jeannerod & J. Grafman (Eds.), *Handbook of Neuropsychology*, Vol. 11: *Action and Cognition* (pp. 111–130). Amsterdam: Elsevier.

Chatrian, G. E. (1976). The mu rythms. In A. Remond (Ed.), *Handbook of Electroencephalography* (pp. 104–114). Amsterdam: Elsevier.

Cochin, S., Barthelemy, C., Lejeune, B., Roux, S., & Martineau, J. (1998). Perception of motion and qEEG activity in human adults. *Electroencephalography and Clinical Neurophysiology, 107*, 287–295.

Cochin, S., Barthelemy, C., Roux, S., & Martineau, J. (1999). Observation and execution of movement: similarities demonstrated by quantified electroencephalograpy. *European Journal of Neuroscience, 11*, 1839–1842.

Cohen-Seat, G., Gastaut, H., Faure, J., & Heuyer, G. (1954). Etudes expérimentales de l'activité nerveuse pendant la projection cinématographique. *Revue Internationale de Filmologie, 5*, 7–64.

Colby, C. L., Duhamel, J.-R., & Goldberg, M. E. (1993). Ventral intraparietal area of the macaque: anatomic location and visual response properties. *Journal of Neurophysiology, 69*, 902–914.

Colby, C. L., & Duhamel, J. R. (1991). Heterogeneity of extrastriate visual areas and multiple parietal areas in the macaque monkeys. *Neuropsychologia, 29*, 517–537.

Colby, C. L., & Goldberg, M. E. (1999). Space and attention in parietal cortex. *Annual Review of Neuroscience, 22*, 319–349.

Craighero, L., Bello, A., Fadiga, L., & Rizzolatti, G. (2002). Hand action preparation influences the processing of hand pictures. *Neuropsychologia, 40*, 492–502.

Darwin, C. (1872). *The Expression of the Emotions in Man and Animals.* London: J. Murray.

De Renzi, E. (1982). *Disorders of Space Exploration and Cognition.* New York: Wiley.

Decety, J., Perani, D., Jeannerod, M., Bettinardi, V., Tadary, B., Woods, R., Mazziotta, J. C., & Fazio, F. (1994). Mapping motor representations with positron emission tomography. *Nature, 371*, 600–602.

Decety, J., Grezes, J., Costes, N., Perani, D., Jeannerod, M., Procyk, E., Grassi, F., & Fazio, F. (1997). Brain activity during observation of actions. Influence of action content and subject's strategy. *Brain, 120,* 1763–1777.

Dimberg, U., Thunberg, M., & Elmehed, K. (2000). Unconscious facial reactions to emotional facial expressions. *Psychological Science, 11,* 86–89.

Duhamel, J.-R., Bremmer, F., Ben Hamed, S., & Graf, W. (1997). Spatial invariance of visual receptive fields in parietal cortex neurons. *Nature, 389,* 845–848.

Duhamel, J.-R., Colby, C. L. & Goldberg, M. E. (1998). Ventral intraparietal area of the macaque: congruent visual and somatic response properties. *Journal of Neurophysiology, 79,* 126–136.

Fadiga, L. Fogassi, L., Pavesi, G., & Rizzolatti, G. (1995). Motor facilitation during action observation: A magnetic stimulation study. *Journal of Neurophysiology, 73,* 2608–2611.

Foerster, O. (1936). The motor cortex in man in the light of Hughlings Jackson's doctrines. *Brain, 59,* 135–159.

Fogassi, L., Gallese, V., Fadiga, L., & Rizzolatti, G. (1998). Neurons responding to the sight of goal directed hand/arm actions in the parietal area PF (7b) of the macaque monkey. *Society of Neuroscience Abstracts, 24,* 257.5.

Gallese, V., Fadiga, L., Fogassi, L., & Rizzolatti, G. (1996). Action recognition in the premotor cortex. *Brain, 119,* 593–609.

Gallese, V., Fogassi, L., Fadiga, L., & Rizzolatti G. (2002). Action representation and the inferior parietal lobule. In W. Prinz, & B. Hommel (Eds.), *Attention & Performance XIX. Common mechanisms in perception and action* (pp. 334–345). Oxford: Oxford University Press.

Gastaut, H. J., & Bert, J. (1954). EEG changes during cinematographic presentation. *Electroencephalography and Clinical Neurophysiology, 6,* 433–444.

Gentilucci, M., Fogassi, L., Luppino, G., Matelli, M., Camarda, R., & Rizzolatti, G. (1988). Functional organization of inferior area 6 in the macaque monkey. I. Somatotopy and the control of proximal movements. *Experimental Brain Research, 71,* 475–490.

Grafton, S. T., Arbib, M. A., Fadiga, L., & Rizzolatti, G. (1996). Localization of grasp representations in humans by PET: 2. Observation compared with imagination. *Experimental Brain Research, 112,* 103–111.

Greenwald, A. G. (1970). Sensory feedback mechanisms in performance control: With special reference to the ideo-motor mechanism. *Psychological Review, 77,* 73–99.

Grèzes, J., Costes, N., & Decety, J. (1998). Top-down effect of strategy on the perception of human biological motion: a PET investigation. *Cognitive Neuropsychology, 15,* 553–582.

Hari, R., & Salmelin, R. (1997). Human cortical oscillations: a neuromagnetic view through the skull. *Trends in Neurosciences, 20,* 44–49.

Hoshi, E., & Tanji, J. (2000). Integration of target and body-part information in the premotor cortex when planning action. *Nature, 408,* 466–470.

Hyvärinen, J., & Poranen, A. (1974). Function of the parietal associative area 7 as revealed from cellular discharges in alert monkeys. *Brain, 97,* 673–692.

Iacoboni, M., Woods, R. P., Brass, M., Bekkering, H., Mazziotta, J. C., & Rizzolatti, G. (1999). Cortical mechanisms of human imitation. *Science, 286,* 2526–2528.

James, W. (1890). *Principles of Psychology,* Vol. 2. New York: Holt, Dover edition.

Jasper, H. H., & Penfield, W. (1949). Electro-corticograms in man: Effect of voluntary movement upon the electrical activity of the precentral gyrus. *Archives of Psychiatry, 183*, 163–174.

Jeannerod, M. (1986). The formation of finger grip during prehension. A cortically mediated visuomotor pattern. *Behavioural and Brain Research, 19*, 99–116.

Leinonen, L., Hyvarinen, J., Nyman, G., & Linnankoski, I. (1979). Functional properties of neurons in lateral part of associative area 7 in awake monkeys. *Experimental Brain Research, 34*, 299–320.

Leinonen, L., & Nyman, G. (1979). Functional properties of cells in anterolateral part of area 7 associative face area of awake monkeys. *Experimental Brain Research, 34*, 321–333.

MacKay, W. A. (1992). Properties of reach-related neuronal activity in cortical area 7a. *Journal of Neurophysiology, 67*, 1335–1345.

Meltzoff, A. N., & Moore, M. K. (1977). Imitation of facial and manual gestures by human neonates. *Science, 198*, 75–78.

Mountcastle, V. B., Lynch, J. C., Georgopoulos, A., Sakata, H., & Acuna, C. (1975). Posterior parietal cortex of the monkey: command functions for operation within extrapersonal space. *Journal of Neurophysiology, 38*, 871–908.

Nishitani, N., & Hari, R. (2000). Temporal dynamics of cortical representation for action. *Proceedings of National Academy of Sciences, 97*, 913–918.

Pandya, D. N., & Seltzer, B. (1982). Intrinsic connections and architectonics of posterior parietal cortex in the rhesus monkey. *Journal of Comparative Neurology, 204*, 196–210.

Pause, M., Kunesch, E., Binkofski, F., & Freund, H.-J. (1989). Sensorimotor disturbances in patients with lesions of the parietal cortex. *Brain, 112*, 1599–1625.

Perrett, D. I., Harries, M. H., Bevan, R., Thomas, S., Benson, P. J., Mistlin, A. J., Chitty, A. J., Hietanen, J. K., & Ortega, J. E. (1989). Frameworks of analysis for the neural representation of animate objects and actions. *Journal of Experimental Biology, 146*, 87–113.

Perrett, D. I., Mistlin, A. J., Harries, M. H., & Chitty, A. J. (1990). Understanding the visual appearance and consequence of hand actions. In M. A. Goodale (Ed.), *Vision and Action: The control of grasping* (pp. 163–342). Norwood, NJ: Ablex.

Prinz, W. (1997). Perception and action planning. *European Journal of Cognitive Psychology, 9*, 129–154.

Rizzolatti, G., Fadiga, L., Fogassi, L., & Gallese, V. (1996a). Premotor cortex and the recognition of motor actions. *Cognitive Brain Research, 3*, 131–141.

Rizzolatti, G., Fadiga, L., Matelli, M., Bettinardi, V., Paulesu, E., Perani, D., & Fazio, F. (1996b). Localization of grasp representation in humans by PET: 1. Observation versus execution. *Experimental Brain Research, 111*, 246–252.

Rizzolatti, G., Camarda, R., Fogassi, L., Gentilucci, M., Luppino, G., & Matelli, M. (1988). Functional organization of inferior area 6 in the macaque monkey: II. Area F5 and the control of distal movements. *Experimental Brain Research, 71*, 491–507.

Rizzolatti, G., Luppino, G., & Matelli, M. (1998). The organization of the cortical motor system: new concepts. *Electroencephalography and Clinical Neurophysiology, 106*, 283–296.

Rizzolatti, G., Fadiga, L., Fogassi, L., & Gallese V. (1999). Resonance behaviors and mirror neurons. *Archives Italiennes de Biologie, 137*, 85–100.

Sakata, H., Taira, M., Murata, A., & Mine, S. (1995). Neural mechanisms of visual guidance of hand action in the parietal cortex of the monkey. *Cerebral Cortex, 5*, 429–438.

Salenius, S., Schnitzler, A., Salmelin, R., Jousmaki, V., & Hari, R. (1997). Modulation of human cortical rolandic rhythms during natural sensorimotor tasks. *Neuroimage, 5*, 221–228.

Salmelin, R., & Hari, R. (1994). Spatiotemporal characteristics of sensorimotor neuro-magnetic rhythms related to thumb movement. *Neuroscience, 60*, 537–550.

Schnitzler, A., Salenius, S., Salmelin, R., Jousmaki, V., & Hari, R. (1997). Involvement of primary motor cortex in motor imagery: a neuromagnetic study. *Neuroimage, 6*, 201–208.

Seitz, R. J., Roland, P. E., & Bohm, C., Greitz, T., & Stone-Elander, S. (1991). Somatosensory discrimination of shape: tactile exploration and cerebral activation. *European Journal of Neurosciences, 3*, 481–492.

Seltzer, B., & Pandya, D. N. (1994). Parietal, temporal, and occipital projections to cortex of the superior temporal sulcus in the rhesus monkey: a retrograde tracer study. *Journal of Comparative Neurology, 15*, 445–463.

Snyder, L. H., Batista, A. P., & Andersen, R. A. (1997). Coding of intention in the posterior parietal cortex. *Nature, 386*, 167–169.

Spence, K. W. (1937). Experimental studies of learning and higher mental processes in infra-human primates. *Psychological Bulletin, 34*, 806–850.

Strafella, A. P., & Paus, T. (2000). Modulation of cortical excitability during action observation: a transcranial magnetic stimulation study. *NeuroReport, 11*, 2289–2292.

Thorpe, W. H. (1956). *Learning and Instinct in Animals*. London: Methuen and Co. Ltd.

Thorpe, W. H. (1963). *Learning and Instinct in Animals* (2nd edition). London: Methuen and Co. Ltd.

Tinbergen, N. (1953). *The Herring Gull's World*. London: Collins.

Vogt, C., & Vogt, O. (1926) Die vergleichend-architektonische und vergleichend-reizphy-siologische Felderung der Grosshirnrinde unter besonderer Beruecksichtigung der menschlichen. *Naturwissenschaften, 14*, 1190–1194.

Von Bonin, G., & Bailey, P. (1947). *The Neocortex of Macaca Mulatta*. Urbana: University of Illinois Press.

Von Economo, C. (1929). *The Cytoarchitectonics of the Human Cerebral Cortex*. London: Oxford University Press.

# Part II

# Further developments in the study of mirror neurons system and interpretations of its functions

# Is the human brain unique?

Gerhard Roth

Brain Research Institute, University of Bremen /
Hanse Institute for Advanced Study, Delmenhorst, Germany

Humans are proud of their brain and their cognitive abilities, and many of us including many neuroscientists believe that the alleged uniqueness of human nature is due to the uniqueness of the human brain. In the following, I will briefly discuss some popular claims about the human brain that can be found even in the scientific literature. These are: (1) The human brain in general is anatomically unique; (2) Humans have the largest brain in absolute terms; (3) Humans have the largest brain relative to body size; (4) Humans have the largest cerebral cortex, particularly prefrontal cortex; (5) Humans have some brain centers or functions not found in other animals.

*First claim*:  The human brain in general is anatomically unique. This is completely wrong. All tetrapod vertebrates (amphibians, reptiles, birds, mammals) have brains that – despite enormous differences in outer appearance, overall size and relative size of major parts of the brain – are very similar in their general organization and even in many details (Wullimann 2000). More specifically, all tetrapod brains possess a median, medial and lateral reticular formation inside the medulla oblongata, pons and ventral mesencephalon, including a noradrenergic locus coeruleus, serotonergic raphe nuclei and a medial ascending reticular activating system. There is a corpus striatum, a globus pallidus, a nucleus accumbens, a substantia nigra, a basal forebrain/septum and an amygdala within the ventral telencephalon, a lateral pallium, homologous to the olfactory cortex of mammals, and a medial pallium, homologous to the hippocampal formation (at least Ammon's horn and subiculum). This means that all structures required for attention, declarative memory (or its equivalents in animals), emotions, motivation, guidance of voluntary actions and evaluation of actions are present in the tetrapod brain. These structures essentially have the same connectivity and distribution of transmitters, neuromodulators and neuropeptides in the different groups of tetrapods.

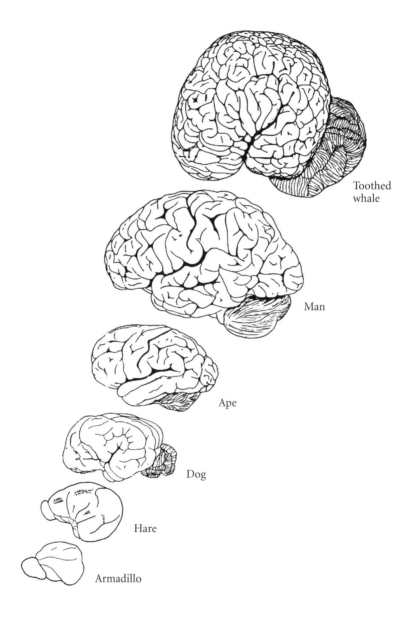

Toothed
whale

Man

Ape

Dog

Hare

Armadillo

**Figure 1.**  Series of mammalian brains, all drawn to the same scale. Evidently, man has
neither the largest brain nor the most convoluted cortex. Convolution of the cortex as
well as of the cerebellum increases monotonically with an increase in brain size.

A more difficult problem is the presence of structures homologous to the mammalian isocortex in the telencephalon of other tetrapods. Amphibians possess a dorsal pallium, turtles and diapsid reptiles have a dorsal cortex plus a dorsal ventricular ridge (DVR), birds have a wulst and a DVR, and these structures are believed by many comparative neurobiologists to be homologous to the isocortex – and not to the basal ganglia – of mammals (Karten 1991; Northcutt & Kaas 1995; MacPhail 2000; Shimizu 2000). However, major differences exist between these structures with regard to cytoarchitecture and size. In amphibians, the dorsal pallium is small and unlaminated; in lizards it is relatively larger, and in turtles and some diapsid reptiles it shows a three-layered structure. In birds, those parts assumed to be homologous to the mammalian cortex (i.e., DVR and wulst) are large, but unlaminated. In mammals – with the exception of insectivores and cetaceans – the dorsal pallium or isocortex shows the characteristic six-layered structure. Despite these differences it is safe to assume that the dorsal pallium and cortex of amphibians and reptiles is at least homologous to the limbic and associative cortex of mammals, while a primary sensory and motor cortex appears to be absent. When we compare birds such as pigeons or parrots with roughly equally intelligent mammals such as dogs, then it becomes apparent that the same or very similar cognitive functions are performed by anatomically very different kinds of pallium/cortex.

*Second claim*: Humans have the largest brain in absolute terms. This is definitely wrong, as can be seen from List 1. Humans have large brains (1.3 kg average weight), which is the largest among extant primates (the extinct *Homo neandertalensis* had a somewhat larger brain), but by far not the largest one among mammals. The largest mammalian brains (and of all animals) are found in elephants (up to 5.7 kg) and whales (up to 10 kg).

**List 1.** List of brain weights in mammals.

| Brain weight in mammals [gram] | | | |
| --- | --- | --- | --- |
| Sperm whale | 8,500 | Chimpanzee | 400 |
| Elephant | 5,000 | Lion | 220 |
| Man | 1,300 | Dog | 135 |
| Horse | 590 | Cat | 30 |
| Gorilla | 550 | Rat | 2 |
| Cow | 540 | Mouse | 0.4 |

*Third claim*: Humans have the largest brain relative to body size. This is wrong, too. While the human brain occupies about 2% of body mass, in very small rodents relative brain size goes up to 10%. However, again among primates, humans have the largest relative brain size.

The relationship between brain size and body size is being discussed for more than hundred years (cf. Jerison 1973). It appears that body size is the single most important factor influencing brain size, i.e., large animals generally have large brains in absolute terms. However, increase in brain size does not strictly parallel the increase in body size, but follows only to the power of 0.66–0.75 (i.e., 2/3 or 3/4, depending on the statistics used; Jerison 1991), – a phenomenon called *negative brain allometry* (Jerison 1973); (Figures 2 and 3). Consequently, small animals of a given taxon have *relatively* larger brains and large animals of this group *relatively* smaller brains. Among mammals, this is reflected by the fact that in very small rodents brains occupy up to 10% of body mass, in pigs 0.1% and in the blue whale, the largest living anima less than 0.01% (Figure 4).

In addition, the different groups of vertebrates, while satisfying the principle of negative brain allometry, exhibit considerable differences in their fundamental brain-body relationship (Figure 5). Among tetrapods, mammals and birds generally have larger brains relative to body volume/weight than amphibians and reptiles, and among mammals, cetaceans and primates have relatively larger brains than other orders. Thus, during the evolution of birds and mammals and more specifically of cetaceans and primates, genetic and epigenetic systems controlling

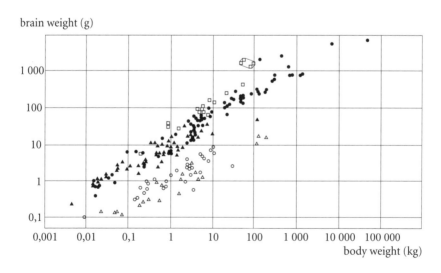

**Figure 2.** The relationship between brain size and body size in vertebrates. Double-logarithmic graph. Open circles: bony fishes; open triangles: reptiles; filled triangles: birds; filled circles: mammals except primates; open squares: primates; encircled open squares: *Homo sapiens*. After Jerison (1973).

brain weight (g)

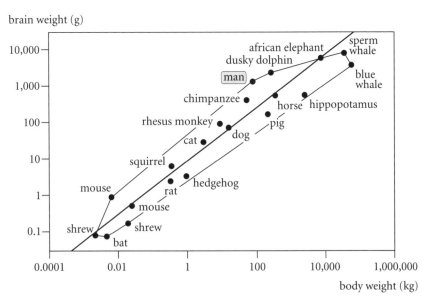

**Figure 3.** The relationship between brain size and body size in mammals. Data from 20 mammalian species. From Nieuwenhuys et al. (1998), modified.

relative brain weight (percentage of body weight)

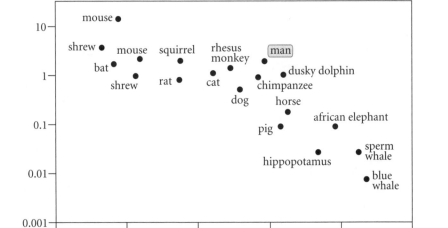

**Figure 4.** Brain weight as a percentage of body weight for the same 20 mammalian species. Double-logarithmic graph. From Nieuwenhuys et al. (1998), modified.

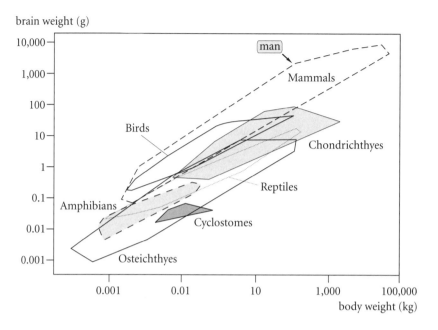

**Figure 5.** Diagrams showing the relationship between body weight and brain weight in a double-logarithmic graph.

brain size have undergone substantial changes in favor of relatively larger brains. These changes resulted in enlargements of brains beyond that associated with body size (Jerison 1991, 2000).

Thus, contrary to a common belief, humans do not have the largest brain either in absolute or relative terms. Unless we accept that cetaceans and elephants are more intelligent than humans and/or have states of consciousness not present in humans, the absolute or relative size of the human brain *per se* cannot account for our factual or alleged superior cognitive abilities. However, among relatively large animals man stands out with a brain that constitutes 2% of body mass. We can quantify this fact by determining the so-called encephalization quotient (EQ) which indicates the ratio between the actual relative brain size of a group of animals to the relative brain size as expected on the basis of brain allometry determined by body size alone (List 2). Calculating the EQ for the human brain, it turns out that it is about seven times larger than that of an average mammal and about 3 times larger than that of a chimpanzee, if they had the size of a human being (Jerison 1973, 1991).

While man stands out in this respect among primates, similar processes must have taken place among cetaceans. Toothed whales, particularly members of the family Delphinidae, exhibit EQs that are far superior to all primates except *Homo*

**List 2.** Encephalization in mammals.

| Encephalization quotient in mammals | | | | | |
|---|---|---|---|---|---|
| Man | 7.4 | Marmot | 1.7 | Cat | 1.0 |
| Dolphin | 5.3 | Fox | 1.6 | Horse | 0.9 |
| Chimpanzee | 2.5 | Walrus | 1.2 | Sheep | 0.8 |
| Monkey | 2.1 | Camel | 1.2 | Mouse | 0.5 |
| Elephant | 1.9 | Dog | 1.2 | Rat | 0.4 |
| Whale | 1.8 | Squirrel | 1.1 | Rabbit | 0.4 |

(After Jerison; Blinkov & Glesner)

*sapiens* (Marino 1998). While man has an EQ of about 7, the dolphins *Sotalia fluviatilis*, *Delphinus delphis* and *Tursiops truncatus* have EQs of 3.2, and the great apes (except man) have EQs around 2. Thus, humans have a much larger brain than expected among primates, but even in this respect their brain is by no means unique, as the example of dolphins shows.

*Fourth claim*:  Humans have the largest cerebral cortex, particularly prefrontal cortex. There are enormous differences both in absolute and relative brain and pallial/cortical size among tetrapods and among mammals in particular. For example, man has a brain and a cortex that are roughly 3,000 times larger in volume than those of a mouse. This implies that changes in *relative* size of cortex are inconspicuous, because in mammals cortical size rather strictly follows changes in brain size, but, again, there are differences within mammalian groups. Apes (including man) have somewhat larger isocortices than other primates and other mammals, because their forebrains (telencephalon plus diencephalon) are generally somewhat larger constituting 74% of the entire brain as opposed to about 60% in other mammals including mice. At 40% of brain mass the human cortex has the size expected in a Great ape (Jerison 1991).

The enormous increase in cortical volume is partly the result of an increase in brain volume and consequently in cortical surface (which is related to an increase in brain volume by exactly the power of 2/3; Jerison 1973), and partly the result of an increase in the thickness of the cortex. The cortex is about 0.8 mm thick in mice and 2.5 mm in man. However, the number of neurons per unit cortical volume decreases with an increase in cortical thickness and brain size. While about 100,000 (or more) neurons are found in one $mm^3$ of motor cortex in mice, "only" 10,000 neurons are found in the motor cortex of man (Jerison 1991). This decrease in the number of cortical neurons per unit volume is a consequence of a roughly equal increase in the length of axonal and dendritic appendages of neurons, in the number of glial cells and in the number of small blood vessels. Without such an increase in glial cells and blood vessels, large isocortices would probably be both architecturally and metabolically impossible.

Thus, the dramatic decrease in nerve cell packing density is at least partly compensated for by an increase in cortical thickness. This could explain why all mammals have a roughly equal number of neurons contained in a cortical column below a given surface area (e.g., 1 mm$^2$) (Rockel et al. 1980). Furthermore, as explained above, what should count for the performance of neuronal networks, is not so much the number of neurons *per se*, but the number of synapses their axons and dendrites form or carry, plus the degree of plasticity of synapses. An increase in length of axons and dendrites paralleling a decrease in nerve cell packing density, should lead to more synapses, and such an increase in the number of synapses, could compensate for the strong decrease in nerve cell packing density as well. It has been estimated that the mouse cortex contains about 10 million ($10^7$) neurons and 80 billion ($8 \times 10^{10}$) synapses and the human cortex about 100 billion ($10^{11}$) neurons and a quadrillion ($10^{15}$) synapses, ten thousand times more than the mouse cortex (Jerison 1991; Schüz 2000; Schüz & Palm 1989). These differences certainly have important consequences for differences in the performance of the respective cortices.

What about animals with brains and cortices that are much larger than those of man, e.g., elephants or most cetaceans? Shouldn't they be much more intelligent than man or have some superior states of consciousness (a popular assumption for whales and dolphins)? As to cetaceans, there is currently a debate on how many neurons their cortices really contain. Their cortex is unusually thin compared to large-sized land mammals and shows a different cytoarchitecture (e.g., lacking a distinct cortical layer IV). Accordingly, experts report a lower number of nerve cells contained in a standard cortical column than in land mammals.

While Garey and Leuba (1986) report that in dolphins the number of cortical neurons per standard column is ⅔ that of land mammals, recently Güntürkün and von Fersen (1998), after examining the brains of three species of dolphins reported that this value amounted only to ¼. Accepting this latter lower value, then – given a cortical surface of about 6,000 cm$^2$ in dolphins (three times that of man) – the cortex of the bottlenose dolphin (*Tursiops truncatus*) should contain ¾ the corresponding number of neurons found in humans, i.e., 6 x $10^{10}$, which is about equal to the number of cortical neurons estimated for chimpanzees. Calculations of the number of cortical neurons in cetaceans with much larger brains and cortices, e.g. in the sperm whale with a cortical surface of more than 10,000 cm$^2$, are difficult, because precise data on cortical nerve cell number per standard cortical column are lacking. However, even assuming that – due to enormous expansion of the cortex and consequent "thinning out" of neurons – the respective value is only 1/8 of that found in land mammals, a sperm whale cortex should contain approximately the same number of cortical neurons as dolphins. Based on these calculations we should expect cetaceans to be roughly as intelligent as non-human great apes, which is what cognitive behaviorists have found out about these animals.

The case of elephants remains, with a similarly enormously large brain (around 4 kg) and cortex of about 8,000 cm$^2$, which at the same time is thicker than that of cetaceans, but also possesses a typical six-layered structure. Assuming that the number of cortical neurons is 2/3 the value found in primates, elephants should have at least as many cortical neurons and cortical synapses as humans. Again, we do not know enough about the organization of the elephant cortex, but elephants should come close to the cognitive and mental capabilities of man, if it were only the number of cortical neurons and synapses that count.

Perhaps it might be safer to restrict our consideration to the size of the associative cortex, because – as I mentioned at the outset – different kinds of consciousness are necessarily bound to the activity of specific parts of the associative cortex. There is a common belief that the associative cortex had increased dramatically both in absolute and relative terms during hominid brain evolution and that this was the basis for the uniqueness of human mind. However, such an increase is difficult to assess, as there are no precise criteria for distinguishing primary and secondary sensory cortical areas from true association areas. Recently, Kaas (1995) argued that the number of cortical areas increased dramatically from about 20 such areas in the hypothetical insectivore-like ancestor to more than 60 in primates. However, what has increased – according to Kaas – was the number of functionally intermediate areas (such as V3 or MT), but neither the primary nor the highly associative areas. Kaas is right to warn about the danger of greatly underestimating the number of functionally different cortical areas in small-brained mammals.

Available data suggest that – contrary to common belief – the associative cortex has increased roughly in proportion to an increase in brain and cortical size. This apparently is the case for the prefrontal cortex, which is regarded by many neuroscientists and neurophilosophers as the true seat of consciousness. Anatomically, the prefrontal cortex is defined as the cortical area with major (though not exclusive) subcortical input from the mediodorsal thalamic nucleus (Uylings et al. 1990; Roberts et al. 1998). Using this definition, it turns out that the PFC has increased isometrically with an increase in cortical and overall brain volume within groups of mammals, but here again we find an additional increase in relative PFC size with an increase in absolute brain size across mammalian orders: in rats, PFC constitutes 6.5%, in dogs, 8.7%, in cows 9.8% and in man 10.6% of brain mass (Jerison 1997). What follows is that the human PFC has exactly the size expected according to primate brain allometry. Of course, cetaceans as well as elephants have prefrontal cortices which are much larger in absolute terms than the human PFC, but what they do with this massive "highest" brain center, remains a mystery so far.

We have not yet found anything in brain anatomy that would explain the factual or alleged uniqueness of the human brain and of humans regarding cognition

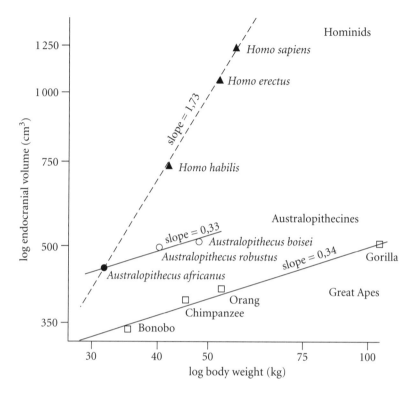

**Figure 6.** Increase in relative brain size in the great apes and in hominids. After Pilbeam and Gould (1974), modified.

and consciousness. Given the fact that *Homo sapiens* has an absolutely and relatively large brain and cortex, it appears to be the animal with the highest number of cortical neurons and/or synapses, probably with the exception of the elephant. Thus, in this respect humans are not truly exceptional. What is highly remarkable, however, is the strong increase in relative (and absolute) brain size in hominid evolution during the last 3–4 million years. While in non-human primates as well as in hominids that did not represent our ancestors, brain size increases with body size to a power of 0.33–0.34, in the lineage leading to *Homo sapiens* it increased to a power of 1.73, i.e. in a *positively allometric fashion*, which means that brain size increased faster than body size (Figure 6). However, the reasons for this phenomenon are completely unclear.

What remains is the question whether there are any anatomical or physiological specializations in the human cortex that could be correlated with the unique cognitive abilities attributed to man. As to the general cytoarchitecture of the human cortex, it is indistinguishable from that of other primates and most other mammals. Likewise, no differences have been discovered so far between humans and non-human mammals with respect to short-term or long-term plasticity of cortical neurons, the action of neuromodulators etc. Only two traits have been discovered that could drastically distinguish the human cortex from that of other primates, viz., (1) differences in growth rate and length of growth period and (2) the presence of the Broca speech center.

As to (1), maturation of the brain is more or less completed at 2 years after birth in prosimians and 6–7 years in monkeys and and non-human apes, but the human brain still continues to mature until the age of 20, which is much longer than in any other primate (Pilbeam & Gould 1974; Hofman 2000). A critical phase in the development of the human brain seems to occur around the age of 2.5 years. At this time, major anatomical rearrangements in the associative cortex have come to a stop, and the period of fine-wiring appears to start, particularly in layer 3 of the prefrontal cortex (Mrzljak et al. 1990). As mentioned above, at this time, human children "take off" cognitively compared to non-human primates. Without any doubt, the drastically prolonged period of brain development constitutes one important basis for an increased capability of learning and memory formation.

The other trait concerns the presence of the Broca speech center in the frontal lobe responsible for temporal aspects of language including syntax, along with the Wernicke speech center in the temporal lobe which is responsible for the meaning of words and sentences (although meaning is likewise dependent on syntax and grammar). It is to date unclear whether these speech centers are true evolutionary novelties. All mammals studied so far have a center for intraspecific communication within the temporal lobe (mostly left side) which may be homologous to the Wernicke center for semantics. It has been reported that destruction of these areas leads to deficits in intraspecific vocal communication. In addition, it has long been argued that the posterior part (A44) of the Broca speech center in humans and the ventral premotor area of non-human primates probably are homologous (Preuss 1995). The ventral premotor area controls the movement of forelimbs, face and mouth, which is likewise the case for the posterior portion of the Broca area.

According to a number of primatologists, non-human primates lack a direct connection between the motor cortex and the nucleus ambiguus, where the laryngeal motor neurons are situated. In man, bilateral destruction of the facial motor cortex abolishes the capacity to produce learned vocalization including speech or humming a melody, while a similar destruction in monkeys has no such consequences (Jürgens 1995). According to a number of experts, the evolutionary basis for human language was an emotionally driven stereotyped language typical of non-human primates. During hominid evolution, the cortex gained control over this system such that beyond the initiation of hard-wired, innate sounds a flexible production of sounds and their sequences became possible (Deacon 1990; Jürgens

1995). Such an interpretation, however, contrasts with recent evidence of a high degree of sound learning in monkeys (Zimmermann 1995) and the mentioned consequences of destruction of left-hemispheric, Wernicke-like temporal areas in all mammals.

Be that as it may, non-human primates including the great apes are strongly limited even in non-vocal speech based on the use of sign language or symbols, and these limitations seem to concern mostly syntax. Accordingly, anything concerning language in the human brain developed relatively recently or underwent substantial modifications, it was probably the Broca center rather than the Wernicke center. Such an assumption is consistent with the fact that the most clear-cut differences between humans and non-human primates concern syntactical complexity of language. Thus, during hominid evolution a reorganization of the frontal-prefrontal cortex appears to have been organized such that the facial and oral motor cortices and the related subcortical speech centers came under the control of a kind of cortex that is specialized in all aspects of temporal sequence of events including the sequence of action (Deacon 1990).

## Acknowledgment

I am grateful to Prof. Harry Jerison, UCLA/Hanse Institute for Advanced Study, for helpful criticism.

## References

Deacon, T. W. (1990). Rethinking mammalian brain evolution. *American Zoologist, 30,* 629–705.

Garey, L. J., & Leuba, G. (1986). A quantitative study of neuronal and glial numerical density in the visual cortex of the bottlenose dolphin: Evidence for a specialized subarea and changes with age. *Journal of Comparative Neurology, 247,* 491–496.

Güntürkün, O., & von Fersen, L. (1998). Of whales and myths. Numerics of cetacean cortex. In N. Elsner & R. Wehner (Eds.), *New Neuroethology on the Move*. Proceedings of the 26th Göttingen Neurobiology Conference (Vol. II: 493). Stuttgart: Thieme.

Hofman, M. A. (2000). Evolution and complexity of the human brain: Some organizing principles. In G. Roth & M. F. Wullimann (Eds.), *Brain Evolution and Cognition* (pp. 501–521). New York, Heidelberg: Wiley-Spektrum Akademischer Verlag.

Jerison, H. J. (1973). *Evolution of the Brain and Intelligence*. New York: Academic Press.

Jerison, H. J. (1991). *Brain Size and the Evolution of Mind*. New York: American Museum of Natural History.

Jerison, H. J. (1997). Evolution of prefrontal cortex. In N. A. Krasnegor, G. R. Lyon, & P. S. Goldman-Rakic (Eds.), *Development of the Prefrontal Cortex: Evolution, Neurobiology, and Behavior* (pp. 9–26). Baltimore, London, Toronto, Sydney: Brookes Publ. Company.

Jerison, H. J. (2000). The evolution of neuronal and behavioral complexity. In G. Roth, & M. F. Wullimann (Eds.), *Brain Evolution and Cognition* (pp. 523–553). New York, Heidelberg: Wiley-Spektrum Akademischer Verlag.

Jürgens, U. (1995). Neuronal control of vocal production in non-human and human primates. In E. Zimmermann, J. D. Newman & U. Jürgens (Eds.), *Current Topics in Primate Vocal Communication* (pp. 199–206). New York, London: Plenum Press.

Kaas, J. H. (1995). The evolution of isocortex. *Brain, Behavior and Evolution, 46*, 187–196.

Karten, H. J. (1991). Homology and evolutionary origins of the "neocortex". *Brain, Behavior and Evolution, 38*, 264–272.

Marino, L. (1998). A comparison of encephalization between odontocete cetaceans and anthropoid primates. *Brain, Behavior and Evolution, 51*, 230–238.

MacPhail, E. M. (2000). Conservation in the neurology and psychology of cognition. In G. Roth & M. F. Wullimann (Eds.), *Brain Evolution and Cognition* (pp. 401–430). New York, Heidelberg: Wiley-Spektrum Akademischer Verlag.

Mrzljak, L., Uylings, H. B. M., van Eden, C. G., & Judás, M. (1990). Neuronal development in human prefrontal cortex in prenatal and postnatal stages. In H. B. M. Uylings, C. G. van Eden, J. P. C. de Bruin, M. A. Corner, & M. G. P. Feenstra (Eds.), *The Prefrontal Cortex. Its structure, function and pathology* (pp. 185–222). Amsterdam, New York, Oxford: Elsevier.

Nieuwenhuys, R., Donkelaar, H. J. ten, & Nicholson, C. (1998). *The Central Nervous System of Vertebrates*, Vol. 3. Berlin: Springer.

Northcutt, R. G., & Kaas, J. H. (1995). The emergence and evolution of mammalian isocortex. *Trends in Neurosciences, 18*, 373–379.

Pilbeam, D., & Gould, S. J. (1974). Size and scaling in human evolution. *Science, 186*, 892–901.

Preuss, T. M. (1995). Do rats have a prefrontal cortex? The Rose-Woolsey-Akert program reconsidered. *Journal of Cognitive Neuroscience, 7*, 1–24.

Roberts, A. C., Robbins, T. W., & Weiskrantz, L. (1998). *The Prefrontal Cortex. Executive and cognitive functions.* Oxford, New York, Tokyo: Oxford University Press.

Rockel, A. J., Hiorns, W., & Powell, T. P. S. (1980). The basic uniformity in structure of the neocortex. *Brain, 103*, 221–244.

Shimizu, T. (2000). Evolution of the forebrain in tetrapods. In G. Roth & M. F. Wullimann (Eds.), *Brain Evolution and Cognition* (pp. 135–184). New York, Heidelberg: Wiley-Spektrum Akademischer Verlag.

Schüz, A. (2000). What can the cerebral cortex do better than other parts of the brain? In G. Roth & M. F. Wullimann (Eds.), *Brain Evolution and Cognition* (pp. 491–500). New York, Heidelberg: Wiley-Spektrum Akademischer Verlag.

Schüz, A., & Palm, G. (1989). Density of neurons and synapses in the cerebral cortex of the mouse. *Journal of Comparative Neurology, 286*, 442–455.

Uylings, H. B. M., & van Eden, C. G. (1990). Qualitative and quantitative comparison of the prefrontal cortex in rat and in primates, including humans. In H. B. M. Uylings, C. G. van Eden, J. P. C. de Bruin, M. A. Corner & M. G. P. Feenstra (Eds.), *The Prefrontal Cortex. Its structure, function and pathology* (pp. 31–62). Amsterdam, New York, Oxford: Elsevier.

Wullimann, M. F. (2000). Brain phenotypes and early regulatory genes: The Bauplan of the metazoan central nervous system. In G. Roth & M. F. Wullimann (Eds.), *Brain Evolution and Cognition* (pp. 11–40). New York, Heidelberg: Wiley-Spektrum Akademischer Verlag.

Zimmermann, E. (1995). Loud calls in nocturnal prosimians: Structure, evolution and ontogeny. In E. Zimmermann, J. D. Newman, & U. Jürgens (Eds.), *Current Topics in Primate Vocal Communication* (pp. 47–72). New York, London: Plenum Press.

# The co-evolution of language and working memory capacity in the human brain

Oliver Gruber

Max Planck Institute of Cognitive Neuroscience, Leipzig, Germany

## 1.  Introduction

The detection of so-called mirror neurons in area F5 of the ventral premotor cortex in monkeys has led to a vivid discussion on the evolution and the functions of premotor cortices. Mirror neurons discharge both when the monkey grasps or manipulates objects and when it observes the experimenter making similar actions. Hence, these neurons have been proposed to underlie a fundamental mechanism for gesture recognition. Furthermore, on the basis of comparative cytoarchitectonical data it has been suggested that it may be Broca's area that is the human homologue of area F5 in monkeys. Recent neuroimaging studies in humans seem to support this view that a similar mirror system in humans may be located in or at least near Broca's area (Rizzolatti et al. 1996; Grafton et al. 1996; Iacoboni et al. 1999). Consequently, it has been hypothesized that mirror neurons in premotor cortices may have played a pivotal role in the evolution of human language and communication (e.g., Rizzolatti & Arbib 1998). In the present contribution, I will extend this view by providing evidence that the evolution of premotor cortices during human phylogeny not only formed the neuronal basis for language functions, but also strongly affected working memory capacity and, presumably, other higher cognitive functions.

## 2.  Conflicting functional-neuroanatomical models of working memory

In fact, Broca's area as well as other parts of the human premotor cortex have been repeatedly demonstrated to be critical for the performance of working memory tasks, in particular in the verbal domain. Various neuroimaging studies have established the view that these premotor brain areas subserve the verbal rehearsal mechanism which, together with the phonological store presumably located in the left inferior parietal lobe, constitutes the so-called phonological loop (e.g., Paulesu et al. 1993; Awh et al. 1996). According to the influential model proposed by Baddeley and Hitch (1974), the phonological loop represents the verbal component of working memory in humans, whereas the visuospatial sketchpad, on the other hand, is regarded as a counterpart which is specialized for the storage of visual and spatial material. Both components are considered to be subsidiary systems dedicated to pure storage. These so-called "slave" systems are supervised by the central executive, a hypothetical attentional-controlling system.

Other, conflicting functional-neuroanatomical models of working memory have been derived from studies of non-human primates using single-cell recordings and anatomical tract-tracing techniques (see Becker & Morris 1999, for a recent discussion). One of the most prominent of these models claims that working memory is topographically organized along parallel prefronto-parietal circuits according to different informational domains (Goldman-Rakic 1996). According to this model, the dorsolateral prefrontal cortex subserves the online-maintenance of visuospatial information, whereas the ventrolateral prefrontal cortex is involved in the maintenance of information about the features of visual objects. While some functional neuroimaging studies of working memory in human subjects produced results consistent with these findings in non-human primates (e.g., Courtney et al. 1998; Haxby et al. 2000), others failed to confirm the suggested organizational principle (e.g., Nystrom et al. 2000; Postle et al. 2000). In sum, the comparability between empirical data derived from studies of either humans or non-human primates appears to be compromised by the special endowment of humans with language. Obviously, the development of language has led to changes in the functional implementation of working memory in the human brain, which so far have been widely neglected by researchers in this field.

### 3.   Articulatory suppression – A method to reduce human working memory capacity to a level comparable to that of non-human primates?

The functional magnetic resonance imaging (fMRI) studies, which will be presented here, reinvestigated the functional neuroanatomy of working memory in humans by making use of articulatory suppression, a classical domain-specific interference technique which is well-established in experimental psychology (Baddeley et al. 1984). The articulatory suppression effect refers to the observation that verbal short-term memory is reduced when one has to perform other concurrent articulations. This effect is usually explained by a disruption of the rehearsal mechanism. Thus, memory performance under articulatory suppression has to rely on other, non-articulatory phonological and/or visual storage mechanisms which could be more similar to working memory mechanisms in non-human primates. In the following studies articulatory suppression was used to deprive human subjects of specific verbal strategies and to make thus the results of these studies more comparable to the findings in non-human primates.

During the first fMRI experiment (Gruber 2000; Gruber 2001), 11 pretrained subjects performed blockwise verbal item-recognition tasks in cued alternation with letter case judgment tasks. Each experimental trial began with a 1-s presentation of four letters, which were randomly taken out of a set of eight phonologically similar letters, followed by a 4-s fixation delay, and then a 1-s presentation of a single letter (see Figure 1). Trials were separated by a 1-s fixation period. A cue instructed the subjects to either quickly read and memorize the four target letters, maintain them during the delay and to decide whether the probe letter matched one of these items or not, or, alternatively, to read them without memorizing and

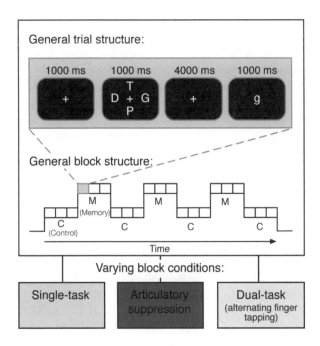

**Figure 1.** Experimental design. Subjects performed blockwise a verbal item-recognition task (M) in cued alternation with a letter case judgment task (C). Different blocks varied with respect to the 4-s delays, which were either unfilled (single-task condition) or filled with silent counting (articulatory suppression) or alternating finger tapping to tones (dual-task condition).

maintaining, and to judge whether the single letter was uppercase or lowercase. Different blocks varied with respect to the 4-s delays, which were either unfilled (single-task condition) or filled with silent counting (articulatory suppression) or alternating finger tapping to tones (alternative dual-task condition). Thus, a 2 × 2 factorial design was employed with one factor being the verbal short-term memory demands during the item-recognition task and the other formed by concurrent dual-task components. The latter, alternative dual-task condition is comparable to articulatory suppression in terms of its general attentional demands and was introduced as a further control in order to differentiate the specific interference effect of articulatory suppression from possible more general dual-task effects. In order to avoid confounding shifts of priorities in the different dual-task situations, the subjects were instructed to consider both silent counting and alternating finger tapping as the respective primary tasks and the item-recognition and letter case judgment tasks as secondary. Furthermore, they were explicitly instructed to rehearse the letters both in the single-task and alternative dual-task condition, and not to use visual memory strategies during articulatory suppression. Instead, they were told to keep the phonologically recoded information in mind although without any rehearsal of it. Finally, letter case was systematically changed between the targets and the probe in the memory conditions (see Figure 1 for an example) in order to preclude a pure visual-matching strategy. A 3.0 Tesla MRI scanner (Bruker Medspec 30/100) with a circularly polarized head coil was used to obtain a high-resolution structural scan for each subject followed by three runs of 518 gradient echo-planar image (EPI) volumes each (TR 2-s, TE 40ms, flip angle 90°; number of slices 16, voxel size $3 \times 3 \times 5\,\text{mm}^3$, distance factor 0.2) that were synchronized with stimulus presentation.

## 4.    Two different brain systems underlie phonological working memory in humans

As expected, silent articulatory suppression led to a significant reduction of memory performance, whereas alternating finger tapping showed no such interference effect (mean percentage of correct responses during single-task condition/silent articulatory suppression/alternating finger tapping: 93.2/77.6/ 91.1%; F = 21.61, p < 0.001). In order to reveal brain areas involved in memory performance under the various secondary conditions, we compared brain activity during each memory condition with activity during the corresponding letter case judgment task, the essential difference between these two tasks being in every case the short-term memory requirements. Verbal working memory performance under both non-interfering conditions activated Broca's area, the left premotor cortex, the cortex along the left intraparietal sulcus and the right cerebellum thus replicating the re-sults from various previous studies. By contrast, no significant memory-related ac-tivation was found in these "classical" areas of verbal working memory when silent articulatory suppression prevented the subjects from rehearsal. Instead, this non-articulatory maintenance of phonological information was associated with en-hanced activity in another prefronto-parietal network, including the cortex along the anterior part of the intermediate frontal sulcus and the inferior parietal lobule (Figure 2, see Appendix).

A straightforward interpretation of these findings is that this network of pre-frontal and parietal areas underlies a brain mechanism by which phonological in-formation can be maintained across a short period of time, in particular if it is not possible to rehearse. Since articulatory suppression is thought to interfere only with the rehearsal mechanism, one may argue that the observed dissociation between the two brain systems corresponds to a dissociation of non-articulatory phono-logical storage from explicit verbal rehearsal. Accordingly, these results suggested that phonological storage may be a function of a complex prefronto-parietal net-work, and not localized in only one, parietal brain region. This assumption re-ceives support from data of another recent brain imaging study which revealed a strikingly similar pattern of brain activation when it was explicitly tested for phonological storage by subtracting a letter match from a letter probe task (Hen-son et al. 2000).

However, although the subjects were explicitly instructed and, in addition, let-ter case was systematically varied to force them to respond on the basis of phono-logical identity and not visual form, it is impossible to rule out by this first study that the memory-related activations during articulatory suppression may have been produced by visual working memory strategies. Therefore, in order to dif-ferentiate the short-term memory system detected under articulatory suppression from prefrontal and parietal areas that are known to underlie visual working mem-ory, we conducted a second fMRI experiment using similar tasks with silent ar-ticulatory suppression, during which colored letters in different fonts were pre-sented and either the letters themselves or their colors or specific forms were to be remembered (Gruber & von Cramon 2001).

## 5.  Similar brain systems for phonological storage and visual working memory are differentially distributed along human prefrontal and parietal cortices

Although both phonological and visual working memory processes (under articu-latory suppression) activated similar prefronto-parietal networks, they were found to be differentially distributed along these cortical structures. In particular, while the phonological task variant yielded strong activations along the anterior parts of the intermediate and superior frontal sulci and in the inferior parietal lobule, working memory for visual letter forms or colors preferentially activated more posterior prefrontal regions along the intermediate and superior frontal sulci as well as the superior parietal lobule (Figure 3, see Appendix). Thus, a prefronto-parietal working memory system presumably subserving non-articulatory maintenance of phonological information could be differentiated in a domain-specific way from the cortical areas subserving visual working memory. On the other hand, the fact that both phonological and visual working memory processes were distributed along identical neuroanatomical structures gives rise to the assumption that these brain structures may represent a multimodal working memory system whose subdivisions deal with different informational domains. Importantly, a very similar anterior-posterior segregation of domain-specific working memory processes appears to exist in the prefrontal cortex of non-human primates as several recent studies indicate a role of the posterior principal sulcus in visuospatial and possibly also in auditory-spatial processing, whereas the anterior part of the principal sulcus may subserve non-spatial auditory and probably also some aspects of species-specific phonetic processing (e.g., Romanski et al. 1999).

## 6.    A new hypothesis regarding the role of premotor cortices in the evolution of human working memory and cognition

Together, these studies suggest that human working memory is supported by two brain systems which fundamentally differ from each other in terms of their evolutionary origin. A phylogenetically older working memory system, which is also present in non-human primates, seems to be topographically organized along parallel prefronto-parietal and prefronto-temporal brain circuits according to different informational domains. A second system, which probably developed later on in the context of the evolution of language, is supported by premotor speech areas and mediates explicit verbal rehearsal. This system represents a flexible, functionally superior and therefore predominant memory mechanism, which operates independently from the original input modality (Schumacher et al. 1996). The two fMRI studies presented here provide empirical support for this plausible theory, firstly by demonstrating a possible functional-neuroanatomical dissociation of the rehearsal mechanism from other working memory mechanisms that are also able to keep phonological information online (Figure 2; Gruber 2000; Gruber 2001) and, secondly, by indicating a domain-specific topographical organization of these latter fronto-parietal working memory circuits in humans (Figure 3; Gruber & von Cramon 2001). Moreover, other studies provide evidence for a functional heterogeneity of the premotor brain areas subserving verbal rehearsal, as the same brain regions seem to be involved in the manipulation of working memory contents (Gruber et al. 1999), and may also support the mnemonic processing of temporal information (Gruber et al. 2000). From these findings one may conclude that Baddeley's formal conception of the phonological loop as the verbal counterpart of the visuospatial sketchpad is inappropriate. Rather, the rehearsal mechanism should be regarded as a functional new acquisition during human phylogeny which interacts with another working memory system concerned with various, and not exclusively visual, submodalities. Based on the presented empirical data and in view of the homologous functional-neuroanatomical organization revealed in monkeys, I have recently introduced an alternative model of human working memory that emphasizes this evolutionary special role of the rehearsal mechanism in human working memory (see Figure 4; Gruber 2000; Gruber & von Cramon 2001). Although additional studies are clearly needed for further validation and refinement, this model

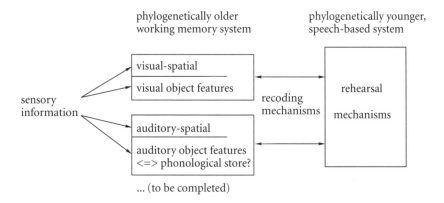

Figure 4. An evolutionary-based model of human working memory. Verbal rehearsal is considered to be the most efficient and predominant working memory mechanism in humans which can be accessed via recoding mechanisms and which operates independently from the original stimulus modality. It is neurally implemented by the brain areas depicted in green in Figure 2. A probably phylogenetically older working memory system is topographically organized along parallel prefronto-parietal circuits according to different informational domains. For instance, the red-colored brain regions in Figure 2 may subserve the maintenance of phonologically coded representations, in particular when the information will not or even cannot be rehearsed. This phonological storage mechanism in the human brain may thus have evolved as a further differentiation from a non-spatial auditory working memory mechanism that is also present in various non-human species.

appears promising in that it may offer new explanations for many behavioral, neuropsychological and neuroimaging findings in human subjects. The model also permits the harmonization of conflicting working memory models derived from human respectively animal research (e.g., Baddeley & Hitch 1974; Goldman-Rakic 1996).

With regard to our discussion on mirror neurons and the role of premotor cortices in human evolution, the findings presented here strongly suggest that Broca's area and other premotor cortices constitute not only a sophisticated language system, but also a very efficient working memory mechanism. The well-known effect of articulatory suppression on memory performance can be taken as an indication for the clearly higher capacity of this memory mechanism as compared to the phylogenetically older working memory mechanism, which human subjects have to rely on when verbal rehearsal is prevented. In this sense, it appears that a co-evolution of language and working memory capacity has taken place in the human brain. Finally, the observable functional heterogeneity of Broca's area that I briefly mentioned, suggests that the evolution of premotor brain areas may also have provided the basis for other higher cognitive functions that make humans in some sense unique.

## References

Awh, E., Jonides, J., Smith, E. E., Schumacher, E. H., Koeppe, R. A., & Katz, S. (1996). Dissociation of storage and rehearsal in verbal working memory. *Psychological Science, 7*, 25–31.

Baddeley, A. D., & Hitch, G. J. (1974). Working memory. In G. Bower (Ed.), *Recent Advances in Learning and Motivation*, Vol. VIII (pp. 47–90). New York: Academic Press.

Baddeley, A., Lewis, V., & Vallar, G. (1984). Exploring the articulatory loop. *Quarterly Journal of Experimental Psychology A, 36*, 233–252.

Becker, J. T., & Morris, R. G. (1999). Working memory(s). *Brain and Cognition, 41*, 1–8.

Courtney, S. M., Petit, L., Maisog, J. M., Ungerleider, L. G., & Haxby, J. V. (1998). An area specialized for spatial working memory in human frontal cortex. *Science, 279*, 1347–1351.

Goldman-Rakic, P. S. (1996). The prefrontal landscape: Implications of functional architecture for understanding human mentation and the central executive. *Philosophical Transactions Royal Society of London B, 351*, 1445–1453.

Grafton, S. T., Arbib, M. A., Fadiga, L., & Rizzolatti, G. (1996). Localization of grasp representations in humans by positron emission tomography. 2. Observation compared with imagination. *Experimental Brain Research, 112*(1), 103–111.

Gruber, O., Bublak, P., Schubert, T., & von Cramon, D. Y. (1999). The neural correlates of working memory components: A functional magnetic resonance imaging study at 3 Tesla. *Journal of Cognitive Neuroscience*, Suppl. S, 32–33.

Gruber, O. (2000). Two different brain systems underlie phonological short-term memory in humans. *Neuroimage, 11*, S407.

Gruber, O. (2001). Effects of domain-specific interference on brain activation associated with verbal working memory task performance. *Cerebral Cortex, 11*, 1047–1055.

Gruber, O., Kleinschmidt, A., Binkofski, F., Steinmetz, H., & von Cramon, D. Y. (2000). Cerebral correlates of working memory for temporal information. *Neuroreport, 11*(8), 1689–1693.

Gruber, O., & von Cramon, D. Y. (2001). Domain-specific distribution of working memory processes along human prefrontal and parietal cortices: A functional magnetic resonance imaging study. *Neuroscience Letters, 297*, 29–32.

Haxby, J. V., Petit, L., Ungerleider, L. G., & Courtney, S. M. (2000). Distinguishing the functional roles of multiple regions in distributed neural systems for visual working memory. *Neuroimage, 11*, 380–391.

Henson, R. N. A., Burgess, N., & Frith, C. D. (2000). Recoding, storage, rehearsal and grouping in verbal short-term memory: An fMRI study. *Neuropsychologia, 38*, 426–440.

Iacoboni, M., Woods, R. P., Brass, M., Bekkering, H., Mazziotta, J. C., & Rizzolatti, G. (1999). Cortical mechanisms of human imitation. *Science, 286*, 2526–2528.

Nystrom, L. E., Braver, T. S., Sabb, F. W., Delgado, M. R., Noll, D. C., & Cohen, J. D. (2000). Working memory for letters, shapes, and locations: fMRI evidence against stimulus-based regional organization in human prefrontal cortex. *Neuroimage, 11*, 424–446.

Paulesu, E., Frith, C. D., & Frackowiak, R. S. J. (1993). The neural correlates of the verbal component of working memory. *Nature, 362*, 342–344.

Postle, B. R., Stern, C. E., Rosen, B. R., & Corkin, S. (2000). An fMRI investigation of cortical contributions to spatial and nonspatial visual working memory. *Neuroimage, 11*, 409–423.

Rizzolatti, G., Fadiga, L., Matelli, M., Bettinardi, V., Paulesu, E., Perani, D., & Fazio, F. (1996). Localization of grasp representations in humans by PET. 1. Observation versus execution. *Experimental Brain Research, 111*(2), 246–252.

Rizzolatti, G., & Arbib, M. A. (1998). Language within our grasp. *Trends in Neurosciences, 21*(5), 188–194.

Romanski, L. M., Tian, B., Fritz, J., Mishkin, M., Goldman-Rakic, P. S., & Rauschecker, J. P. (1999). Dual streams of auditory afferents target multiple domains in the primate prefrontal cortex. *Nature Neuroscience, 2*(12), 1131–1136.

Schumacher, E. H., Lauber, E., Awh, E., Jonides, J., Smith, E. E., & Koeppe, R. A. (1996). PET evidence for an amodal verbal working memory system. *Neuroimage, 3*, 79–88.

**Figure 2.** Brain regions subserving phonological working memory under different conditions. Green indicates memory-related activations that occurred only in absence of articulatory suppression during both single- and non-interfering dual-task (ST/DT) conditions. Red indicates memory-related activations that occurred only under articulatory suppression (AS). Brown indicates memory-related activations that were present in all conditions investigated in this study, i.e. independent from articulatory suppression. Bars in the inserts show the mean percentage of signal changes produced by the memory tasks in relation to the respective control conditions (L, left; R, right; from Gruber 2001).

**Figure 3.** Domain-specific distribution of working memory processes along human prefrontal and parietal cortices. Predominant activation of the cortex along the anterior parts of the intermediate and superior frontal sulci and of the inferior parietal lobule by phonological memory (indicated in yellow and red), and of the cortices along posterior parts of the same frontal sulci and of the superior parietal lobule by visual working memory (indicated in blue and green). Each task was performed under articulatory suppression (from Gruber & von Gramon 2001).

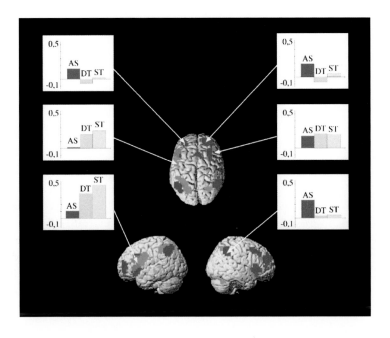

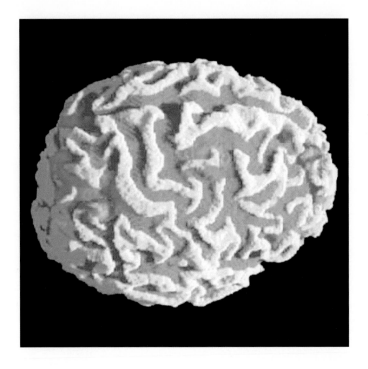

# Episodic action memory

## Characterization of the time course and neural circuitry

Ava J. Senkfor

NMR Center, Massachusetts General Hospital, Harvard Medical School, Charlestown, MA, USA

## 1. Introduction

Memory for one's own past behavior is fundamental for planning new activities and for most social interactions. This aspect of memory has been less thoroughly investigated than memory for passively presented materials, despite the fact that an individual's behavior often plays a critical role in determining the nature of his or her experience.

My colleagues and I have investigated memory for actions by allowing participants to examine real three-dimensional objects, under well controlled study conditions, then recording brain electrical activity during attempts to retrieve information about both the objects and their own activities from the study phase. The experiments use four encoding tasks to manipulate the nature of the episodic memory: (1) *Performing* a typical action with an object; (2) *Imagine* performing an action without touching the object or moving one's hands; (3) *Watching* the experimenter perform an action; or (4) estimating the *Cost* of the object. The last condition serves as a non-action control condition, one that involves cognitive effort but no motoric involvement.

On each trial during the test phase of the experiments, participants are presented with a digital color photo of an object, and asked to indicate whether or not they had studied it, and/or to indicate *how* they had studied it. During the test phases, event-related potentials (ERPs) are recorded from the scalp of healthy young volunteers. The ERP provides a record of synchronous synaptic activity from large populations of neurons (Regan 1987). ERPs have proved sensitive to successful episodic recognition across a variety of stimulus types (Smith & Halgren

1989; Senkfor & Van Petten 1998; Van Petten & Senkfor 1996; Van Petten, Senkfor, & Newberg 2000). In the two action memory experiments reviewed here, four questions are addressed:

1.  Are objects encoded with actions more recognizable than objects encoded without motor involvement?
2.  Is the behavior associated with an object during the study phase retrieved automatically upon re-exposure to the object or only when participants are explicitly queried about the encoding context?
3.  Are different patterns of brain activity associated with retrieval of action memories versus non-action memories?
4.  If memories with an action component prove to be qualitatively different from non-action memories, how closely will retrieval recapitulate the motor activities from encoding?

## 2.  Efficacy of action encoding for object memory

In the first experiment, participants initially study 150 objects, evenly divided between the Perform and Cost estimation encoding tasks (trials randomly intermixed; Senkfor, Van Petten, & Kutas 1999, submitted). These two encoding tasks

| DAY1 | | DAY2 | |
|---|---|---|---|
| Study (Objects) | Item Test (Photos) | Study (Objects) | Source Test (Photos) |
| "Perform" "Cost" "Perform" "Cost" · · · | | "Perform" "Cost" "Perform" "Cost" · · · | |

Figure 1. During the study phases, 150 real objects or toy versions of real objects, are presented one at a time. Preceding each object is an encoding task cue – "Perform" or "Cost". After 7 seconds a tone signals removal of the object and the next cue and object occurs 4 seconds later. On day one, the test phase includes all studied objects (digital photographs) plus an equal number of new photographs of objects (Item Test). Participants make Old/New judgments to each object. On day two, the study phase is identical to day 1, but participants receive a new set of objects. At test, the procedures are the same except participants make "Old-Perform", "Old-Cost", or "New" judgments to each object. EEG is recorded during the test phase only from 28 scalp electrodes.

**Table 1.** Accuracies (standard error) in the item and source memory tests.

|  | Item test | Source test |
|---|---|---|
| Hit | 93 (0.8) | 94 (0.5) |
| Perform | 94 (0.9) | 94 (0.8) |
| Cost | 92 (1.2) | 94 (1.0) |
| Hit/Hit | – | 90 (3.0) |
| Perform | – | 93 (3.1) |
| Cost | – | 87 (3.2) |
| Correct Rejection | 95 (0.8) | 96 (0.6) |

are well matched in several ways: both require that an object be identified, and both require a self-initiated strategy to produce an acceptable response; participants selected their own "typical" action, or the best basis for a cost estimate. However, only the Perform task required analysis of the object's somatomotor properties (size, shape, and weight relative to hand aperture and muscular effort), selection of an appropriate motor program, and execution of the motor program. At test, participants viewed all of the studied objects intermixed with an equal number of unstudied objects, and responded "old" or "new" to each object. Figure 1 shows the experimental design in more detail.

Table 1 shows that both encoding tasks are very effective in promoting high levels of recognition accuracy, but yield no advantage with Perform encoding over Cost encoding. Previous studies have suggested that action encoding is particularly beneficial for memory (Engelkamp 1998), but these have typically used a weak comparison task of passively listening to action commands.

## 3.    Automatic retrieval of motoric information?

In the second session of this experiment, the same participants study another 150 objects, also evenly divided between the Perform and Cost encoding tasks. The test phase of this second session includes a *source memory* test. Here, participants responded "Perform" or "Cost" to indicate how they studied each of the old objects, in addition to "New" responses for unstudied objects. Table 1 indicates that although there was again no advantage of Perform-encoding for episodic object recognition, participants are slightly more accurate in indicating the source of their memories after action encoding than the purely cognitive task of cost estimation.

Figure 2 shows that in both sessions of the experiment, ERPs recording during the memory tests differentiated studied from unstudied objects. Beginning around

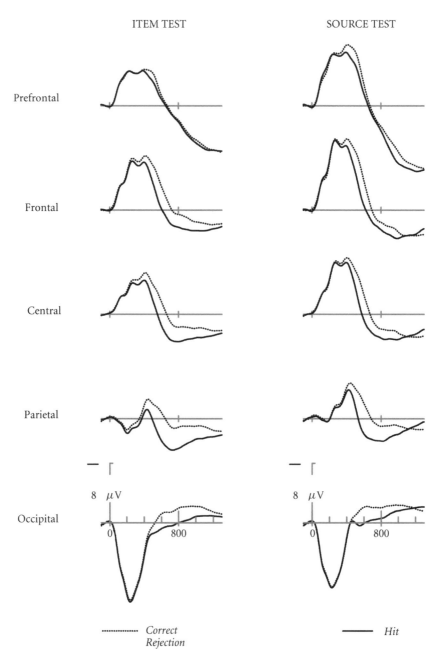

**Figure 2.** Grand average ERPs from left lateral prefrontal, frontal, central, parietal, and occipital sites elicited by correctly identified new trials (Correct Rejection) and studied trials (Hit).

300 ms after the onset of the digital object images, studied objects elicit more positive ERPs than new objects, and this difference continues for the remainder of the 1300 ms epoch in both the item and source memory tests. A generally similar *old/new effect* is observed across a variety of stimulus types in previous studies that have linked the positivity to specifically to successful retrieval (unstudied items receiving incorrect "old" responses, and unrecognized old stimuli elicit different ERPs than hits in recognition tests, Rubin et al. 1999; Van Petten & Senkfor 1996). However, of greatest interest is whether this general similarity of the old/new effect across stimulus types may conceal content-specific retrieval processes. The Perform and Cost encoding tasks were selected to engage different neural processes in the study phases of the experiment, and retrieval of such qualitatively different information may manifest different brain activity during the test phases.

The left side of Figure 3 contrasts the ERPs elicited by the two classes of studied objects during the old/new (item) recognition test. No difference between Perform and Cost-encoded objects is observed. Thus, the brain's response to objects is not automatically altered by the context of the original study episode. The apparent failure to retrieve contextual information when it is not requested stands in contrast to a large number of ERP studies. These studies show that ERP differences between studied and unstudied items (old/new effects) do not require explicit retrieval instructions and are observed during tasks that do not require any overt differentiation between old and new items (see Van Petten & Senkfor 1996 for review).

In contrast, the right side of Figure 3 shows that when participants are asked to retrieve information about the encoding tasks from study during the source memory test, the nature of the original encoding task modulates brain activity during retrieval. Starting around 800 ms post stimulus onset, Perform-encoded objects elicits more positive ERPs than Cost-encoded objects. This Perform-Cost difference is evident over all but prefrontal scalp sites. The delayed onset of the Perform/Cost difference relative to the old/new effect suggests that retrieval of the encoding task information occurs after episodic object recognition (see also Senkfor & Van Petten 1998). In the late portion of the epoch, the spatial distribution of ERPs across the scalp also differs between the two classes of encoding tasks. These two scalp distribution patterns suggest that different cortical areas are engaged during the retrieval of episodes with actions and episodes with cost estimates.

## 4. Content specificity of action memories

The first experiment provides some evidence that action encoding of objects leads to memory traces that are qualitatively distinct from those formed during non-

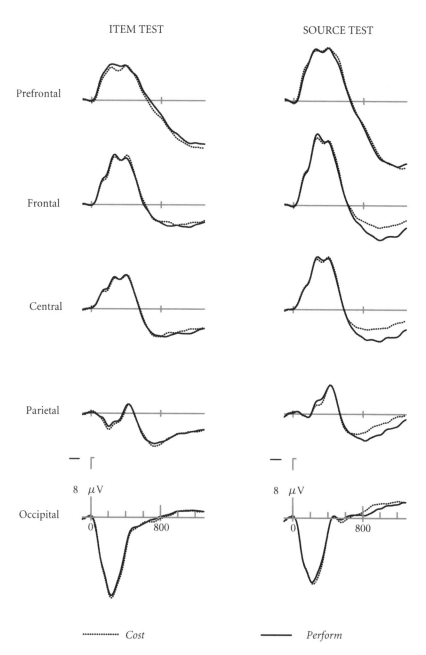

**Figure 3.** Grand average ERPs from left lateral prefrontal, frontal, central, parietal, and occipital sites elicited by correctly remembered Perform and Cost encoded trials.

motoric interaction. It also shows that these differences are apparent only when relevant to one's retrieval goals. A second experiment probes content-specific retrieval processes more closely by including four encoding tasks: Perform, Imagine, Watch, and Cost (Senkfor, Van Petten, & Kutas 1999, 2002).

Performed, watched, and imagined actions engender both common and unique features during encoding. Motor planning and execution, motor imagery alone, and observing someone else's performance are likely to share some neural substrates. Neuroimaging studies report some overlapping activations among these action conditions (Hallet, Feldman, Cohen, Sadato & Pascual-Leone 1994; Roland, Larsen, Lassen, & Skinhoj 1980; Rizzolatti et al. 1996; see Decety 1996a, 1996b for reviews). These similarities during the initial experience predict some commonalities during *memory* for actions, as compared to memory for non-actions. However, an inability to differentiate past performance from imagination from observation would not serve us well, nor would a general category for the three types of actions. Performed actions share perceptual attributes with observed actions (overt motion, but in a mirror-reversed form), but also include features of agency (goal formation, selection and execution of motor programs) and proprioceptive/tactile feedback that are lacking when one only observes. Performed and imagined actions may share some motoric attributes, but motor imagery is bereft of overt movement and somatosensory experience. Comparisons across the three action tasks thus allow a closer examination of what qualitative attributes contribute to the observed difference between retrieval of episodes with and without action in our initial experiment.

During the study phase, real objects are presented together with a cue signaling the encoding task (Perform, Imagine, Watch, or Cost estimation). Object location cues right or left hand, so that presentation to a participant's right side corresponds to right-hand manipulation or imagery on Perform and Imagine trials, but a left-hand action by the experimenter on Watch trials (location had no signal value on Cost trials). At test, participants view images of the studied objects (no new objects) and press one of four buttons to indicate the encoding task (Figure 4).

Behavioral results show a graded benefit of action encoding (see Table 2). Performed actions are better remembered than observed actions, which are better remembered than Imagine or Cost trials. The ERPs at test reveal processing differences beginning around 600 ms poststimulus onset, but the relationship among encoding tasks varies across the scalp (Figure 5). At prefrontal sites, Imagine trials elicit more positive ERPs than all other tasks, beginning around 900 ms. At frontal sites approximately over premotor cortex, the three action conditions are indistinguishable from each other, but elicit more positive ERPs than the non-action Cost condition beginning around 800 ms. A third pattern is seen over occipital, temporal, and posterior parietal sites beginning around 800 ms. At these posterior sites, Perform and Watch trials are differentiated from Imagine and Cost trials – a di-

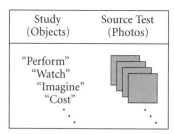

**Figure 4.** 216 real objects, or toy versions of real objects, are presented one at a time, each preceded by a spoken encoding task cue from a tape recorder – "Perform", "Watch", "Imagine", or "Cost". After 7 seconds a tone signals the removal of the object and another cue and object occurs 4 seconds later. At test, digital color photographs of all studied objects are presented as participants determine which encoding task (Perform, Watch, Imagine, or Cost) was conducted with that object via button presses. EEG is recorded during the test phase from 28 scalp electrodes.

vision between conditions that did and did not include visual motion during the study phase. Our interpretation is thus that the prefrontal sites are uniquely sensitive to memory for motor imagery, frontal sites reflects a neural circuit devoted to action, and posterior areas are sensitive to retrieval of at least one visual attribute (motion).

In sum, the nature of the encoding task is reflected differentially across cortical areas. Moreover, brain activity during retrieval shows fairly binary distinctions between episodic memories, suggesting that at least three distinct features are used to differentiate the four classes of trials. We confidently assign these content-specific effects to memory processes because the four patterns of brain activity are elicited by identical sets of stimuli (across participants) and vary only in how they are originally experienced. Such content-specific effects are a rich source of evidence about how events are initially parsed and how the brain disentangles similar episodes during retrieval.

The three attribute circuits we identify here – actions in general, motor imagery, and visual motion – appear to be conducted in parallel. The earliest division – between conditions with and without motion – began around 800 ms over

**Table 2.** Reaction times and accuracies (standard error) in the memory tests.

| Encoding task | Reaction time | Accuracy |
| --- | --- | --- |
| Perform | 1546 (47) | 93 (1.1) |
| Watch | 1651 (58) | 88 (1.7) |
| Imagine | 2072 (99) | 82 (2.5) |
| Cost | 1762 (59) | 78 (2.6) |

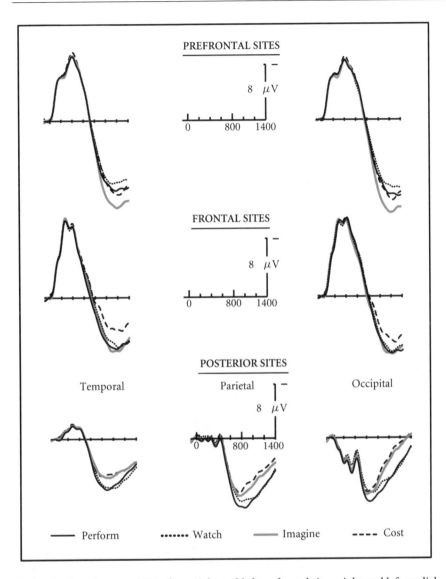

**Figure 5.** Grand average ERPs from right and left prefrontal sites, right and left medial fronto-central sites, and left lateral temporal, parietal, and occipital sites elicited by correctly remembered objects encoded with perform, watch, imagine, and cost tasks.

posterior sites as well as a second distinction between the action and non-action tasks is occurs over the frontal sites. The prefrontal division between Imagine and the other conditions began last, at 900 ms, but also shows temporal overlap with the other two processes. All three attribute circuits, apparent in the ERP, occurs well in advance of the participants overt button press responses (average RT was 1757 ms). We thus hypothesize that the three patterns of ERP responses are causally related to an accurate decision, so that reduction or elimination of any one of the three binary distinctions would be accompanied by distinct patterns of memory confusions.

## 5.    Recapitulation of motor activity during memory retrieval

Fronto-central scalp sites near premotor cortex differentiated the action and non-action memory episodes. However, a stronger form of the motor recapitulation hypothesis predicts lateralization of the memory effects depending on the hand used to conduct the performed, watched, or imagined action. After sorting the retrieval trials according to encoding hand, we find that objects encoded with the right and left hand produced essentially identical accuracy levels and reaction times. The ERP data are quantified as mean amplitude measures from 600–1400 ms poststimulus onset, and lateral pairs of electrode sites closest to the midline are selected for analysis (more ventral pairs were excluded a priori).

*Hand tag effect*
While no difference between objects presented to the participants' left or right side is seen in the non-action task of cost estimation, the three action tasks show more positive ERPs for trials corresponding to the right hand used during study than the left hand (Figure 6). The results thus suggest that a "hand tag" is embedded in the episodic memory trace, and recovered during retrieval of the event.

*Hemisphere and encoding hand*
The encoding hand data is further broken down to examine hemispheric asymmetries. A strong recapitulation hypothesis would incorporate the contralateral organization of motor cortex to predict a hand by hemisphere interaction: asymmetries that reverse direction depending on the hand used during the study phase. At prefrontal and frontal sites, a significant contralateral pattern is observed for the Perform and Watch conditions: a left-greater-than-right asymmetry for objects encoded with the right hand accompanied by a right-greater-than-left asymmetry for objects encoded by the left hand – a motor recapitulation effect (Figure 6). Like the more general hand tag, this pattern follows encoding hand (both participant's and experimenter's) rather than object location during the study phase and thus demonstrates a true "mirror effect". No hand by hemisphere interaction is observed in the Imagine or Cost conditions; its absence in the Imagine condition is puzzling, but all of the hand analyses are based on data with a lower signal-to-noise ratio than the overall analyses of the four encoding conditions.

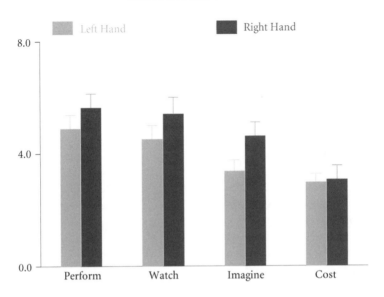

**Figure 6.** Mean amplitude from medial sites for correctly remembered objects encoded with the right versus left hand during study for each of the four encoding tasks – Perform, Watch, Imagine, and Cost.

## 6.   Summary and current directions

Our first experiment shows that although recently experienced objects elicit different brain activity than unstudied objects, mere presentation of an object does not necessarily evoke specific memories of one's prior interactions with it. Both experiments show that when one's prior activities *are* retrieved, different brain circuits are engaged depending on the nature of that prior experience: performing an action, watching the experimenter perform an action, imagining an action, or a cognitive control task of cost estimation. In noninvasive scalp ERP recordings, clear binary distinctions are observed between retrieval of episodes with and without action over premotor cortex, between episodes with and without visual motion over posterior cortex, and between episodes with and without motor imagery over prefrontal cortex. The results further suggested a fair degree of specificity in action memory traces, including information about which hand participated in the action. The motor recapitulation effects in the ERP are also mirror-reversed in the watch condition. Similar effects are observed for the participants' and experimenter's right hands, despite the fact that the two parties faced each other during the study phase of the experiment.

In ongoing work, pursuits include a more detailed view about the cortical circuits underlying the content-specific retrieval effects and the hand-specific mirror effect via fMRI (Senkfor, Busa, & Gabrieli, in preparation; Senkfor, Busa, & Halgren, in preparation), altering encoding features that would affect processing circuits during retrieval (Senkfor, in preparation). A third line of investigation concerns the fate of these effects during normal aging, when confusions between become more prevalent (Senkfor & Kutas 2000).

## Acknowledgments

Funding provided by National Research Service Award (MH12557), National Institute of Mental Health (MH 52893), and National Institute of Aging (AG 14792, AG 08313).

## References

Decety, J. (1996a). The neurophysiological basis of motor imagery. *Experimental Brain Research, 111,* 429–436.

Decety, J. (1996b). Neural representation for action. *Reviews in the Neurosciences, 7,* 285–297.

Engelkamp, J. (1998). *Memory for Actions.* London: Psychology Press/Taylor & Frances.

Hallet, M., Feldman, J., Cohen, L. G., Sadato, N., & Pascual-Leone, A. (1994). Involvement of primary motor cortex in motor imagery and mental practice. *Behavioral and Brain Sciences, 17,* 210.

Regan, D. (1989). *Human Brain Electrophysiology: Evoked potentials and evoked magnetic fields in science and medicine.* New York: Elsevier.

Rizzolatti, G., Fadiga, L., Gallese, V., & Fogassi, L. (1996). Premotor cortex and the recognition of motor actions. *Cognitive Brain Research, 3,* 131–141.

Roland, R. E, Larsen, B., Lassen, N. A., & Skinhoj, E. (1980). Supplementary motor area and other cortical areas in organization of voluntary movements in man. *Journal of Neurophysiology, 43,* 118–136.

Rubin, S., Van Petten, C., Glisky, E., & Newberg, W. (1999). Memory conjunction errors in younger and older adults: Electrophysiological and neuropsychological evidence. *Cognitive Neuropsychology, 16,* 459–488.

Senkfor, A. J. (in preparation). Perceptual influence of "to have and to hold": An ERP analysis of episodic action memories with and without objects.

Senkfor, A. J., Busa, E., & Halgren, E. (in preparation). Encoding task echoes during retrieval: A direct test of the recapitulation hypothesis with fMRI.

Senkfor, A. J., Busa, E., & Gabrieli, J. D. E. (in preparation). Recapitulation of episodic action events revealed by fMRI.

Senkfor, A. J., & Kutas, M. (2000). Effects of aging on episodic action memory. *Journal of Cognitive Neuroscience*, Supplement, 29.

Senkfor, A. J., & Van Petten, C. (1998). Who said what: An event-related potential investigation of source and item memory. *Journal of Experimental Psychology: Learning, Memory & Cognition, 24*, 1005–1025.

Senkfor, A. J., Van Petten, C., & Kutas, M. (1999). Episodic action memory: An ERP analysis. *Journal of Cognitive Neuroscience*, Supplement, 31.

Senkfor, A. J., Van Petten, C., & Kutas, M. (submitted). A source is a source? An ERP analysis of source and item memory.

Senkfor, A. J., Van Petten, C., & Kutas, M. (2002). Episodic action memory for real objects: An ERP investigation with perform, watch, and imagine action encoding tasks versus a non-action task. *Journal of Cognitive Neuroscience, 14*, 402–419.

Smith, M. E., & Halgren, E. (1989). Dissociation of recognition memory components following temporal lobe lesions. *Journal of Experimental Psychology: Learning, Memory & Cognition, 15*, 50–60.

Van Petten, C., & Senkfor, A. J. (1996). Memory for words and novel visual patterns: Repetition, recognition, and encoding effects in the event-related potential. *Psychophysiology, 33*, 491–506.

Van Petten, C., Senkfor, A. J., & Newberg, W. (2000). Memory for drawings in locations: Spatial source memory and event-related potentials. *Psychophysiology, 37*, 551–564.

# The role of objects in imitation

Andreas Wohlschläger and Harold Bekkering

Max-Planck-Institut für psychologische Forschung, München, Germany

## 1. Introduction

Imitation plays an important role in skill acquisition – and not merely because it avoids time-consuming trial-and-error learning. Observing and imitating is also a special case of the translation of sensory information into action. The actor must translate a complex dynamic visual input pattern into motor commands in such a way, that the resulting movement visually matches the model movement. For that reason, imitation is one of the most interesting examples of perceptual-motor co-ordination.

Although humans are very successful in imitating many complex skills, the mechanisms that underlie successful imitation are poorly understood. The translation problem is particularly interesting in children, because they must perform the translation despite the obviously great differences in orientation, body size, limb lengths, and available motor skills. Additionally, these differences result in very different dynamic properties (Meltzoff 1993). Nevertheless, children spontaneously and continuously try to imitate the customs and skills manifested by the adults and peers.

Based on earlier findings (Meltzoff & Moore 1977), Meltzoff and Moore (1994) developed an influential theory – the theory of active inter-modal mapping (AIM) – that assumes a supra-modal representational system that merges the perceptual and the action systems. This supra-modal representational system is thought to match visual information with proprioceptive information. The AIM theory is in line with the common view that – in imitation – perception and action are coupled by means of a direct perceptual-motor mapping (cf. e.g., Butterworth 1990; Gray et al. 1991).

A direct perceptual-motor mapping is also supported by neurophysiological findings. The so-called mirror neurones (di Pellegrino et al. 1992) in the monkey's pre-motor area F5 are potential candidates for a neural implementation of an observation-execution matching system, because they fire both during the observation and during the execution of particular actions. Support for a similar system in humans comes from the finding of a motor facilitation during action observation (Fadiga et al. 1995).

Unfortunately, direct-mapping theories, including AIM, cannot account for certain findings in human imitation behaviour. For example, 18-month-old children do not only re-enact an adult's action, but are also able to infer what the adult intended to do when the model fails to perform a target act (Meltzoff 1995). These findings suggest that young children apprehend the equivalence between acts seen and acts done not only on an inter-modal sensorial level, but also on a higher cognitive, intentional level. While direct mapping can cope with *that* finding by making a few additional assumptions, *other* robust findings are harder to explain using direct-mapping approaches. Imitation movements – especially in children – consistently and systematically deviate from the model movements. First of all, it is well documented that while young children spontaneously imitate adults in a mirror-like fashion, older children sometimes tend to transpose left and right (Swanson & Benton 1955; Wapner & Cirillo 1968). Hence, if direct-mapping is the basic process for imitation, it is either less 'direct' in younger children than in older ones, or it could be better called 'direct-mirroring' in younger children than 'direct-mapping.' Secondly, a hand-to-ear test (originally developed for aphasics by Head in 1920) repeatedly showed that young children prefer to imitate both ipsi-lateral (e.g. left hand touching left ear) and contra-lateral (e.g. left hand to right ear) movements with an ipsi-lateral response (Schofield 1976). Clearly, it is not the movement (ipsi- vs. contra-lateral) that is mapped, because it is mapped inconsistently. However, Bekkering, Wohlschläger and Gattis (2000) found that children consistently reached for the "correct" ear.[1]

The reason for the avoidance of cross-lateral movements in children is not due to an immature bifurcation as Kephart (1960) suggested. Recently, we (Bekkering et al. 2000) showed that bimanual contra-lateral movements (i.e. left hand to right ear and at the same time right hand to left ear) are imitated contra-laterally quite often and more frequently than unimanual contra-lateral movements are, even though the bimanual movements require a double crossing of the body midline. In addition, we were able to show that unimanual contra-lateral movements are imitated contra-laterally if throughout the session only one ear is touched. Based on these findings, we speculated that children probably primarily imitate the goal of the model's action while paying less attention to – or not caring about – the course of the movement. However, if the goal is unambiguous (both ears are touched simultaneously) or if there is only one goal (only one ear is touched), then aspects of the movement come into play. In other words: in imitation it is primarily the goal of an act that is imitated; how that goal is achieved is of only secondary interest. Of course, perceiving the goal of an action would be a prerequisite for such a goal-directed imitation. Indeed, recent research showed that already 6-month-old infants selectively encode the goal object of an observed reaching movement (Woodward 1998). These results demonstrate that children perceive the goals and intentions of others from a very early age on.

We tested our hypothesis of goal-directed imitation by a variation of the hand-to-ear task that allowed the removal of the goal objects of the model's movement. Instead of touching the ears, the model now covered one of two adjacent dots stuck to the surface of a table with either the ipsi- or the contra-lateral hand. Results were similar to those of the hand-to-ear task. Children always covered the correct dot; but they quite often used the ipsi-lateral hand when the model covered the dot contra-laterally. However, when the same hand-movements were performed with the dots removed, children imitated almost perfectly ipsi-lateral with ipsi-lateral and contra-lateral with contra-lateral movements.

Thus, it seems that in imitation the presence or absence of goal object has a decisive influence on imitation behaviour. Goal-oriented movements seem to be imitated correctly with respect to the goal; but the movement itself is frequently ignored. Movements without goal objects or with a single, non-ambiguous goal object are imitated more precisely. It seems that if the goal is clear (or absent), then the course of the movement plays a more central role in imitation. One might also say that then, the movement itself becomes the goal.

## 2.    A goal-directed theory of imitation

Based on these results, we developed a theory of goal-directed orientation that nevertheless does not make a principle differentiation between object-oriented movements and movements lacking a goal object. It rather suggests

a.  *Decomposition.* The perceived act is cognitively decomposed into separate goal aspects;

b.  *Selection of goal aspects.* Due to capacity limitations, only a few goal aspects are selected;

c.  *Hierarchical organisation.* The selected goal aspects are hierarchically ordered. The hierarchy of goals follows the functionality of actions. Ends (objects and treatments) are more important than means (effectors and movement paths);

d.  *Ideo-motor principle.* The selected goals elicit the motor program with which they are most strongly associated. These motor programs are not necessarily leading to matching movements;

e.  *General validity.* There is no principle difference in imitation behaviour between children, adults, and animals. Differences in accuracy are due to differences in working memory capacity.

The goal-directed theory of imitation does not only explain the recent data of imitation research, but also gives imitation a more functional nature. Direct mapping, on the other hand, has a rather automatic taste. The goal-directed theory of imitation allows imitators to learn from models even if the differences in motor skills or in body proportions are so huge that the imitator is physically unable to make the same movement as the model. Whatever movement the imitator uses, the purpose of learning by imitation can be regarded as being fulfilled as soon as he reaches the same goal as the model.

## 3.   Experiments

The series of experiments presented here provides further evidence for the theory of goal-directed imitation. Experiment 1 tests the *ideo-motor principle* in children's imitation behaviour. Experiment 2 and 3 test the *general validity* of our goal-directed theory of imitation by using adult subjects instead of children. Experiment 2 replicates the dot experiment (see above) with adults and thus tests the *general validity* of the *ideo-motor principle*. Experiment 3 tries to clarify the *hierarchical organisation* by investigating the imitation of more complex object-oriented actions in adults and thus tests the *general validity* of *selection of goals aspects* in imitation.

### Experiment 1

According to the *ideo-motor principle*, the movements elicited by the goal of an action are those that are most strongly associated with the achievement of the goal. We already showed (Bekkering et al. 2000) that contra-lateral movements are quite frequently imitated with ipsi-lateral ones (so-called contra-ipsi-error). This finding is in keeping with the *ideo-motor principle*, because it is quite likely that the more direct, ipsi-lateral movement is more strongly associated with reaching for an object than the indirect, contra-lateral one. Experiment 1 tries to show, that the contra-ipsi-error is not the only manifestation of the *ideo-motor principle* in imitation. One way to show that the *ideo-motor principle* is of more general validity in imitation is to use the same spatial relations between effectors and objects (again ipsi- and contra-lateral movements), but varying in addition the treatment of the object. Prehension movements, for example, should be more strongly associated with the use of the more skilled dominant hand, whereas the choice of the effector in pointing to an object should depend more on the spatial relation between effector and object.

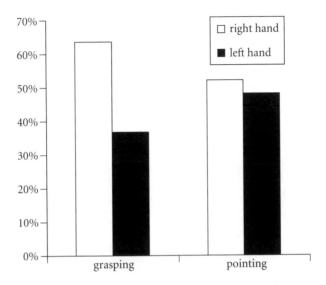

**Figure 1.** Results of Experiment 1. Irrespective of the fact that the model used both hands with equal frequency, children prefer to use the right hand when imitating the grasping of objects. No such hand preference is observed when children imitated pointing towards the objects.

We therefore asked 16 children to imitate contra- and ipsi-lateral movements towards one of two objects (a comb and a pen) on the table. Half of the children were shown grasping movements, whereas the other half observed the experimenter pointing towards the objects. In the grasping condition, the dominant hand was used about twice as often as the left hand, whereas in the pointing condition hand use was balanced (see Figure 1). Interestingly, the preference for the dominant hand in the grasping condition led to 17% ipsi-contra-errors (all of them were made with the right hand), an error that hardly occurred in previous experiments. No such errors occurred in the pointing condition. In summary, the results of Experiment 1 demonstrate the goal-directedness of imitation, the strength of the ideo-motor principle in imitation, and that not only the objects identity (comb vs. pen), but also the treatment of the object (grasping vs. pointing) determines which motor programme is activated.

Experiment 2

The theory of goal-directed imitation is thought to be valid for all individuals, irrespective of age and developmental state. However, up to now, our own evidence stems exclusively from imitation research in children. Of course, in such simple tasks like touching the contra-lateral ear, we don't expect adults to show the same error-prone behaviour that we found in children. Nevertheless, if the goal-directed theory of imitation is generally valid, some (perhaps weaker) effects in adult's imitation behaviour should be detectable. We (Wohlschläger & Bekkering 2002) therefore replicated one of our core experiments – covering dots on a table – in adults, expecting to find a reflection of the children's error pattern at a lower level in adults and in their response times (RT).

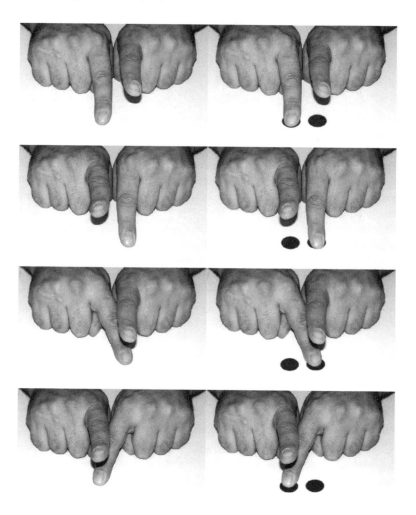

**Figure 2.** Stimuli used in Experiment 2. The adult participants had to put their hands in a position similar to that depicted in the stimuli photographs. Note that the only difference between the left and right column of photographs is in the presence of dots.

In order to be able to measure response times precisely, we slightly modified the task. First, we used finger movements instead of whole hand movements. Second, the model movements were not presented by the experimenter but on a computer screen. Subjects were instructed to put their hands next to each other on the table, just as depicted in the stimuli (see Figure 2), and to imitate the depicted downward finger movement as quickly as possible after the presentation of one of the stimuli. As in the experiment with children, there were two conditions. In one condition, the stimuli contained two dots, one of which was covered by one of the fingers at the end of an either ipsi-lateral or contra-lateral downward movement. In the other condition, the stimuli depicted the same movements, but there were no dots present.

Twelve adult subjects went through both, the dots and the no-dots condition, in blocks. Results showed that although adults almost made no errors (0.6%), these few errors mainly (77.8%) occurred with stimuli depicting contra-lateral movements towards dots (contra-ipsi error). Second, RT were faster for ipsi-lateral movements, but only if dots were present (see Figure 3). These results, that basically replicate the findings in children, show that also in adults dots as action goals are activating the direct, ipsi-lateral motor programme, which leads to faster responses and sometimes even to errors.

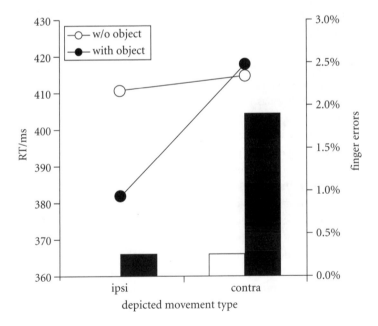

**Figure 3.** Results of Experiment 2. Imitating contra-lateral was slower and more finger errors were made, but only if dots were present.

## Experiment 3

Although adults show the same effects as children in simple actions, more complex actions are needed to investigate the *general validity* and the *hierarchical organisation* of our goal-directed theory of imitation. Currently, we have only data from actions that comprise two variable aspects: the goal object and the effector. In the following experiment, we increased the number to four variable aspects: the goal *object*, the *treatment* of the object, the *effector*, and the *movement path*. Given the higher complexity, we expected also adults to show a substantial number of errors in imitation. Ranking the different error types according to the number of errors should yield insight into the hierarchy of goal aspects. We expected the subjects to show the least number of errors in choosing the object and the treatment, whereas the choice of the effector and the movement path should be quite error-prone.

The action we used was more complex but nevertheless quite simple. It consisted of moving a pen upside down into one of two cups (*object*) or touching the cup's handle with the pen's cap (*treatment*). In either case, the pen had to rotated by 180°. The experimenter served as the model and he either used his right or his left hand (*effector*). In addition, he either turned the pen clockwise or counterclockwise (*movement path*) to bring it into an upside down position at the end of the movement (see Figure 4).

32 adults served as participants in the experiment. They were kept naïve about the purpose of the experiment. Before showing the action to them, they were simply asked "Can you do what I do?" We were interested in "spontaneous imitation" and therefore we ran only one trial for each subject.

The results showed that indeed, adults produced a considerable amount of errors if the action that has to be imitated gets more complex. Actually, only 10 subjects exactly copied the model's movement. As suggested by the theory of goal-directed imitation, most errors were made due to using the wrong *movement path*, followed by the wrong *effector*. The *treatment* was almost always imitated correctly. The error rate for the *object* depended on whether the two cups were of the same or of different colour. If the objects had the same colour, subjects randomly chose one of the objects. However, if the objects had different colours, the cup with colour corresponding to the cup the experimenter used was chosen (see Figure 5). This last finding illustrates the goal-oriented nature of imitation. If an object is uniquely identifiable, it is considered the unique goal of an action. However, if there are several similar ones around, imitation picks out an arbitrary one, ignoring the location of the object. As a consequence, the choice of the effector and the movement path just follow what is necessary to achieve the goal. In fact, subjects almost always used their right hand to grasp the pen and put it into the cup (thus replicating the finding of Experiment 3 in adults).

**Figure 4.** Three frames of an imitation sequence of Experiment 3. The model on the right uses the left hand to put the pen upside down into the right cup by turning it counter-clockwise. The imitator uses the right hand and turns the pen clockwise to put it into the left cup (not shown). In this example, the imitator perfectly mirrored the model movement. However, most subjects failed to do so. See text for details.

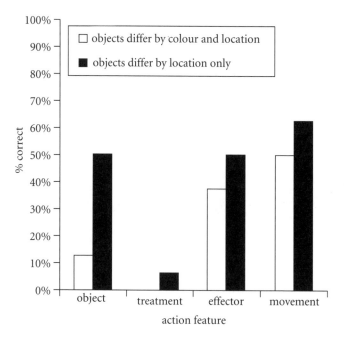

**Figure 5.** Results of Experiment 3. It was mainly the treatment of the object (pen went into the cup vs. pen touched its handle) that was imitated correctly. The correct[1] cups was only used, if the cups differed by colour. Otherwise, the choice of the cup as well as the choice of the effector and the movement path were basically at chance level.

## 4.  Discussion

In this chapter we reported a series of experiments that demonstrate the importance of objects and their treatments in human imitation, both for children and adults. The experiments showed that it is primarily the treatment of an object that is imitated in object-oriented actions, whereas the choice of the effector and the movement path are following the so-called *ideo-motor principle*: The motor programme most strongly associated with the achievement of the goal is activated during the execution of the imitative act and it is probably already executed during the observation of the action that is imitated later on (Fadiga et al. 1995). This motor programme leads in most cases to the most direct and effective movement. In contrast to current theories that explain imitation by a direct mapping of visual input onto motor output, our new goal-directed theory of imitation states that the matching takes place between action goals. Actions involving objects are thus imitated in such a way, that the same treatment is done to the same object, thereby ignoring the motor part of the action. Of course, in most cases the model acts in an efficient and direct way on the object. If the imitator copies the action goal and if this action goal in turn activates the most direct motor programme in the imitator, then both actions resemble each other in all aspects, leading to an impressive, mirror-like behaviour. When there is no object, the movements themselves become the goal and they are also imitated in a mirror-like fashion. It is probably the frequently observed parallelism between the movements of the model and the imitator, that led to direct-mapping theories. However, according to our new theory of goal-directed imitation, this similarity between the movements of the model and the imitator is only superficial and incidental: the underlying similarity is a similarity of goals and intentions.

Imitating goals and/or intentions of course requires that the imitator understands the action of the model. In our view, thus action understanding is a prerequisite for imitation. It is a necessary but not a sufficient condition for imitation to occur: Within a goal-directed theory (as opposed to direct-mapping explanations) it is possible to explain why imitation sometimes occurs and sometimes not. *Because* action understanding precedes imitation the observer can decide whether or not he wants to imitate the goals and intentions of the model. In addition, a goal-directed theory of imitation also gives room to creativity in imitation, because the way the goal is achieved is left to the imitator, whereas direct-mapping approaches have a rather automatic taste. Observing the imitator achieving the same goal in a more efficient way in turn might cause the model to imitate the new movement of the former imitator. This type of creativity, based on the decoupling of ends and means and on mutual imitation, probably plays a very important role in the evolution of culture and technique.

Further evidence for our view that action understanding is the main ingredient for imitation comes from neurobiology. The recently discovered so-called mirror-neurones in the macaque monkey (di Pellegrino et al. 1992) can be considered a neural system for action understanding. These neurones, located in the rostral part of the monkeys pre-motor cortex (area F5), discharge during the observation of object-oriented actions. Each single unit seems to code a particular object manipulation, e.g. a precision grip on a seed. The cells do not fire if a different movement type (e.g. whole hand grip) is shown or if a different object is manipulated (e.g. larger object). Thus monkeys seem to have a neural system for action understanding, at least for object-oriented ones.

Interestingly, a mirror-neurone does not only fire during action observation, but also if the monkey executes the action the neurone is tuned for. Mirror-neurones were thus thought to be a good candidate for playing a role imitation. However, monkeys do not imitate, but their relatives, the great apes and humans do. Recently, a fMRI study (Iacoboni et al. 1999) showed that the human homologue of the monkey's F5 mirror-neurone area is particularly active during the imitation of finger movements. One might speculate that during the evolution of species, first a system for action understanding had to be evolved before the imitation of action goals could evolve.

## Note

1. "Correct" has to be understood in the mirror-sense, because children spontaneously imitate ipsi-lateral movements in a mirror-fashion.

## References

Bekkering, H., Wohlschläger, A., & Gattis, M. (2000). Imitation of gestures in children is goal-directed. *The Quarterly Journal of Experimental Psychology: Section A: Human Experimental Psychology, 53*, 153–164.

Butterworth, G. (1990). On reconceptualising sensori-motor development in dynamic systems terms. In H. Bloch & B. I. Bertenthal (Eds.), *Sensory Motor Organizations and Development in Infancy and Early Childhood* [NATO Advanced Science Institutes series, D: Behavioural and Social Sciences, 56] (pp. 57–73). Dordrecht, Netherlands: Kluwer Academic Publishers.

Fadiga, L., Fogassi, L., Pavesi, G., & Rizzolatti, G. (1995). Motor facilitation during action observation: A magnetic study. *Journal of Neurophysiology, 73*, 2608–2611.

Gray, J. T., Neisser, U., Shapiro, B. A., & Kouns, S. (1991). Observational learning of ballet sequences: The role of kinematic information. *Ecological Psychology, 3*, 121–134.

Head, H. (1920). Aphasia and kindred disorders of speech. *Brain, 43*, 87–165.

Iacoboni, M., Woods, R. P., Brass, M., Bekkering, H., Mazziotta, J. C., & Rizzolatti, G. (1999). Cortical mechanisms of human imitation. *Science, 286*, 2526–2528

Kephart, N. C. (1960). *The Slow Learner in the Classroom*. Columbus, OH: Charles E. Merrill.

Meltzoff, A. N. (1993). The centrality of motor coordination and proprioception in social and cognitive development: From shared actions to shared minds. In J. Geert & P. Savelsbergh et al. (Eds.), *The Development of Coordination in Infancy* [Advances in Psychology, 97] (pp. 463–496). Amsterdam, Netherlands: North-Holland/Elsevier Science Publishers.

Meltzoff, A. N., & Moore, M. K. (1977). Imitation of facial and manual gestures by human neonates. *Science, 198*, 75–78.

Meltzoff, A. N., & Moore, M. K. (1994). Imitation, memory and the representation of persons. *Infant Behavior and Development, 17*, 83–99.

Pellegrino G. di, Fadiga, L., Fogassi, L., Gallese, V., & Rizzolatti, G. (1992). Understanding motor events: A neurophysiological study. *Experimental Brain Research, 91*, 176–180.

Schofield, W. N. (1976). Do children find movements which cross the body midline difficult? *Quarterly Journal of Experimental Psychology, 28*, 571–582.

Swanson, R., & Benton, A. L. (1955). Some aspects of the genetic development of right-left discrimination. *Child Development, 26*, 123–133.

Wapner, S., & Cirillo, L. (1968). Imitation of a model's hand movements: Age changes in transpositions of left-right relations. *Child Development, 39,* 887–895.

Wohlschläger, A., & Bekkering, H. (2002). Is human imitation based on a mirror-neurone system? Some behavioural evidence. *Experimental Brain Research, 143,* 335–341.

Woodward, A. L. (1998). Infants selectively encode the goal object of an actor's reach. *Cognition, 69,* 1–34.

# The mirror system and joint action

Günther Knoblich and Jerome Scott Jordan
Max Planck Institute for Psychological Research, Munich, Germany /
Illinois State University, Normal, USA

## 1. Introduction

The most exciting aspect of the discovery of mirror neurons is that parts of the cognitive system are entirely devoted to the processing of social information (Fadiga, Fogassi, Pavesi, & Rizzolatti 1995; Gallese, Fadiga, Fogassi, & Rizzolatti 1996; Rizzolatti & Arbib 1998). For a single being like a monkey on mars, the mirror system would be useless because it is specialized in detecting the action of its peers and matching them to the monkey's own action repertoire. However, the existence of a mirror system in macaque monkeys is also puzzling in a sense. Why are the same monkeys who possess a special system to process social information quite poor in coordinating their actions with other monkeys? What could the mirror system be good for if not for action coordination? Why does this system not allow the monkeys to speak to each other or to trade shares on the stock market? Is it only because they did not develop organs to produce language yet? To be sure, monkeys show some behaviors that allow for coordinated group action. But these behaviors tend to be quite inflexible. A particularly puzzling finding is that the mirror system does not even enable them to imitate the actions of their peers, although the mirror metaphor strongly suggests that they should be able to.

The issue of imitation is treated elsewhere in this book (cf. Wohlschläger & Bekkering, this volume; see also Bekkering, Wohlschlaeger, & Gattis 2000; Iacoboni et al. 1999). We will focus on the issue of coordination of self- and other-generated actions in joint action, that is, in situations where neither member of a group can achieve a common goal on his own but only with the help of the other member. To treat this issue, we first provide some principled arguments to make the general case that a system providing a perception-action match is not *sufficient* for successful coordination of self- and other-generated actions. Nevertheless, the perception-action match provided by this system may well be a *necessary* condition for certain forms of action coordination. We then propose a functional mechanism for action coordination that extends the functionality of the mirror system by modulating the planning of one's own action in response to perceiving the outcomes of somebody else's actions. Although this mechanism is inherently non-linguistic, it may also be a bridging element in the transition from a system that produces and understands manual actions to a sophisticated language faculty (Calvin & Bickerton 2000; Rizzolatti & Arbib 1998). Finally, we will provide some empirical for the existence of such a mechanism in humans.

## 2.    Ego-centered and group-centered action understanding

The most obvious interpretation of the functioning of the mirror system is that it is specialized in processing information about object-directed actions of others that match actions in the observer's own repertoire. At a closer look, this interpretation is quite fuzzy. What exactly is matched? Is the match based on the kinematics of movements, anatomical cues, object features, or the object-actor relationship (Barresi & Moore 1996; Gallese 1998)? We would like to suggest that it is not the observed movement per se that is matched with the observer's own action. Rather, it is the perceived effect the action exerts on the object that is matched to a possible effect that could be also exerted by one of the observer's own actions. As a consequence, the informational content that is processed by the mirror system is best described in terms of perceivable action effects, as suggested by the functional notion of a common coding system for perception and action (Hommel, Müsseler, Aschersleben, & Prinz, in press; Prinz 1997). This notion has been quite successful in explaining different phenomena in the area of human action perception and action planning. An empirical finding supporting the action effect interpretation is that the mirror system is silent when the monkey observes movements in the absence of an object.

One implication of the action effect notion is that the kind of action understanding the mirror system provides is ego-centered and does not necessarily include an explicit representation of another agent. As a consequence, organisms that are equipped with a mirror system may have the ability to understand that objects are affected in a way in which they could also affect them, but they may not understand that the peer who is producing the action is an agent like themselves. If this is true, the obvious function of the system would then be to notice that something is exerting an influence on objects that the organism itself is interested in (for instance, "a banana becoming vanished"). It is easy to see that such a system may help one avoid loosing interesting objects, especially when the environment is crowded by other beings interested in the same objects. Hence, the type of action understanding provided by the mirror system may be purely egocentric, and its main function to grasp interesting (eatable) objects before they are gone.

However, there are situations in which a purely egocentric perspective is not sufficient to successfully attain the desired effects on the environment. A subset of these effects can be obtained when joining efforts with a peer. In other words, in certain situations joint action (Clark 1997) allows one to achieve goals that could not be achieved otherwise. For instance, a shared goal could be to remove a heavy object that is blocking the way, in order to proceed. If one individual tries to move the object, the effect obtained might be rather small. If two individuals join their efforts and push the object simultaneously, the effect obtained will be much larger. Hence, the joint effect will be much larger than the sum of the individual effects. Alternatively, if the goal is to jointly steer a canoe towards a certain location, a group is more likely to achieve this, when paddling in turn. Otherwise, it is very likely that the actions of one individual counteract the actions of the other individual, thereby keeping the group from achieving the common goal. Independently of whether the situation requires a group to act simultaneously or in turns, coordination of self- and other-generated actions is required.

## 3.  A mechanism for action coordination

A simple perception-execution match will not allow an individual to successfully coordinate his/her own actions with somebody else's. As mentioned above, purely egocentric forms of action understanding prevent rather than support coordinated action of two individuals. Nevertheless, the ability to match observed actions to one's own action repertoire is a likely prerequisite for action coordination. What is lacking is a coordination mechanism that modulates one's own actions in response to perceiving the effects of a peer's actions. Moreover, such a coordination mechanism requires separate representations of joint action effects and individual action effects. In other words, organisms engaging in joint action have to keep apart what the group achieves as a whole from what they achieve themselves. Hence, joint action requires group-centered action understanding in addition to egocentric action understanding. If one thinks of the mirror system as coding individual action effects (observable changes in the environment that might happen as the consequence of one's own actions), only two steps are needed to add a mechanism for action coordination (see Figure 1).

The first step is to add codes that represent joint action effects (observable changes in the environment that are partly but not fully controlled by one's own actions). The second step is to couple the joint and individual effect level by excitatory or inhibitory connections. Whether these connections will be excitatory

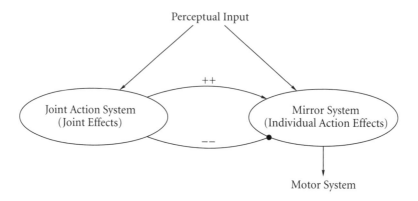

Figure 1. Model for modulation of the mirror system by joint effects.

or inhibitory depends on whether a certain code on the mirror level might trigger actions that increase or decrease the probability with which the desired joint effect occurs. If activation of the codes on the mirror level increases the probability with which a desired joint effect occurs, as in the push-together example, an excitatory connection between the joint effect code and the individual effect code will arise. As a consequence, observing the joint effect will increase one's own tendency to carry out certain actions that are apt to increase the effect, irrespective of whether the organism produced the effect itself or another organism produced it. Alternatively, if one's own actions decrease the probability with which a certain joint effect occurs, as in the canoe example, observing the joint effect will decrease the tendency to carry out the action.

## 4.  Empirical evidence from action coordination in humans

One implication of the proposed mechanism is that action coordination in groups should be difficult when individual and joint effects are not easily distinguishable, especially when action conflicts arise, that is, when one group members' actions can interfere with the other group member's actions that could produce the desired joint effect. Alternatively, action coordination should be successful whenever individual (self- and other-generated) effects and joint effects are easily distinguishable, because an inhibitory connection between joint and individual effect level will be acquired whenever one's own actions are not adequate to achieve the desired joint effect. Therefore, groups should learn to coordinate conflicting actions when joint and individual effects are easily distinguishable, but not otherwise. The situation is different for individuals who are able to resolve an action conflict on their own. In this case, performance can be optimized without the additional representation of joint effects. Therefore, the individual condition provides a baseline for the group conditions, because the action conflict can be resolved without the processing of social information (without taking the other's actions into account). Our prediction is that groups learn to coordinate their actions as well as individuals if individual and joint effects can be clearly distinguished, but perform worse if that is not the case.

We used a simple tracking task in all experiments (see Figure 2). A target moved across the computer screen horizontally with constant velocity. As soon as it reached a screen border it changed its direction abruptly and moved back towards the other border, changed its direction again, and so on. The task was to keep a tracker on the target by controlling its velocity with two keys. When the tracker was moving to the right, hitting the right key accelerated it by a constant amount and hitting the left key decelerated it by the same amount. When the tracker was moving to the left, hitting the left key accelerated it and hitting the right key decelerated it. Within the border regions, a conflict arose between two alternative strategies. The first alternative, i.e., trying to stay on target as long as possible, minimized the immediate error up until the point at which the target changed its direction. Afterwards, a large error arose because tracker velocity could only be changed gradually. Several key-presses were needed to stop the tracker and more were needed to gain velocity in the opposite direction. During this interval, the target continued moving in the opposite direction, constantly increasing the distance between itself and the tracker. Thus, trying to minimize immediate error created a large future error.

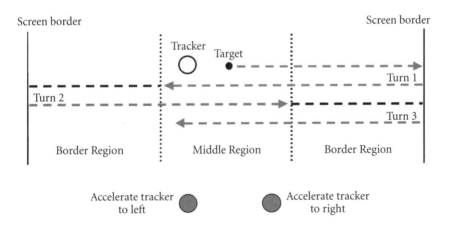

**Figure 2.** Illustration of tracking task.

The second alternative was to slow down the tracker before the target turned. In this case, the immediate error was increased to prevent future error. This is the case because the target continued to move toward the border as the tracker decelerated with each key-press. Using the latter strategy is the only way to improve performance within the border region. We refer to key-presses that decreased immediate error as compensatory presses, and those that increased immediate error in order to reduce future error, as anticipatory brakes (see Figure 3).

We used two versions of the task that differed in one single aspect. In one version each key-press triggered an acoustical signal, e.g., a left key triggered a high tone, and a right key triggered a low tone. Hence, joint effects (tracker movement) and individual effects (tones) were clearly distinguishable. In the other version, there were no tones and therefore joint and individual effects were hard to distinguish. We investigated performance of individuals and groups for each version of the task. In the individual condition each person controlled both keys, in the group condition each person controlled one key. Twenty individuals and twenty groups were asked to optimize performance without the acoustical signal, fifteen individuals and fifteen groups were asked to optimize performance with the acoustical signal. In both experiments, participants in the group condition were divided by a partition. They could neither see nor talk to one another. However, each partic-

before                          after

a) Effect of a compensatory button press.

before                          after

b) Effect of an anticipatory button press.

**Figure 3.** Effects of compensatory and anticipatory button presses. The small solid circles stand for the target, the larger transparent circles for the tracker. The length of the arrows indicates the target and tracker velocity. In the actual paradigm the tracker moved on a level with the target.

ipant was provided with a separate computer monitor, and all events taking place during the experiment (e.g. the movements of the tracker and the movements of the target) were presented simultaneously on both monitors.

We will focus on two measures, a performance measure and one for the extent to which the anticipatory strategy was used. As a performance measure, we computed the absolute distance between tracker and target in the border regions. We restricted the evaluation to the border regions, because these are the regions where individuals and groups had to coordinate conflicting actions.

Figure 4 shows the results for performance. Different points on the x-axis refer to different blocks. Hence, the progressions of each line illustrate learning effects. Each line in the graph stands for one experimental condition, that is, they show the performance of individuals who received (Individual +) or did not receive (Individual –) auditory feedback, and the performance of groups who received (Joint +) and did not receive (Joint –) auditory feedback. We will only describe and discuss the results that turned out to be statistically significant in appropriate analyses.

During the initial trials (Block 1) groups performed worse than individuals irrespective of whether the acoustical signal was present or not. However, while the

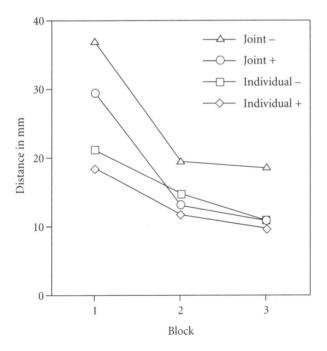

**Figure 4.** Tracker-target-distance for individuals and groups with and without acoustical signal across blocks.

acoustical did not make a difference in the individual condition, it helped groups to perform better right from start. Individual as well as group performance improved in later blocks. The decrease was almost linear in the individual conditions, that is, individual performance improved gradually. Group performance increased more than individual performance across consecutive blocks, especially from Block 1 to Block 2. However, while groups receiving the additional acoustical information reached the performance level of individuals after the first block, groups who did not receive such a signal did not reach that level. In fact, the latter group never reached a performance level that was better than the initial individual performance.

The performance data indicate that groups could use the acoustical signal to better coordinate their actions to implement a joint anticipatory strategy. Looking at a further dependent variable allows one to assess more directly whether individuals and groups got better by such a strategy. The extent to which the anticipatory strategy was employed within the boundary regions can be defined as the proportion of anticipatory button presses (see Figure 2, panel b) occurring within that region.

Figure 5 shows the result of the analysis of the anticipatory brake rate. Again, each line stands for one experimental condition, that is, individuals who received

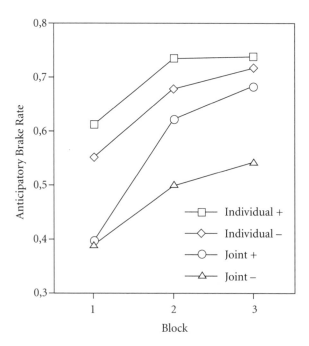

**Figure 5.** Anticipatory brake rate for individuals and groups with and without acoustical signal across blocks.

(Individual +) or did not receive (Individual –) auditory feedback, and groups who received (Joint +) and did not receive (Joint –) auditory feedback. The progression of each line reflects changes in anticipatory brake rate across blocks.

During the initial trials (Block 1), groups and individuals were clearly different and the presence of the acoustical signal did not make a real difference for either of them. Individuals started out at a relatively high level right from start and the initial anticipatory brake rate was clearly lower in the group conditions. In later blocks, the anticipatory brake rate increased for individuals as well as groups. However, while the presence of the acoustical signal did not affect the increase in the anticipatory brake rate in the individual condition, it did affect the increase in the group condition. The anticipatory brake rate increased sharply from the first to the second block in the group condition in which the acoustical signal was present. In the final block it was almost as high in the individual conditions. In contrast, the anticipatory brake rate increased only slightly for groups who did not receive the acoustical signal, the extent of the increase being comparable to the individual conditions.

The results for the performance and the strategy measure are consistent with the predictions. Groups learned to coordinate their actions as well as individuals when individual and joint effects could be clearly distinguished, but performed worse if that was not the case, whereas individuals did not benefit from the additional acoustical action effect. At present, there are probably alternative ways of explaining the results. Further experiments are needed to unambiguously assess whether the proposed mechanism (modulation of activity in the mirror system by joint effects) exists.

## 5.   Conclusions

The notion of joint action (Clark 1997) might prove useful in understanding how a sophisticated language faculty developed from an earlier system for action understanding (Rizzolatti & Arbib 1998). This notion suggests that the successful coordination of self- and other-generated actions might have provided an evolutionary advantage because coordinated action allows creatures achieving effects in the environment that cannot be achieved by individual action alone. Although the mirror system is not sufficient for successful action coordination, it is easy to see how an additional system that codes joint action effects might modulate activation of codes on the mirror level in order to coordinate self- and other-generated actions. Naturally, these evolutionary considerations are largely speculative (but not more so than other evolutionary considerations). Empirical evidence from action coordination in humans is also consistent with the assumption that the mirror system might be modulated by representation of joint effects. A further implication of the present study is that it might be useful to look at other forms of coordinated action than imitation to get a grip on how language developed from manual actions.

## Acknowledgements

Günther Knoblich, Cognition and Action, Max Planck Institute for Psychological Research, Amalienstrasse 33, 80799 Munich, Germany. Scott Jordan, Department of Psychology, Saint-Xavier-University, 3700 West 103rd Street, Chicago, IL 60655, USA. We thank Rüdiger Flach for helpful comments and Irmgard Hagen, Patrick Back, and Lucia Kypcke for their help in collecting the data. Correspondence concerning this article should be addressed to Günther Knoblich, Max-Planck-Institut für psychologische Forschung, Amalienstrasse 33, 80799 Munich, Germany. Electronic mail may be sent via Internet to knoblich@mpipf-muenchen.mpg.de.

## References

Barresi, J., & Moore, C. (1996). Intentional relations and social understanding. *Behavioral and Brain Sciences, 19*(1), 107–154.

Bekkering, H., Wohlschlaeger, A., & Gattis, M. (2000). Imitation of gestures in children is goal-directed. *Quarterly Journal of Experimental Psychology. A, Human Experimental Psychology, 53A*(1), 153–164.

Calvin, W. H., & Bickerton, D. (2000). *Lingua ex Machina: Reconciling Darwin and Chomsky with the human brain.* Cambridge, MA: MIT Press.

Clark, H. H. (1997). *Using Language.* Cambridge, U.K.: Cambridge University Press.

Fadiga, L., Fogassi, L., Pavesi, G., & Rizzolatti, G. (1995). Motor facilitation during action observation: A magnetic stimulation study. *Journal of Neurophysiology, 73*(6), 2608–2611.

Gallese, V. (1998). Mirror neurons and the simulation theory of mind-reading. *Trends in Cognitive Sciences, 2*(12), 493–501.

Gallese, V., Fadiga, L., Fogassi, L., & Rizzolatti, G. (1996). Action recognition in the premotor cortex. *Brain, 119,* 593–609.

Hommel, B., Müsseler, J., Aschersleben, G., & Prinz, W. (in press). The theory of event coding: A framework for perception and action. *Behavioral and Brain Sciences.*

Iacoboni, M., Woods, R. P., Brass, M., Bekkering, H., Mazziotta, J. C., & Rizzolatti, G. (1999). Cortical mechanisms of human imitation. *Science, 286*(5449), 2526–2528.

Prinz, W. (1997). Perception and action planning. *European Journal of Cognitive Psychology, 9*(2), 129–154.

Rizzolatti, G., & Arbib, M. A. (1998). Language within our grasp. *Trends in Neurosciences, 21*(5), 188–194.

# Brain activation to passive observation of grasping actions

Francis McGlone, Matthew Howard, and Neil Roberts
Centre for Cognitive Neuroscience, University of Wales, Bangor and
Unilever Research, Wirral, U.K. / Magnetic Resonance and Image
Analysis Research Centre, Liverpool University, U.K.

## 1. Introduction

The recent description of a group of neurones located in areas of monkey premotor cortex (F5) that fire not only when the animal passively observes stereotypical behaviours (grasping, eating), but also if the animal performs similar actions (Rizzolatti & Arbib 1998), has aroused a great deal of interest across many science disciplines. Neuroscientists, linguists, evolutionary and experimental psychologists, functional neuroanatomists and ethologists have all recognised that the discharge characteristics of these neurones provide a tool by which we can probe the basis of conspecific, non-verbal communication, and possibly the evolution of language itself (Corballis 1999). This ecumenical fervour is due in part to the fact that the human equivalent of this area is hypothesised to lie in Broca's area (Passingham 1993), a region known to be involved in speech production. Several lines of evidence point to this region playing a role in communication, and Gallese and Goldman (1998) have proposed that this cortical region operates as an observation/execution matching system (OEMS). Based on evidence provided by single unit recordings from monkey area F5, two classes of neurones have been described: one class, the canonical neurones, respond when the monkey observes graspable objects, and the second class, termed 'mirror neurones' (MN) (Rizzolatti & Gentilucci 1988) respond to both the observation of an action and its self-generated action. The types of motor acts that can provoke these responses in MN's are highly specific and their repertoires have so far been limited to studies of reaching, grasping, hand rotation and eating.

The existence for a similar system in man could have been inferred from the pioneering work of one of the first electromyographers, E. Jacobson (the first scientist to record eye movements during sleep), who in the early part of the last century recorded "minute" voltages from somatic muscles when his subjects imagined performing specific arm movements. Only the muscle groups that would have generated these movements produced the signals. More recently, employing transcranial magnetic stimulation (TMS) of the motor cortex, coupled with electromyographic recording from arm muscles, Fadiga et al. (1995) demonstrated an enhancement of motor evoked potentials when subjects were observing an actor grasping an object. This enhancement was only found in recordings from those muscles that would normally be recruited if the subjects performed the observed task themselves. These studies provide evidence that a 'mirror' system exists in man, but give no clue as to its anatomical localisation. However, in two recent PET studies (Rizzolatti et al. 1996b; Grafton et al. 1996), brain areas activated by the observation of grasping movements *and* their performance, or imagined performance, have been found in the inferior frontal gyrus (BA 44/45, monkey F5), as well as pre-motor and SMA (BA6). Neurones that respond to the observation of stereotypical hand movements have been found in the superior temporal sulcus (STS) of the macaque by Perrett et al. (1985 and 1989), and share many of the properties of MN's: they code similar actions; generalise to differences within the action and do not generalise to mimicked actions or goal alterations. However, the properties of the cells in monkey F5 differ from those found in STS in one fundamental respect – in F5 the same neural pattern of activation is found during observation *and* activation.

The high degree of stereospecific tuning found in MN's has been elegantly demonstrated by Rizzolatti et al. (1996a), where single neurones responding to the observation of the experimenters hand grasping a grape fail to respond if the object is grasped by a pair of pliers held by the experimenter. Goal directed actions, such as the experimenter eating a grape, have also been shown to correlate with the firing of single cells in this region, but that same cell will not fire if the goal directed action is meaningless i. e. if the grape is placed on the perioral area of the face. This degree of specificity would be required for a visuomotor encoding system that provided not only an interpretation of the meaning of observed actions, but also their meaningful replication i.e. a communication capability. In order to determine if this defining characteristic of monkey MN's is also found in the humans, we employed functional magnetic resonance imaging (fMRI), in a paradigm where we sought to identify homologous regions in the human brain. We tested two stereotypical actions; picking up a pile of coins or picking up a pen. However, the latter action had a confounded meaning.

## 2.  Methods

*Subjects.* Written consent was obtained for 13 right-handed healthy subjects (5 female; 8 male; mean age 28 years). All subjects had normal vision or wore contact lenses, and were paid for their time.

*Experimental Design.* For each experimental condition, 100 EPI brain volumes were collected over 300 seconds. This acquisition period was divided into 20 epochs, each of fifteen seconds in length. Subjects performed a baseline (OFF event) in the first epoch, followed by the task of interest (ON event). This ON/OFF or 'boxcar' design was repeated in subsequent epochs.

Each subject underwent fMRI scanning of two different experimental conditions. In the ON event of each condition, subjects passively viewed a 15 second QuickTime movie (Apple Computer, USA) of actors performing a stereotypical, goal directed motor behaviour. For the ON events, subjects viewed a close up of an actor's hand performing a precision grip task, picking up a pen from a desktop, or a number of small coins. In all conditions the OFF event was a static picture of the actor and object (see Figure 1).

*fMRI Stimulus Presentation.* All stimuli were generated by an Apple PowerMacintosh G3 running Psyscope software. An LCD projector (Epson LMP7300) was used to back-project these stimuli onto a screen positioned at the feet of the subject. Subjects viewed the screen down the bore of the magnet via an arrangement of mirrors.

*MR Data Acquisition.* Magnetic Resonance Imaging of the brain was performed using a SIGNA LX/Nvi Neuro-Optimised Imaging System (General Electric, Milwaukee, USA). A multi-slice localising sequence was used to prescribe 22 contiguous T2$^*$-weighted gradient-echo Echo-Planar Images (EPI) (TE = 40 ms, TR = 3 s, Flip angle = 90°, 5 mm thickness, FOV = $24 \times 24$ cm, $128 \times 128$ matrix), in an axial orientation. The EPI volume encompassed the entire frontal, parietal, and occipital lobes, and the superior aspect of the temporal lobe. Finally, a high resolution, T1-weighted 3D IR-PREP sequence was acquired (TR = 12.3 msec, TE = 5.4 msec, TI = 450 msec, 1.6 mm thickness, matrix = $256 \times 192$, FOV = $20 \times 20$ cm), which encompassed the whole brain. This scan provides anatomical detail and is used for spatial normalisation which is a prerequisite for SPM group analysis (detailed below).

*Data Processing.* Functional Group Analyses were performed using Statistical Parametric Mapping software (SPM99b, Friston et al. 1995). Several processing stages were necessary prior to statistical analysis. These were: (1) Motion Correc-

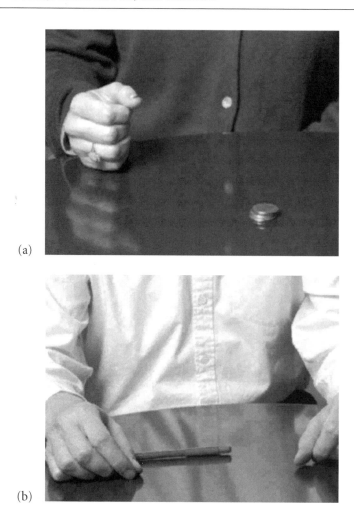

(a)

(b)

**Figure 1.** Still image of coins (a) and pen (b) shown during OFF period. During ON period objects were picked up in a 15 second video clip back-projected on to a screen placed at the end of the scanner. Subjects viewed the images through prism spectacles.

tion: Each of the functional volumes is realigned with the first one in the series, using a cubic spline interpolation technique. The algorithm also incorporates spin-history correction to minimise motion correlated intensity variations. (2) Spatial normalisation: The functional data were spatially transformed to a standardised canonical co-ordinate frame of reference based upon the bicommisural co-ordinate system of Talairach and Tournoux (1988). This is achieved by transforming the 3D high-resolution volume to a T1-weighted template, initially using a 12 parameter affine transformation, followed by a multi-parameter discrete cosine transform basis function to describe the deformations. (3) Spatial smoothing: The stereotactically normalised functional volumes were smoothed using an isotropic Gaussian kernel of 6mm full width at half maximum. This allowed for inter-individual variation in the location of functional activity, and conditioned the data such that they conformed to the random Gaussian field model (Worsley et al. 1996) underlying the statistical analysis. Finally, Gaussian smoothing increased the signal to noise ratio of the data, in a manner consistent with a haemodynamic response anticipated at a spatial extent of several mm. (4) Temporal Smoothing: High-pass filtering was used with a cutoff of 60 seconds to minimise slow temporal artifacts (e.g. respiration, scanner baseline drift). Low pass filtering employed a smoothing kernel modelled on the canonical haemodynamic response function.

*Group Data Analysis.* Statistical group analysis was performed within SPM99b using a multivariate General Linear Model (GLM), in which the main covariate of interest was a square wave depicting the ON/OFF profile of the design, convolved with the haemodynamic response function. After the elements of the GLM have been specified, SPM99b calculated a statistical t-score map showing brain regions that preferentially respond to the ON task compared to the OFF task. These brain regions, or *clusters* of voxels are assigned corrected significance probabilities according to the theory of Gaussian fields. In these group analyses, a corrected significance threshold of $p < 0.05$ was used.

## 3.   Results

### 3.1  Activations common to both conditions

Three distinct regions were activated in both conditions (Table 1a). A bilateral cluster of activation in the lateral occipital cortex, visual area V5/MT (BA39/37), a broad band of activation in the superior [LPs] (BA7) and inferior parietal lobules [LPi] (BA 40/2/7) and a further small cluster of activation, bilaterally, at the border of the middle frontal gyrus [MFg] and the precentral gyrus [GPrC] (BA6/8). More activation was found in the right rather than left hemisphere. The superior frontal gyrus [GFs] (BA6) was also activated in both conditions, however, in this region laterality differed as a function of condition. In the coin grasping condition, activation was seen in the right hemisphere, whereas the opposing hemisphere was activated in the pen grasping condition.

### 3.2  Condition-specific activations

1.  *Coin grasping condition.* Two distinct bands of activation extending from dorsal to ventral were present in this condition (Figure 2a). Several distinct clusters of activation were seen in the precentral gyrus [BA4], conjoined with activations in the parietal lobules. More anteriorly, a second band of activation was observed in the premotor cortices [BA6], extending ventrally from the superior frontal sulcus [BA6] to the dorsal aspect of the inferior frontal gyrus [BA44 & 9].

2.  *Pen grasping condition.* Activations in this condition were reduced in comparison to the coin grasping condition (Figure 2b). Only several additional small precentral clusters anterior to the sulcus of Roland (BA4) and a discrete cluster in the premotor cortex [BA6] were additionally activated.

### 3.3  Fixed effects analysis (regions with a cluster size greater than one voxel considered)

A fixed effects analysis was undertaken to identify voxels which were preferentially activated in the coin, as opposed to pen condition, and vice versa (Table 1b). Voxels

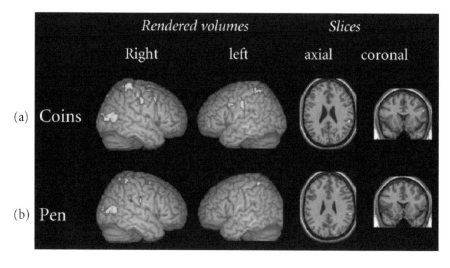

**Figure 2.** Averaged activation patterns found from watching coins (top row) and pen (bottom row) being picked up. The volume rendered images show the global activation produced by observing these actions, with more activation being generated by the coin condition. The coronal sections demonstrate clearly the lack of activity found in the pen grasping condition in IFG. Data thresholded at $p < 0.05$.

**Table 1a.** Areas of significant activation (p < 0.05, corrected) for all conditions.

| Region | R/L | BA | x | y | z | Z-score |
|---|---|---|---|---|---|---|
| Visual Area V5 | R | 19 | 45 | −66 | 6 | 6.84 |
| Visual Area V5 | L | 39 | −42 | −66 | 3 | Inf |
| Superior Parietal Lobule | R | 7 | 39 | −45 | 57 | Inf |
| Superior Parietal Lobule | L | 7 | −36 | −51 | 57 | Inf |
| Inferior Parietal Lobule | L | 40 | −60 | −27 | 21 | 5.92 |
| Inferior Parietal Lobule | R | 40 | 51 | −39 | 42 | 5.4 |
| Middle Frontal Gyrus/Sulcus | L | 44/6 | −57 | 6 | 36 | 5.92 |
| Middle Frontal Gyrus | R | 6/8 | 48 | 0 | 54 | 6.16 |
| Superior Frontal Gyrus * | R | 6 | 33 | −6 | 57 | 5.77 |
| Superior/Middle Frontal Gyrus + | L | 6 | −33 | −9 | 63 | 7.75 |
| Inferior Frontal Gyrus * | R | 44 | 57 | 15 | 27 | 5.16 |

(* = Coin condition only, + = pen condition only)

**Table 1b.** Fixed effects analysis (Voxels uniquely activated in coin condition).

| Region | R/L | BA | x | y | z | Z-score |
|---|---|---|---|---|---|---|
| Visual Area V5 | R | 19 | 48 | −75 | 18 | 6.69 |
| Visual Area V5 | L | 40/39 | −45 | −57 | 6 | 6.04 |
| Superior Parietal Lobule | R | 19 | 24 | −78 | 45 | 6.56 |
| Superior Parietal Lobule | R | 7/40 | 42 | −39 | 57 | 5.2 |
| Inferior Parietal Lobule | L | 40 | 63 | −48 | 33 | 5.75 |
| Middle/Inferior Frontal Gyrus | R | 44/6 | 51 | 6 | 39 | 5.92 |
| Precentral Gyrus | R | 4 | 63 | −15 | 36 | 5.84 |
| Extrastriate Cortex | R | 39 | 48 | −63 | 9 | 5.28 |

uniquely activated in the coin condition were found in the middle occipital gyrus, visual area V5, precentral gyrus, superior and inferior parietal lobules and inferior frontal gyrus. By contrast, only a small cluster in the superior temporal gyrus was activated in the pen as opposed to coin condition.

## 4.   Discussion

The data presented here shows that the passive viewing of behaviourally relevant actions performed by another human subject activates a network of prefrontal brain regions normally associated with direct control over the generation and implementation of movement (BA6, 9) and the generation of speech (BA44). The expectation that the former would be expected was first noted by Jacobson in the early

1930's (Jacobson 1930, 1932) where he describes, in a series of papers, the electro-physiological recording of "microscopic" contractions in voluntary muscles concordant with subjects' imagining or observing stereotypical movements, such as serving a tennis ball. The more recent observations of enhanced motor evoked potentials (MEP's) recorded during Transcranial Magnetic Stimulation of motor cortex while the subject observed the experimenter grasping objects provides further evidence that a 'mirror system' exists in humans (Fadiga et al. 1995).

Our results are in general agreement with the PET data (Fadiga, Rizzolatti), but pose further questions relating to: (1) lateralisation of the response, and (2) its dependence on precise matching of biologically relevant actions. Subjects in this study observed right handed manipulations that resulted in activation of right hemisphere BA44/9, reflecting a possible ipsilateral representation of observed actions. These results contrast with those found by previous authors. Most recently Buccino et al. (2001), who were primarily interested in the somatotopic representation of actions (performed by the mouth, hand or foot) in premotor cortex, reported bilateral activation with fMRI in BA44 in their hand grasping observation condition – the hand used was not mentioned, but we presume it was the right. All their subjects (right handed) observed videotaped images in the same manner as this study. The PET studies (Rizzolatti et al. 1996b; Grafton et al. 1996) reported only unilateral activation in left BA45 (inferior frontal gyrus), to their right hand object grasping conditions. Their subjects observed actions made from an actor entering the 'stage' from the subject's right side. Recent findings from our group (in preparation) have confirmed an ipsilateral representation in a separate study in which we employed the same stimulus as used here (only the coin picking), and imaged 10 more subjects with the projected video reversed (the right hand now being on the left) – IFG activation was significant only in the left hemisphere. These confounds will need to be addressed in future studies but one observation may be pertinent. With all fMRI experiments subject's view images via mirrors, whereas with PET, subjects are able to see the world normally. There has been a recent interest in mirror reflections of self, and the coding or recoding of peripersonal space (Tipper et al. 1998; Tipper et al. 2001; Maravita et al. 2000), where it has been shown that stimuli presented in a mirror are treated as if they were in peripersonal space and not in physical space. This higher order re-representation of allocentric space into egocentric space, purported to be due learning based, may well explain some of the confounded findings from these imaging studies.

The lack of significant activation in the 'pen' condition is interpreted as demonstrating the specificity of the observed actions as identified by Rizzolatti's observation that actions with meaningless goals (the grape missing the mouth) did not activate neurones in F5. The pen did not afford normal use as it was not only picked up in an inappropriate manner (see Figure 1b) but was also not useable as it had a cap on. If high degrees of subtlety are indeed coded by MN's, and these observations are at a very preliminary stage, then we should be able to demonstrate this by investigating the effects of manipulating egocentric and allocentric space and the dependence of mirror activity on observer relevant actions. We carried out a further series of experiments in which we used subjects over-trained in a specific manual task, namely professional musicians (violinists), to establish the degree to which observer familiarity is represented in MN's (Howard 2001).

In humans, MN activity could be covert, as well as overt, the former providing a feeling of empathy or familiarity in the observer as the observed action has to be by definition one that is in the observers behavioural repertoire – or imagined repertoire (Grafton 1996). This might underpin the attraction of watching football matches, boxing, or evocative films in that the observer is 'on the pitch', 'in the ring', or 'emotively engaged' whilst remaining motorically passive. It has been postulated (Lhermitte et al. 1986) that this system normally inhibits motor cortex (BA6, 4), suppressing the expression of the observed action by the observer, and clinical evidence of the 'imitation behaviours' (echopraxia) found in patients with prefrontal damage supports this view. There may also be developmental links with the immature brains of infants not having developed the inhibitory circuitry so they can learn by mimicry – echolalia.

In summary, this study has confirmed an MN area in the human brain that is specific for meaningful actions, and added a further dimension to this area's function in that simple manual actions are represented ipsilaterally.

## References

Buccino, G., Binkofski, F., Fink, G. R., Fadiga, L., Fogassi, V., Gallese, V., Seitz, R. J., Zilles, K., Rizzollatti, G., & Freund, H-J. (2001). Action observation activates premotor and parietal areas in a somatotopic manner: an fMRI study. *European Journal of Neuroscience, 13*, 400–404.

Corballis, M. C. (1999). The gestural origin of language. *American Scientist, 87*, 138–145.

Fadiga, L., Fogassi, L., Pavesi, G., & Rizzolatti, G. (1995). Motor facilitation during actions observation: a magnetic stimulation study. *Journal of Neurophysiology, 73*, 2608–2611.

Friston, K. J., Holmes, A. P., Worsley, K. J., Poline, J. P., Frith, C. D., & Frackowiak, R. S. J. (1995). Statistical parametric maps in functional imaging: a general linear approach. *Human Brain Mapping, 2*, 189–210.

Gallese, V., & Goldman, A. (1998). Mirror neurons and the simulation theory of mind-reading. *Trends in Cognitive Sciences, 2*(12), 493–501.

Grafton, S. T., Arbib, M. A., Fadiga, L., & Rizzolatti, G. (1996). Localization of grasp representations in humans by positron emission tomography: 2. Observation compared with imagination. *Experimental Brain Research, 112*, 103–111.

Howard, M. A., McGlone, F., Tipper, S., Brooks, J. C. W., Sluming, V., Phillips, N., & Roberts, N. (2001). Action observation in orchestral string players using eFMRI. *NeuroImage, 13*(6), 1189.

Jacobson, E. (1930). Electrical measurements of neuromuscular states during mental activities. *American Journal of Physiology, 92,* 567–608.

Jacobson, E. (1932). Electrophysiology of mental activities. *American Journal of Psychology, 44,* 677–694.

Lhermitte. F., Pillon, B., & Serdaru, M. (1986). Human autonomy and the frontal lobes 1. Imitation and utilization behavior – a neuropsychological study of 75 patients. *Annals of Neurology, 19*(4), 326–334.

Maravita, A., Spence, C., Clarke, K., Husain, M., & Driver, J. (2000). Vision and touch through the looking glass in a case of crossmodal extinction. *Neuroreport, 11,* 3521–3526.

Passingham, R. (1993). *The frontal lobes and voluntary action.* Oxford: Oxford University Press.

Perrett, D. I., Smith, P. A. J., Mistlin, A. J., Chitty, A. J., Head, A. S., Potter, D. D., Broennimann, R., Milner, A. D., & Jeeves, M. A. (1985). Visual analysis of body movements by neurons in the temporal cortex of the macaque monkey: A preliminary report. *Behavioural Brain Research, 16,* 153–170.

Perrett, D. I., Harries, M. H., Bevan, R., Thomas, S., Benson, P. J., Mistlin, A. J., Chitty, A. J., Hietanen, J. K., & Ortega, J. E. (1989). Frameworks of analysis for the neural representation of animate objects and actions. *Journal of Experimental Biology, 146,* 87–113.

Rizzolatti, G., & Gentilucci, M. (1988). Motor and visual-motor functions of the premotor cortex. In P. Rakic & W. Singer (Eds.), *Neurobiology of Neocortex* (pp. 269–284). New York: John Wiley.

Rizzolatti, G., Fadiga, L., Gallese, V., & Fogassi, L. (1996a). Premotor cortex and the recognition of motor actions. *Cognitive Brain Research, 3,* 131–141.

Rizzolatti, G., Fadiga, L., Matelli, M., Bettinardi, V., Paulesu, E., Perani, D., & Fazio, F. (1996b). Localization of grasp representations in humans by PET: 1. Observation versus execution. *Experimental Brain Research, 111,* 246–252.

Rizzolatti, G., & Arbib, A. (1998). Language within our grasp. *Trends in Neurosciences, 21*(5), 188–194.

Talairach, J., & Tournoux, P. (1988). *Coplanar stereotactic atlas of the human brain. Three-dimensional proportional system; an approach to cerebral imaging.* New York: Thieme.

Tipper, S. P., Lloyd, D., Shorland, B., Dancer, C., Howard, L. A., & McGlone, F. (1998). Vision influences tactile perception without proprioceptive orienting. *NeuroReport, 9,* 1741–1744.

Tipper, S., Philips, N., Dancer, C., Lloyd, D., Howard, L. A., & McGlone, F. (2001). Vision influences tactile perception at body sites that cannot be directly viewed. *Experimental Brain Research,* in press.

Worsley, K. J., Marrett, S., Neelin, P., Vandal, A. C., Friston, K. J., & Evans, A. C. (1996). A unified statistical approach for determining significant signals in images of cerebral activation. *Human Brain Mapping, 4*(1), 58–73.

# Mirror neurons and the self construct

Kai Vogeley and Albert Newen
Institute of Medicine, Research Center Jülich, Germany /
Department of Philosophy, University of Bonn, Germany

## 1. Introduction

The concept of mirror neurons postulates a neuronal network that represents both observation and execution of goal-directed behavior and is taken as evidence for the validity of the simulation theory, according to which human subjects use their own mental states to predict or explain mental processes of others. However, the concept of mirror neurons does not address the question, whether there is a specific difference between the other individual observed and myself, between first-person- and third-person-perspective. Addressing this issue, a functional magnetic resonance imaging study is presented that varies first-person- and third-person-perspective systematically. A classical theory of mind paradigm was employed and extended to include first-person- and third-person-perspective stimuli. During the involvement of third-person-perspective increased neural activity in the anterior cingulate cortex and left temporopolar cortex was observed. During the involvement of first-person-perspective increased neural activity in the right temporoparietal cortex and in the anterior cingulate cortex was found. A significant interaction of both perspectives activated the right prefrontal cortex. These data suggest that these different perspectives are implemented at least in part in distinct brain regions. With respect to the debate on simulation theory, this result rejects the exclusive validity of simulation theory.

## 2. Mirror neurons and simulation theory

Gallese et al. (1996) identified a neuronal mechanism matching observation and execution of goal-related motor actions in the macaque brain in the inferior area 6 (corresponding to F5). This neuronal system was found to respond not only to the observation of goal-directed behavioral sequences of other animals. In addition, this network also responded to the execution of the same movements of the experimental animal studied. This led the authors coin the term "mirror neurons" (MN) for this group of neurons. Closely related to the finding of the MN system are PET studies on human subjects in which the neural representations of grasp movements were studied. These studies demonstrated predominantly left hemisphere activations during grasping observation in the region of the superior temporal sulcus and the human analogue of inferior area 6 (Rizzolatti et al. 1996) and the left inferior frontal cortex, the left rostral inferior parietal cortex, the left supplementary motor area and the right dorsal premotor cortex (Grafton et al. 1996). These studies provide convincing evidence for this specific neuronal system, that responds both to execution and observation of certain actions. These papers are of eminent importance and have already become canonical papers describing crucial experiments in the field.

This MN concept is so important because it has enormously stimulated the discussion on the mechanisms of prediction and/or explanation of behaviors or, more generally, mental states of others. In this respect the MN system was taken as an argument in favor of the so-called "simulation theory" (ST). According to ST, human subjects use their own mental states to predict or explain mental processes of others (Gallese & Goldman 1998). The opponent of ST is the so-called "theory theory" (TT), according to which subjects performing theory of mind (TOM) use a specific body of knowledge to predict or explain the behavior or mental states of others, that is independent from own mental states. In favor of ST, Gallese & Goldman (1998) proposed in a very stimulating and important contribution to this debate, that "mirror neurons (MNs) represent a primitive version, or possibly a precursor in phylogeny, of a simulation heuristic that might underlie mind-reading" (Gallese & Goldman 1998:498). Furthermore, they speculated that "a cognitive continuity exists within the domain of intentional-state attribution from non-human primates to humans, and that MNs represent its neural correlate" (Gallese & Goldman 1998:500). The existence of the MN system is taken as an argument in favor of ST. It is speculated that human subjects modeling mental states of others in order to predict and/or explain the behavior of others use their own mental states to predict or explain mental processes of others (Gallese & Goldman 1998:496), thus following the general line of ST in contrast to the TT concept, according to which subjects performing TOM use a specific commonsense psychological theory, also referred to as "folk psychology".

In fact, these studies provide a strong empirical argument that a simulation component is involved during observation of movements. However, this particular finding is not a proof for the *exclusive* validity of ST. This aspect was also emphasized by the authors themselves, in fact, they do not claim that ST is a "full-scale realization of the simulation heuristic" (Gallese & Goldman 1998:498). So they are not able to completely reject the conceptual counterpart TT. In the case that ST was exclusively true, all mental states requiring the modeling of mental states of others, irrespective of whether they are attributed to someone else or to oneself, should involve the same neuronal system, as there is no functional difference between attributing mental states to others and oneself. Following this concept, the attribution of mental states to others would be entirely based on the simulation of own mental states and subsequent projection onto other persons. In contrast to this, one would assume two distinctly implemented neural mechanisms if TT was true. In this case the attribution of a mental state to someone else would refer to a particular "theory", a specific and independent body of knowledge, whereas the attribution of a mental state to oneself would be something completely different involving a distinct neural mechanism.

Especially with respect to the debate on ST and TT, one central question remains unanswered by the MN concept, that tells us, what both processes, observation and execution of motor actions, have in common. The question is: what makes the specific difference between the execution and observation of actions? Obviously, it makes a big difference both at a behavioral as well as phenomenal level of subjective experience, whether I observe a motor act of another individual or whether I perform a motor act myself. The essential difference between observation and execution of motor acts is obviously the involvement of myself as generator of these motor actions. The specific class of motor representations generated by MNs, lets say "mirror" motor representations, do not allow the involved agent to distinguish, whether these motor act representations are generated by someone else or by him/herself. To put it in more formal terms, representation in this context can be defined as a relational process, that provides an internal description (e.g. a certain MN activation pattern) of an external event to be represented (e.g. a certain motor action) for an agent. In the specific case of representations provided by MNs this external event might be an observed motor action or an executed motor action.

That means, that a specific property of the agent (either "being an observer" or "being an executor") is crucially involved in this specific class of representations, but this property is not represented by MNs. The mirror neuron concept as such already intrinsically implies the involvement of at least one other neuronal network, that provides this additional information, whether I am involved as a generator or as an observer of this specific mirror neuron representation.

## 3.   Self-consciousness and self construct

One of the focuses of the recent debate in cognitive neurosciences is the concept of the human self as a matter of empirical neuroscience. If empirical indicators for different domains of the human self model can be found, then an operationalization and a mapping to neuronal structures becomes possible. "Classical" features of the self dealt with in the philosophical as well as psychological tradition may be then addressed empirically with respect to their implementation in specific neuronal network architectures.

Consciousness in general may be defined as the integrated internal representation of the outer world and our organism based on actual experiences, perceptions and memories providing reflected responses to the needs of our environment. Consciousness is a fundamental tool for our orientation in the world and relies upon the integrative, supramodal, sensory-independent, holistic representation of the world. This world model refers to different coordinate systems, both object- and viewer-centered perspectives in space representation, both physical and subjective time scales in time representation. These reference frames are in turn based on data of the different sensory systems. Self-consciousness includes consciousness of ones own mental states, such as perceptions, attitudes, opinions, intentions to act, and so forth. Representing such mental states into one combined framework that allows us to maintain the integrity of our own mind is a meta-representational cognitive capacity.

Essential for such a teleological and functionalistic view on self-consciousness are specific experiences that reflect the involvement of a specific "sense of self". For this group of features or properties, that are constitutive for human self-consciousness, the term self construct is used to indicate a collection of properties that are potentially accessible by adequate operationalizations without strong a priori implications. The following essential features of human self-consciousness can be identified (Vogeley et al. 1999). Firstly, the experience of ownership (with respect to perceptions, judgements etc.) or agency (with respect to actions, thoughts etc.), secondly, the experience of perceptivity with conscious states being "centered" around myself, and thirdly, the experience of unity forming a long term coherent whole of beliefs and attitudes. The experience of ownership is reflected by the use of a pronominal syntax in language and the experiential quality of agency, that I am performing my movements for myself, having my own perceptions, memories, and thoughts. The experience of perspectivity refers to the centeredness of my memory, perceptions, and thoughts around my own body and thus to the experience of a literally spatial, body-centered perspective. The experience of unity is associated with long term coherent wholes of beliefs and attitudes, that are consistent with preexisting autobiographical contexts.

It was postulated that these basic properties are integrated in a postulated so-called "self model" as an episodically active complex neural activation pattern in the human brain, possibly based on an innate and "hard-wired" model (Metzinger 1993, 1995; Damasio 1994; Melzack et al. 1997). This self model could then plausibly serve as a continuous source of a specific kind of milieu information on the own body and organism, that is activated whenever conscious experiences including properties of ownership, perspectivity and unity occur (Vogeley et al. 1999).

A special aspect related to the experience of self perspectivity is the body image. It was hypothesized, that the above mentioned self model creates a literally spatial model of one's own, around which the experiential space is centered (Berlucchi & Aglioti 1997). As Damasio worked out in his "somatic marker hypothesis", the representation of this body image probably involves activation of the right parietal region and of the prefrontal cortex (PFC), especially in its ventromedial parts, which "establishes a simple linkage ... between the disposition for a certain aspect of a situation ..., and the disposition for the type of emotion that in past experience has been associated with the situation" (Damasio 1996: 1415). This linkage then serves judging situations on the basis of former emotional reactions to similar situations to "constrain the decision-making space by making that space manageable for logic-based, cost-benefit analyses" (Damasio 1996: 1415). The rapid and repetitive re-instantiation of the body image is assumed to be based on a prefronto-parietal network, which is unconscious as such as it is continuously reconstituted in its process (Damasio 1994, 1996; Metzinger 1995).

## 4.   Theory of mind and self perspective

An important empirical approach to access self perspective is provided by so-called "theory of mind" paradigms. When Premack and Woodruff (1978) introduced the concept of "theory of mind" (TOM), it referred to the attribution of mental states to both oneself and others. This ability of "mindreading" (Baron-Cohen 1995) is an important component in social interaction and communication and can be tested in TOM paradigms, originally designed in primates and further developed in developmental psychology of humans. In a typical TOM paradigm, a subject has to model the knowledge, attitudes or beliefs of another person. On the basis of a cartoon or a short story, the behavior of another person has to be modeled prospectively by the test person. The capacity of mindreading or TOM appears to be related to the ability to assign and maintain a self perspective (hereafter: SELF). Whereas in classical TOM paradigms (e.g. Fletcher et al. 1995), in which mental states or propositional attitudes of an agent with regard to a particular set of information or propositions need to be modeled (e.g. "Person A knows, believes, etc., that p" with p being a physical event), SELF in this context refers to the special situation, in which I am the agent myself (e.g. "I know, believe, etc., that p") and to the subjective experiential multi-dimensional space centered around one's own person. In this basic sense, SELF is a constituent of a "minimal self" defined as "consciousness of oneself as an immediate subject of experience, unextended in time" (Gallagher 2000). The correct assignment and involvement of the SELF is reflected by the use of personal pronouns ("I", "my" e.g. perception, opinion, and so forth).

According to ST, the capacity of TOM is based on taking someone else's perspective, and projecting one's own attitudes on someone else (Harris 1992). Thus, the capacity to develop a SELF is reduced to a subcomponent of a more general TOM capacity. Both capacities would then be functionally closely related and should employ the same neural mechanisms. By contrast, according to TT, the TOM capacity is a distinct body of theoretical knowledge acquired during ontogeny different from SELF (Gopnik & Wellman 1992; Perner & Howes 1992). On a purely behavioral level, an independent cerebral implementation of the two capacities could only be inferred on the basis of a double dissociation. Arguments based on information of simultaneous or subsequent development of the two differential cognitive capacities are non-conclusive with regard to their putative differential cerebral implementation as reflected by the current controversial debate (for more detail see e.g. Gopnik & Wellman 1992; Gopnik 1993; Carruthers 1996).

To empirically address the issue, as to what extent taking the SELF is involved in modeling someone else's states of mind, and whether taking the SELF or modeling the mind of someone else (TOM) employ the same or differential neural mechanisms, an fMRI study was performed presenting TOM and SELF stimulus material in a two-way factorial design (Vogeley et al. 2001).

For this purpose, a well-characterized collection of short stories (Fletcher et al. 1995; Happé et al. 1996, 1999; Gallagher et al. 2000), which comprised "unlinked sentences", "physical stories" and "TOM stories" was used. Two newly developed groups of stories were introduced that allowed subjects to engage SELF with and without engaging TOM at the same time. This enabled us to study both cognitive capacities of TOM and SELF in a fully factorial design (Figure 1). In the "physical story" condition (T−S−), short consistent texts with no perspective taking were shown presenting a short story on a certain physical event. In the "TOM story" condition (T+S−) stories were presented in which agents play a particular role, to which a mental state (e.g. perception, judgement) had to be ascribed. Two newly developed conditions which engaged the capacity of SELF in the presence or absence of TOM were added. These latter conditions incorporated the study participant as one of the agents in the story. In the "self and other ascription stories" participants had to ascribe adequate behavior, attitudes, or perceptions to themselves in a given plot, similar to "TOM stories". In the "self ascription stories", persons were asked to report their behavior, attitudes, or perceptions in inherently

| Theory of Mind | | |
|---|---|---|
| | TOM + | TOM − |
| SELF − | Condition 3<br><br>TOM Stories<br>(T+S−) | Condition 2<br><br>Physical Stories<br>(T−S−) |
| SELF + | Condition 4<br><br>Self and Other<br>Ascription Stories<br>(T+S+) | Condition 5<br><br>Self Ascription Stories<br>(T−S+) |

*Self Perspective* is indicated along the left side of the table.

**Figure 1.** Two-way factorial design of the study. This schema demonstrates the two-way factorial experimental design applied, in which both factors TOM and SELF were varied systematically.

ambiguous situations. The correct assignment of another person's mental state in the TOM conditions was tested by asking the participants to infer a specific behavior or attitude of another person in the given context of the story, judged as adequate or inadequate according to Fletcher et al. (1995) and Happé et al. (1996). Correct assignment of SELF was monitored by the use of personal pronouns in the documented answer of the particular story.

*Example of "physical stories" (T−S−)*
A burglar is about to break into a jeweller's shop. He skillfully picks the lock on the shop door. Carefully he crawls under the electronic detector beam. If he breaks this beam it will set off the alarm. Quietly he opens the door of the store-room and sees the gems glittering. As he reaches out, however, he steps on something soft. He hears a screech and something small and furry runs out past him, towards the shop door. Immediately the alarm sounds.

Question: Why did the alarm go off?

*Example of "TOM stories" (T+S–)*
A burglar who has just robbed a shop is making his getaway. As he is running home, a policeman on his beat sees him drop his glove. He doesn't know the man is a burglar, he just wants to tell him he dropped his glove. But when the policeman shouts out to the burglar, "Hey, you! Stop!", the burglar turns round, sees the policeman and gives himself up. He puts his hands up and admits that he did the break-in at the local shop.

Question: Why did the burglar do that?

*Example of "self and other ascription stories" (T+S+)*
A burglar who has just robbed a shop is making his getaway. He has robbed your store. But you can not stop him. He is running away. A policeman who comes along sees the robber as he is running away. The policeman thinks that he is running fast to catch the bus nearby. He does not know that the man is a robber who has just robbed your store. You can talk quickly to the policeman before the robber can enter the bus.

Question: What do you say to the policeman?

*Example of "self ascription stories" (T–S+)*
You went to London for a weekend trip and you would like to visit some museums and different parks around London. In the morning, when you leave the hotel, the sky is blue and the sun is shining. So you do not expect it to start raining. However, walking around in a big park later, the sky becomes grey and it starts to rain heavily. You forgot your umbrella.

Question: What do you think?

Stories were presented during the fMRI BOLD contrast EPI measurements for 25 seconds on a display, with the question being presented subsequently for 15 seconds. Subjects were instructed to read the story carefully and to read and answer the subsequent question silently (covertly). Volumes were acquired continuously every 5 seconds over the whole period of 40 seconds while subjects performed the experimental tasks. After each presentation subjects were asked to give the answers overtly. In each of the four experimental conditions and the baseline, 8 trials were presented. Eight right handed, healthy male volunteers (age 25 to 36 years) with no history of neurological or psychiatric illness were studied. Functional magnetic resonance (fMRI) was performed (echo planar imaging on a 1.5 T MRI system, SIEMENS Magnetom VISION, TR = 5000 ms, TE = 66 ms, FOV = $200 \times 200$ mm$^2$, $\alpha = 90°$, matrix size = $64 \times 64$, voxel size = $3.125 \times 3.125 \times 4.4$ mm$^3$). The scanning procedure was performed continuously over one trial and was re-started, after the test person answered. The entire image analysis including realignment, normal-ization, and statistical analysis was performed using Statistical Parametrical Mapping (SPM99, Wellcome Department of Cognitive Neurology, London, UK). For the fMRI group data analysis, all images of all subjects were analyzed in one design matrix in a fixed-effect model. The data were analyzed both with respect to the specific effects of each condition against the baseline ("unlinked sentences" condition) and with respect to the main effects of TOM and SELF. In addition, the contrast of SELF relative to TOM was calculated to assess the significance of the specific differences between TOM and SELF. Finally, we assessed whether the neural mechanisms underlying TOM and SELF interacted with each other. Throughout, we report activations significant at $p < 0.05$ corrected for multiple comparisons at an extent threshold of a minimum of 17 pixels (Figure 2).

The brain activation pattern under the main effect of TOM ([T+S+ plus T+S–] relative to [T–S+ plus T–S–]) demonstrated increases in neural activity predominantly in the right anterior cingulate cortex and left temporopolar cortex (Figure 2a). The main effect of SELF ([T–S+ plus T+S+] relative to [T+S– plus T–S–]) resulted in increased neural activity predominantly in the right temporoparietal cortex and in the anterior cingulate cortex. Further significant increases in neural activity associated with SELF were observed in the right premotor and motor cortex and in the precuneus bilaterally (Figure 2b). When contrasting SELF with TOM directly (T–S+ relative to T+S–), activation of the right temporoparietal cortex and bilateral precuneus was found, thus corroborating the specific difference between SELF and TOM (Figure 2c). The interaction of TOM and SELF ([T+S+ relative to T+S–] relative to [T–S+ relative to T–S–]) was calculated to identify those areas activated specifically as a result of the presence of both TOM and SELF. This calculation revealed an increase in brain activity in the area of the right lateral prefrontal cortex (Figure 2d).

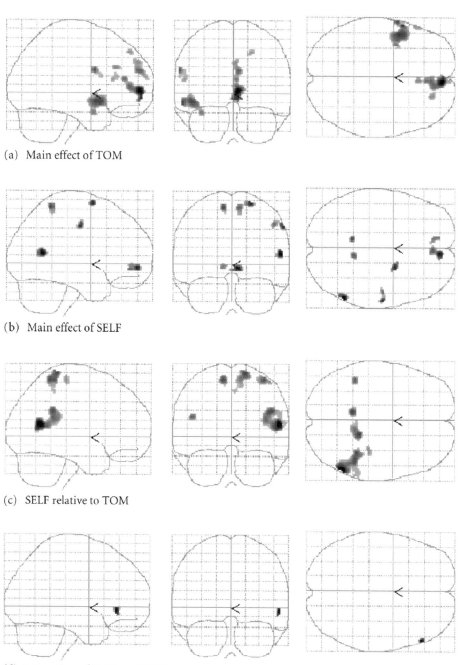

(a)   Main effect of TOM

(b)   Main effect of SELF

(c)   SELF relative to TOM

(d)   Interaction of TOM and SELF

## 5.   Neuronal implementation of the self model

Our results demonstrate that the ability to attribute opinions, perceptions or attitudes to others, often referred to as TOM or "mind-reading" and the ability to apply SELF rely on both common and differential neural mechanisms. The cerebral implementation of TOM capacity is located predominantly in the anterior cingulate cortex. This part of the described experiment replicates previous studies using this particular paradigm employed in our study (Fletcher et al. 1995; Happé et al. 1996; Gallagher et al. 2000). The right hemisphere dominance for TOM is in good accordance with right hemispheric activations under pragmatic language tasks (Brownell et al. 1990; Bottini et al. 1994). Patients with right hemispheric lesions demonstrate difficulties with verbal and non-verbal communication, understanding of metaphors, non-conventional or indirect meaning, indirect questions or the emotional-prosodic quality of expressions, and TOM (Brookshire & Nicholas 1984; Foldi 1987; Bryan 1988; Weylman et al. 1989; Brownell et al. 1994; Happé et al. 1999).

The main finding of our study was that taking SELF leads to additional neural activations in the right temporoparietal cortex and the precuneus bilaterally (Vogeley et al. 2001). The fact of this differential brain activation suggests that these components are implemented at least in part in different brain modules and thus constitute distinct cognitive processes. This view is supported by the observation of a significant interaction between TOM and SELF in the right prefrontal cortex. Interestingly, this region has previously been implicated in "supervisory attentional" mechanisms (Shallice & Burgess 1996) or monitoring situations that involve conflict of senses (Fink et al. 1999).

While the anterior cingulate cortex seems to be the key structure for assigning a mental state to someone else, irrespective of whether SELF is involved or not, our results also imply that activation of this brain region is not sufficient when the

**Figure 2.**   Main effects of TOM and SELF and their interaction. (2a) (T+S+ plus T+S–) relative to (T–S+ plus T–S–). Under the main effect of TOM there is significant activation in the right anterior cingulate cortex, and left superior temporal cortex. (2b) (T+S+ plus T–S+) relative to (T+S– plus T–S–). Under the main factor SELF there is still considerable activation at the anterior cingulate cortex and significant activation in the right temporoparietal cortex and the precuneus bilaterally. (2c) (T–S+ relative to T+S–). The direct contrast of SELF versus TOM corroborates activation of the right temporoparietal junction and bilateral precuneus. (2d) ([T+S+ relative to T+S–] relative to [T–S+ relative to T–S–]). During interaction of TOM and SELF an isolated area with increased neural activity in the right lateral prefrontal cortex was found.

ability to apply SELF is required. Taking SELF appears to activate the right inferior temporoparietal cortex in addition. This activation is independent from the need to assign TOM at the same time. Interestingly, lesions in this region lead to visuo-spatial neglect (Vallar et al. 1999). This conjecture in turn is in good accordance with reports on increased neural activity of right inferior parietal cortex involving visuo-spatial attention e.g. navigation through virtual reality scenes (Maguire et al. 1998) or assessment of the subjective mid-sagittal plane (Vallar et al. 1999). The activation of the temporoparietal cortex during SELF is also compatible with evidence for the implementation of our body image in this region (Berlucchi & Aglioti 1997), thus suggesting, that taking SELF may draw on a body representation as the center of an ego-centric experiential space. These data imply that the temporoparietal cortex is involved in computing an egocentric reference frame. However, our data strongly suggest a more general role for this region which goes beyond visuo-spatial judgements: Increased neural activity in this region was also evoked by the use of personal pronouns in our language-based stimulus material.

The interaction of TOM capacity and SELF involves the right prefrontal cortex suggesting that this region is specifically activated when an integration of TOM and SELF is needed. Previous studies suggested an involvement of right prefrontal cortex in the segregation and integration of different cognitive capacities including situations with increased monitoring demand (Fink et al. 1999) and self-recognition (Keenan et al. 2000). However, it must be clearly stated, that Keenan et al. (2000) studied self-recognition paradigms in which the own face appeared as an object (in the sense of "me"), that had to be identified. Our approach deals with the self as the subject of an experience, and not as object (in the sense of "I"). The prefrontal cortex, especially its dorsolateral parts, is a constitutive component of a complex neural network architecture comprising various sites, to generate experiences of ownership, perspectivity and unity on the phenomenal level. It does so by integrating multimodal perceptions and proprioceptive body image informations (Vogeley et al. 1999). Behavioral adaptation to challenging new situations is provided by monitoring ongoing elaborated programs and previously established automatic programs. The function of the prefrontal cortex may thus be defined as "active, transformational process in which sensory data are synthesized into the simplest possible representation for the purpose of maximizing behavioral efficiency" (Shobris 1996). Symptoms of a dysfunction of the prefrontal cortex, may be the result of a disturbance of crosstemporal contingencies (Fuster 1991). Crosstemporal contingencies are responsible for keeping contents "on line" in working memory (Fuster 1997). Deficits in the experience of ownership could well be due to a complex dysconnection syndrome between the prefrontal and other association cortex areas. Disturbances in the prefronto-parietal network as putative source of continuously generated input about internal milieu data may result in the loss of experience of body-centered perspectivity. If this continuous re-actualization of current experiences and proprioceptive information is disturbed, the result would be the loss of the experiential perspectivity.

## 6.   Conclusion

The results of our study demonstrate that the ability to attribute opinions, perceptions or attitudes to others and the ability to apply SELF rely in part on differential neural mechanisms. Whereas TOM is predominantly associated with increased activity in the anterior cingulate cortex, the capacity to take SELF is predominantly located in the right temporo-parietal cortex. However, there is a marked overlap with shared activity increases in both SELF and TOM in the anterior cingulate cortex.

With respect to the debate on ST and TT, one can state the following. The MN system described by Gallese et al. (1996) responding both to observation of other animals performing goal-directed actions as well as to execution of the same movements of the experimental animal studied provide valid evidence in favor for ST. It might well be, that this MN system represents a primitive version of a simulation algorithm that is essential for mind-reading (Gallese & Goldman 1998). However, this particular system can not prove the exclusive validity of ST, as in the case of exclusive validity of ST, all mental states requiring TOM, irrespective of whether they were attributed to someone else or to oneself, should activate the same brain region. As our experiment shows, this is not the case. That TOM and SELF involve at least in part distinct neural mechanisms is demonstrated by the different activation patterns of the main effects (Figure 2) and is further corroborated by the finding of a significant interaction between both factors. Thus, our data reject both ST and TT in a pure form and are in favor of a mixture of both concepts. On the basis of our study, the TT component appears to be based on the anterior cingulate cortex activation, whereas the ST component is primarily associated with increased brain activity in the area of the right temporoparietal cortex. This is compatible with the view that "knowledge of another's subjectivity is going to have to involve one's own" (Bolton & Hill 1996: 135).

Allowing a differential induction of a SELF or self-related experiences such as the experiences of ownership, agency or unity are necessary requisites to evaluate the theoretical concepts of ST and TT. Expansions of classical TOM paradigms could become useful tools for the further study of the interdependency of the first-person- and third-person-perspective. Our study design or related paradigms to be developed may become useful as an experimental tool in cognitive sciences and clinical applications especially with regard to possible disorders of TOM (e.g. autism and schizophrenia). The findings provide experimental evidence for the cerebral implementation of an important feature of self-consciousness and have important significance for cognitive and neurophilosophical theories of consciousness.

## References

Baron-Cohen, S. (1995). *Mindblindness*. Cambridge, MA: MIT Press.

Berlucchi, G., & Aglioti, S. (1997). The body in the brain: Neural bases of corporeal awareness. *Trends in Neuroscience, 20*(12), 560–564.

Bolton, D., & Hill, J. (1996). *Mind, Meaning and Mental Disorder. The nature of causal explanation in psychology and psychiatry*. Oxford: Oxford University Press.

Bottini, G., Corcoran, R., Sterzi, R., Paulesu, E., Schenone, P., Scarpa, P., Frackowiak, R. S. J., & Frith, C. D. (1994). The role of the right hemisphere in the interpretation of figurative aspects of language. *Brain, 117*, 1241–1253.

Brookshire, R. H., & Nicholas, C. E. (1984). Comprehension of directly and indirectly stated main ideas and details in discourse by brain-damaged and non-brain-damaged listeners. *Brain and Language, 21*, 21–36.

Brownell, H. H., Simpson, T. L., Bihrle, A. M., Potter, H. H., & Gardner, H. (1990). Appreciation of metaphorical alternative word meanings by left and right brain-damaged patients. *Neuropsychologia, 28*, 375–383.

Bryan, K. L. (1988). Assessment of language disorders after right hemisphere damage. *British Journal of Disorders of Communication, 23*, 111–125.

Carruthers, P. (1996). Simulation and self-knowledge: A defence of theory-theory. In P. Carruthers, & P. K. Smith (Eds.), *Theories of Theories of Mind* (pp. 22–38). Cambridge: Cambridge University Press.

Damasio, A. R. (1994). *Descartes' Error. Emotion, reason and the human brain*. New York: G. P. Putnam's Son.

Damasio, A. R. (1996). The somatic marker hypothesis and the possible functions of the prefrontal cortex. *Philosophical Transactions of the Royal Society: Biologic Sciences, 351*, 1413–1420.

Fink, G. R., Marshall, J. C., Halligan, P. W., Frith, C. D., Driver, J., Frackowiak, R. S., & Dolan, R. J. (1999). The neural consequences of conflict between intention and the senses. *Brain, 122*, 497–512.

Fletcher, P., Happé, F., Frith, U., Baker, S. C., Dolan, R. J., Frackowiak, R. S. J., & Frith, C. D. (1995). Other minds in the brain: A functional imaging study of "theory of mind" in story comprehension. *Cognition, 57*, 109–128.

Foldi, N. S. (1987). Appreciation of pragmatic interpretation of indirect commands: comparison of right and left hemisphere brain-damaged patients. *Brain and Language, 31*, 88–108.

Fuster, J. M. (1991). The prefrontal cortex and its relation to behavior. *Progress in Brain Research, 87*, 201–211.

Fuster, J. M. (1997). *The Prefrontal Cortex. Anatomy, physiology, and neuropsychology of the frontal lobe*. New York: Raven Press.

Gallagher, H. L., Happé, F., Brunswick, N., Fletcher, P. C., Frith, U., & Frith, C. D. (2000). Reading the mind in cartoons and stories: An fMRI study of 'theory of mind' in verbal and nonverbal tasks. *Neuropsychologia, 38*, 11–21.

Gallagher, I. (2000). Philosophical conceptions of the self: implications for cognitive science. *Trends in Cognitive Science, 4*, 14–21.

Gallese, V., & Goldman, A. (1998). Mirror neurons and the simulation theory of mind-reading. *Trends in Cognitive Science, 2*, 493–501.

Gallese, V., Fadiga, L., Fogassi, L., & Rizzolatti, G. (1996). Action recognition in the premotor cortex. *Brain, 119*, 593–609.

Gopnik, A. (1993). How we know our minds: The illusion of first-person-knowledge of intentionality. *Behavioral and Brain Sciences, 16*, 1–14.

Gopnik, A., & Wellmann, H. (1992). Why the child's theory of mind really is a theory. *Mind and Language, 7*, 145–151.

Grafton, S. T., Arbib, M. A., Fadiga, L., & Rizzolatti, G. (1996). Localization of grasp representation in humans by positron emission tomography: 2. Observation compared with imagination. *Experimental Brain Research, 112*, 103–111.

Happé, F., Ehlers, S., Fletcher, P., Frith, U., Johansson, M., Gillberg, C., Dolan, R., Frackowiak, R., & Frith, C. (1996). "Theory of mind" in the brain. Evidence from a PET scan study of Asperger syndrome. *Neuroreport, 8*, 197–201.

Happé, F. G. E., Brownell, H., & Winner, E. (1999). Acquired "theory of mind" impairments following stroke. *Cognition, 70*, 211–240.

Keenan, J. P., Wheeler, M. A., Gallup, G. G. Jr., & Pascual-Leone, A. (2000). Self-recognition and the right prefrontal cortex. *Trends in Cognitive Science, 4*, 338–344.

Maguire, E. A., Burgess, N., Donnett, J. G., Frackowiak, R. S., Frith, C. D., & O'Keefe, J. (1998). Knowing where and getting there: A human navigation network. *Science, 280*, 921–924.

Melzack, R., Israel, R., Lacroix, R., & Schultz, G. (1997). Phantom limbs in people with congenital limb deficiency or amputation in early childhood. *Brain, 120*, 1603–1620.

Metzinger, T. (1993). *Subjekt und Selbstmodell*. Paderborn: Schöningh.

Metzinger, T. (1995). Faster than thought. Holism, homogeneity and temporal coding. In T. Metzinger (Ed.), *Conscious Experience* (pp. 425–461). Thorverton: Imprint Academic.

Perner, J., & Howes, D. (1992). "He thinks he knows": And more developmental evidence against simulation (role taking) theory. *Mind and Language, 7*, 72–86.

Premack, D., & Woodruff, D. (1978). Does the chimpanzee have a "theory of mind"? *Behavioral and Brain Sciences, 4*, 515–526.

Rizzolatti, G., Fadiga, L., Matella, M., Bettinardi, V., Paulesu, E., Perani, D., & Fazio, F. (1996). Localization of grasp representation in humans by positron emission tomography: 1. Observation versus execution. *Experimental Brain Research, 111*, 246–252.

Shallice, T. & Burgess, P. (1996). The domain of supervisory processes and temporal organization of behaviour. *Philosophical Transactions of the Royal Society: Biologic Sciences, 351*, 1405–1411.

Shobris, J. G. (1996). The anatomy of intelligence. *Genetic, Social, and General Psychology Monographs, 122*(2), 133–158.

Vallar, G., Lobel, E., Galati, G., Berthoz, A., Pizzamiglio, L., & Le Bihan, D. (1999). A fronto-parietal system for computing the egocentric spatial frame of reference in humans. *Experimental Brain Research, 124*, 281–286.

Vogeley, K., Kurthen, M., Falkai, P., & Maier, W. (1999). The prefrontal cortex generates the basic constituents of the self. *Consciousness and Cognition, 8*, 343–363.

Vogeley, K., Bussfeld, P., Newen, A., Herrmann, S., Happé, F., Falkai, P., Maier, W., Shah, N. J., Fink, G. R., & Zilles, K. (2001). Mind Reading: Neural Mechanisms of Theory of Mind and Self-Perspective. *NeuroImage, 14*, 170–181.

Weylman, S. T., Brownell, H. H., Roman, M., & Gardner, H. (1989). Appreciation of indirect requests by left- and right-brain-damaged patients: The effects of verbal context and conventionality of wording. *Brain and Language, 36*, 580–591.

# Behavioral synchronization in human conversational interaction

Jennifer L. Rotondo and Steven M. Boker
Augustana College, Sioux Falls, South Dakota, USA /
University of Notre Dame, USA

## 1. Introduction

In their 1998 article, Rizzolatti and Arbib noted that neurons in the F6 region of the ventral premotor cortex are activated in primates when they observe or mimic a simple action, such as grasping a grape. The authors suggested that there is, perhaps, something communicative about the repetition of another's actions that indicates understanding and facilitates communication. It is such an action-reaction response cycle that may have led to the beginnings of formalized language.

Such activity is not limited to primates. Postural and gestural mirroring is common in a variety of human activities, including conversational interaction. Quite often, individuals mirror the behaviors of their conversational partners without ever having conscious intention of doing so (Condon & Ogston 1967; Kendon 1970). In an informal group, people may cross their legs at similar angles, hold their arms in similar positions, even simultaneously perform head or hand motions. To a point, such mirrored or matched positions can be of benefit to the continuity of the conversation; however, if people see such displays as deliberate attempts to mimic, it can bring about feelings of discomfort and a desire to break the established symmetry.

Our studies of behavioral coordination address how humans utilize kinesthetic patterns of matching and mirroring in face-to-face activities, such as dancing (see Boker & Rotondo, this volume) and conversational interaction. In both areas, we address issues of symmetry breaking and symmetry formation with respect to time. Symmetry formation refers to the process by which individuals adopt and maintain similar postures over time. Symmetry breaking refers to occasions when behavioral coordination between individuals becomes low, such that they begin to desynchronize or decouple themselves from one another. We wish to determine what personal or situational characteristics influence symmetry breaking as well as the length of duration before which symmetry is reestablished.

Head motion is a particularly rich area for the study of interactive nonverbal behaviors. Many of these movements have direct correlates to speech (i.e., nods and shakes of the head convey assent and disagreement, respectively, in Western cultures; Ekman & Friesen 1969). In addition, they may also symbolize expectancy or interest or may be a way of communicating attentiveness when displayed in synchrony with one's own or a partner's speech stream (Condon & Ogston 1967; Kendon 1970). In addition, movements of the head are some of the most frequent in conversation; tilts, nods, shakes, and turns are commonly displayed throughout our interactions without a great deal of premeditated thought. Therefore, since such a great many of our spontaneous nonverbal behaviors are produced by the head, and since we use its motion in such a wide variety of ways, finding evidence of symmetry in conversation may suggest one of two things. First, the participants may be intentionally attempting to match or mirror a partner's positions during conversation. Alternatively, there may be mechanisms in the human perceptual system (such as mirror neurons) that influence us unconsciously to align our behaviors with those of our partners. This may be an indication of continual interest and understanding between partners, thus facilitating communication.

## 1.1   Determinants of symmetry formation in conversation

Determining whether we mirror or match the behavior of others can be related to the underlying composition of a group or dyad. Differences in age, race, gender, and acquaintanceship have all been hypothesized to influence nonverbal behavior production (Benjamin & Creider 1975; Bull 1987; Burgoon, Buller, & Woodall 1996). If nonverbal behaviors really do vary according to such contextual details, then no preset or fixed actions are expected to arise in every instance of social interaction (Bull 1987). Rather, nonverbal gestures will be evoked by the nature of the relationship between the individuals. As such, the present work was performed to determine how such contextual factors might influence the symmetry formation and breaking process, and dyad composition was dependent upon two factors: gender and dominance.

## 1.2   Gender differences in nonverbal behavior

Gender has been proposed to influence nonverbal behavior in a variety of ways. Women use more closed postures, more facial expressivity (including smiling and gazing more at other individuals), and prefer and/or tolerate closer proximities to their conversational partners than men (Dovidio, Ellyson, Keating, Heltman, & Brown 1988; Hall 1984; Williams & Best 1986). Women are usually the recipients of some of these behaviors also, such as being gazed at more often and being approached more closely by others (Hall 1984). In contrast, men stare away more, are less expressive, more expansive in their posture and sometimes more restless, making a greater number of gross motor movements than women (Duncan & Fiske 1977; Hall 1984; Mehrabian 1981; Williams & Best 1986).

Men and women differ in the types of nonverbal responses they display based on their partner's gender. For instance, Weitz (1976) found that women in mixed-sex interactions adapted their nonverbal reactions to those of their male partners, but they did not make a similar adaptation when in same-sex interactions. Men did not adapt their movements in either situation. In their examination of mixed-sex dyads, Bente, Donaghy and Suwelack (1998) noted that physical motion was more abundant and more complex when one's partner was looking at him or her. However, males and females responded differently to their partner's gazing behaviors. The visual gaze of a male partner triggered body movement in his female partner almost twice as often as the visual gaze of a female partner produced in her male partner. They also noted a difference in follow-up delay time, or the amount of time it took partners to meet one another's gaze after a period when neither was looking at the other. When the male partner glanced at the female partner, the female partner returned his gaze in a little over half a second. However, when female partners were the first to establish visual contact, it took the males over two and one-half times longer to respond.

## 1.3 Dominance and gender as a part of nonverbal behavior production

The concept of dominance can be viewed in a variety of ways. When paired with power and status as an environmental construct, dominance refers to being in charge of resources, making decisions, or having a leading role. Having a dominant personality type refers to one's dispositional tendency to take control of ambiguous situations with some type of social leadership (Burgoon & Dillman 1995). Researchers have disagreed as to whether dominance should be viewed as an individual trait (Weisfeld & Linkey 1985) or as an emergent part of one's social relations (Mitchell & Maple 1985). The current work proposes a mixed view of personality and context in attributing dominance, thus estimating it as both an internal and external characteristic.

A number of researchers have empirically and theoretically connected gender with dominance (Bargh, Raymond, Pryor, & Strack 1995; Henley 1977; Lakoff 1975; Mulac 1989). It is commonly purported that traits associated with dominance are the same traits ascribed to men or described as characteristic of masculinity. In turn, many of these traits are found to be highly valued in American society and have been theoretically linked to the established social structure (Henley 1977; Thorne & Henley 1975).

## 1.4  The present study

Of interest in the present work is how gender and dominance influences on dyad interaction affect symmetry formation and symmetry breaking in patterns of head movement throughout conversation. If traditional gender roles are used as cues, then it is reasonable to believe that men may initiate more symmetry breaking behaviors and take longer to establish symmetry formation than females (see Bente, Donaghy, & Suwelack 1998). Women, attempting to be affiliative, may attempt and engage in more behaviors that establish symmetry formation. Alternatively, if dominance cues are the overriding factors, then individuals high on this factor may be in charge of the symmetry flow of the interaction. Individuals low on this factor may attempt to remain affiliative by mirroring or matching their partners, unconsciously "following the leader" by maintaining symmetry and not initiating any breaking behaviors.

## 2.  Methods

### 2.1  Participants

One hundred twenty-eight collegiate males and females were recruited as participants in the study. Participants were recruited based on gender and their responses to a prescreening measure of dominance (25 items taken from the California Psychological Inventory, Dominance Subscale; Gough 1956). Individuals scoring in the top and bottom third of all dominance scores were recruited. Those in the top third were considered "high dominant", and those in the bottom third were considered "low dominant".

All participants were scheduled in groups of four, consisting of two females and two males each, resulting in 32 groups ("quads") of individuals. Each quad contained two high and two low dominant individuals, so that each gender and personality dimension was represented by one person (high dominant male, high dominant female, low dominant male, low dominant female). The quads were counterbalanced so that conversation order (i.e., mixed- then same-sex vs. same- then mixed-sex) would not introduce spurious effects. None of the participants were previously acquainted with one another.

## 2.2 Apparatus

Physical motion was recorded using Ascension Technologies MotionStar 16 sensor magnetic motion tracking device (for further details, see Boker & Rotondo, this volume). Via transmitter and receivers, this six degree of freedom device computed position and orientation information that was then encoded in a digitized data stream. A single transmitter and 16 receivers were used, producing 96 columns of position and orientation information at each measurement, with a sampling rate of 80 samples per second. For position and orientation of the receivers, see Boker and Rotondo (this volume).

## 2.3 Procedure

Participants were told that they were going to engage in two separate conversations, both resembling a job interview. High dominant individuals were always the interviewers; low dominant individuals were the interviewees. To enhance feelings of dominance and submission, high dominant individuals were given a great deal of information about the study and encouraged to take control of the situation; low dominant individuals were provided with fewer details and encouraged to follow along. None of the participants were informed of the true purpose of the study; rather, they were told that our interest was in monitoring the magnetic fields generated in different conversational settings. During debriefing, none of the participants reported having detected the true nature of the experiment; most were completely surprised when they were informed.

Participants were dressed with eight motion tracking receivers and brought to the conversation room: a wooden platform surrounded by free-standing walls of soundproof foam and curtain closure. The conversation room was monitored by three video cameras, and speech was recorded through lavaliere microphones worn by each participant. They were seated on wooden stools, approximately one meter apart. Participants were instructed to conduct each interview for seven minutes. Upon completion of their second interview, the participants were fully debriefed and thanked for their time.

## 3.   Results

The degree of coordination between head rotation of our dyads was performed using cross-correlation. This method allowed us to determine not only the strength of symmetry between the individuals' behaviors at temporal synchrony, but also how behaviors at slight temporal asynchronies are related. These temporally disparate correlations provide information pertaining to how one individual's actions influence another's, giving a sense of nonverbal conversational flow. Using these cross-correlations, we wished to determine if symmetry breaking behaviors, which would be shifts in the head motions of one person, were followed by symmetry formation, whereby behavioral coordination would be reestablished, and the duration of time required for this to occur. The specifics of this cross-correlation technique can be found in Boker and Rotondo (this volume).

An example of a cross-correlation plot for head motion in conversation can be found in Figure 1. This can be used to illustrate the concepts of lead changes and oscillation. The lag time is along the x-axis, and progression of time during the conversation is along the y-axis. The lag time represents two seconds of time about synchrony (i.e., where lag = 0). Values to the left of synchrony indicate correlations obtained when the window of observations was preceding the fixed interval, and values to the right of synchrony indicate correlations obtained when the window was ahead of synchrony. Because of this structure, these plots are able to show the progression of strongly and weakly correlated movement across time as well as which of the participants initiates the movement to which they are both strongly or weakly responding. When high correlations are on one side of synchrony, person 1

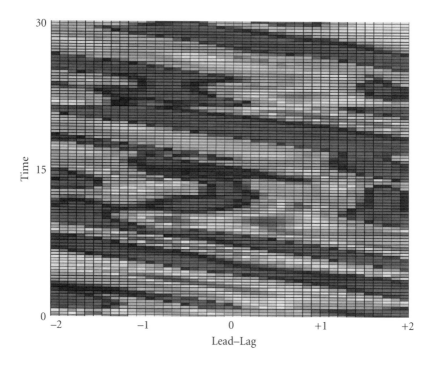

**Figure 1.** Cross correlations of head motion between conversing individuals

is said to be leading the movement; when high correlations are found on the other side of synchrony, person 2 is said to be leading the movement.

Such graphs suggest that behaviors initiated by one individual are subsequently matched or mirrored by the other. There is also evidence of sustained behavioral coordination in the areas of vertical high correlation, which usually are immediately preceding a symmetry break and after which a new pattern of symmetry formation occurs. A peak picking algorithm was devised to differentiate between individuals producing symmetry breaking and symmetry forming behaviors (see Boker & Rotondo, this volume). From the perspective of the present study, we were interested in whether symmetry breaking and formation would be better related to the gender or the personality construction of the dyad.

As a baseline measure of coordination within each dyad, we created a second set of cross correlations that were desynchronized by several minutes. To compare correlations occurring in real time with those created at a temporal asynchrony, we squared the correlations and aggregated them into bins that referred to a particular range of correlation magnitude. The bins were equally spaced between 0 and +1. To determine how many times a value fell into a particular range, we took histogram counts of each bin for both the real time and surrogate matrices. This aggregation technique retained all of our lag information while providing us with a more smoothly distributed set of correlation values. The resultant distributions were then subtracted from one another, producing a Real-Surrogate difference matrix for each conversation.

To assess patterns of symmetry formation and breaking for our 4 conversation types (HDF/LDM, HDM/LDM, HDF/LDF, HDM/LDF), we thought of each lag and bin in the difference matrices as a cell and calculated mean and standard error values for them. T-values for each of these cells were computed to determine if the behavioral coordination patterns were at better than chance levels. The t-values plots can be found in Figure 2, where lighter cyan colors indicate higher t-values.

Near temporal synchrony, symmetry breaking beyond chance levels appears for nearly all of the dyad types. There is strong evidence of this in the HDM/LDF pairing, where both partners appear to be contributing to the elimination of symmetry at better than expected levels. There is little consistent pattern of symmetry breaking in the HDM/LDM pairing beyond chance levels. Perhaps status differences were not as pronounced for this group since gender was constant between the individuals.

In terms of symmetry formation, highly coordinated behavior appeared between .5 and 1.5 seconds after temporal synchrony for most groups. An interesting result appeared for high dominant females, suggesting that high dominant females match or mirror the behaviors of their low dominant partners, whether the partners are male or female. For example, a patch of highly coordinated behavior (symmetry formation) can be found when a low dominant female's behavior precedes

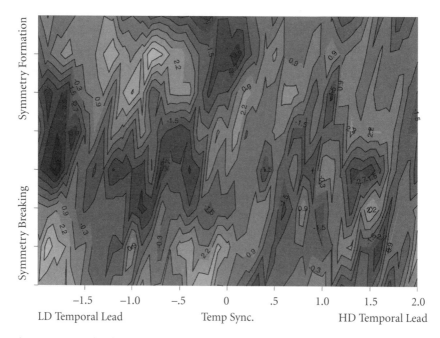

**Figure 2a.** Female–female

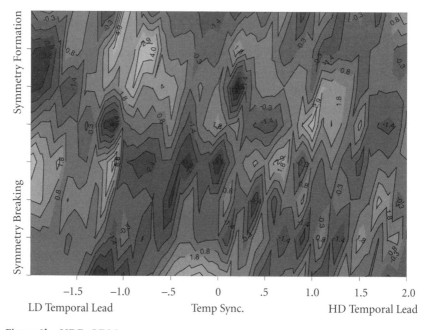

**Figure 2b.** HDF–LDM

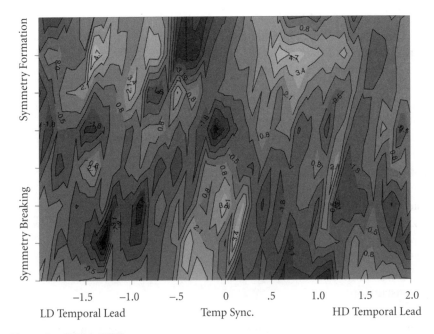

**Figure 2c.** HDM–LDF

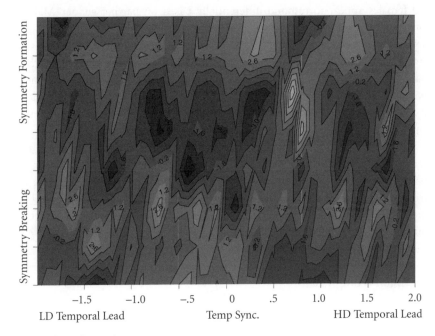

**Figure 2d.** Male–male

a high dominant female's behavior by approximately three-quarters of a second. The same is true when a low dominant male's behavior is displayed first. For this particular pairing there is evidence of greater behavioral synchrony for head motions initiated by the low dominant male, something contrary to the personality perspective.

Another interesting result was found for the HDM/LDF dyads, which demonstrate periodicity in leading and following. The results suggest that head motions performed by low dominant females were matched or mirrored by their high dominant male partners about 1.5 seconds later; in addition, high dominant male head motions were matched or mirrored by the low dominant females about a half-second after they were initiated. This periodicity was not duplicated by any other dyad composition and suggests that further analysis of this particular dyad type might lead to interesting insights about the nonverbal communication patterns displayed by these "traditional" males and females.

## 4.  Discussion

Such indicators suggest that symmetry formation and breaking in conversation follow influences from contextual cues, specifically gender and personality. Males and females appear to respond differentially based on partner gender. Males in conversation with one another appear unaffected by coordinated movement; if viewed alone, one might conclude that nonverbal coordination is an unnecessary component of their conversational style. However, when in conversation with females, males adapt to their conversational partners as well as initiate movements that are indicative of symmetry breaking. High dominant females show evidence of following partner behaviors; low dominant females both lead and follow in conversational interaction.

More experiments and analyses need to be performed to determine if such leading and lagging behaviors might be the result of mirror neuron activity in humans. Evidence of highly coordinated movement during dance suggests that humans are capable of matching and mirroring the behaviors produced by others, often with very little time between action and reaction. In terms of conversation, our perceptual systems may view certain characteristics as important for our continued success in communicating with others. Determining such characteristics would therefore be an important enterprise for furthering our knowledge of such perception-action systems.

# References

Bargh, J. A., Raymond, P., Pryor, J. B., & Strack, F. (1995). Attractiveness of the underling: An automatic power-sex association and its consequences for sexual harassment and aggression. *Journal of Personality and Social Psychology, 68*(5), 768–781.

Benjamin, G. R., & Creider, C. A. (1975). Social distinctions in nonverbal behavior. *Semiotica, 14,* 52–60.

Bente, G., Donaghy, W. C., & Suwelack, D. (1998). Sex differences in body movement and visual attention: An integrated analysis of movement and gaze in mixed-sex dyads. *Journal of Nonverbal Behavior, 22*(1), 31–58.

Bilous, F. R., & Krauss, R. M. (1988). Dominance and accommodation in the conversational behaviors of same- and mixed-gender dyads. *Language & Communication, 8*(3–4), 183–194.

Bull, P. E. (1987). *Posture and Gesture.* New York: Pergamon Press.

Burgoon, J. K., Buller, D. B., & Woodall, W. G. (1996). *Nonverbal Communication: The unspoken dialogue* (2nd ed.). New York: Harper Collins.

Burgoon, J. K., & Dillman, L. (1995). Gender, immediacy, and nonverbal communication. In P. J. Kalbfleisch & M. J. Cody (Eds.), *Gender, Power, and Communication in Human Relationships* (pp. 63–81). Hillsdale, NJ: Erlbaum.

Condon, W. S., & Ogston, W. D. (1967). A segmentation of behavior. *Journal of Psychiatric Research, 5,* 221–235.

Dovidio, J. F., Ellyson, S. L., Keating, C. F., Heltman, K., & Brown, C. E. (1988). The relationship of social power to visual displays of dominance between men and women. *Journal of Personality and Social Psychology, 54,* 233–242.

Duncan, S. D., & Fiske, D. W. (1977). *Face-to-face Interaction: Research, methods, and theory.* Hillsdale, NJ: Erlbaum.

Ekman, P., & Friesen, W. V. (1969). The repertoire of nonverbal behavior: Categories, usage, and coding. *Semiotica, 1,* 49–98.

Gough, H. G. (1956). *Manual for the California Psychological Inventory.* Palo Alto, CA: Consulting Psychologist's Press.

Hall, J. A. (1984). *Nonverbal Sex Differences: Communication accuracy and expressive style.* Baltimore: The Johns Hopkins University Press.

Henley, N. M. (1977). *Body Politics: Power, sex, and nonverbal communication.* Englewood Cliffs, NJ: Prentice Hall.

Kendon, A. (1970). Movement coordination in social interaction: Some examples described. *Acta Psychologica, 32,* 1–25.

Lakoff, R. (1975). *Language and Women's Place.* New York: Harper and Row.

Mehrabian, A. (1981). *Silent Messages: Implicit communication of emotions and attitudes* (2nd ed.). Belmont, CA: Wadsworth.

Mitchell, G., & Maple, T. L. (1985). Dominance in nonhuman primates. In S. L. Ellyson & J. F. Dovidio (Eds.), *Power, Dominance, and Nonverbal Behavior* (pp. 49–66). New York: Springer-Verlag.

Mulac, A. (1989). Men's and women's talk in same-gender and mixed-gender dyads: Power or polemic? *Journal of Language and Social Psychology, 8*(3–4), 249–270.

Rizzolatti, G., & Arbib, M. A. (1998). Language within our grasp. *Trends in Neuroscience,* *21*(5), 188–194.

Thorne, B., & Henley, N. (1975). *Language and Sex: Difference and Dominance.* Rowley, MA: Newbury House.

Weisfeld, G. E., & Linkey, H. E. (1985). Dominance displays as indicators of a social success motive. In S. L. Ellyson & J. F. Dovidio (Eds.), *Power, Dominance, and Nonverbal Behavior* (pp. 109–128). New York: Springer-Verlag.

Weitz, S. (1976). Sex differences in nonverbal communication. *Sex Roles, 2,* 175–184.

Williams, J. E., & Best, D. L. (1986). Sex stereotypes and intergroup relations. In S. Worchel & W. G. Austin (Eds.), *Psychology of Intergroup Relations* (pp. 244–259). Chicago: Nelson-Hall.

# Symmetry building and symmetry breaking in synchronized movement

Steven M. Boker and Jennifer L. Rotondo

Department of Psychology, the University of Notre Dame, USA /
Augustana College, Sioux Falls, South Dakota, USA

## 1. Introduction

Symmetry has many forms and has been shown to be a powerful organizing mechanism in perception (Bertamini, Friedenberg, & Kubovy 1997; Kubovy 1994; Palmer & Hemenway 1978). Types of *spatial symmetry* in natural objects include the bilateral symmetry of most multicelled animals, other classes of repeating patterns exemplified by the symmetry classes of crystalline structures (Senechal 1990), and algorithmic symmetry found in plants (Lindenmayer & Prusinkiewicz 1990). In general, spatial symmetry can be considered to be a form of self-similarity across spatial intervals. When two objects exhibit bilateral symmetry that occurs about a plane between them, then they exhibit *mirror symmetry*. That is, the objects are spatially translated on one axis and flipped about another axis exactly as is one's image in a mirror.

The same logic can be applied to temporal structures. *Temporal symmetry* is considered to be a form of self-similarity over intervals of time. Thus, a simple repeating auditory rhythm such as would be produced by a drum machine can be considered to have a form of symmetry called *translational symmetry*. In this case the rhythm is repeated, thus it is self-similar across time. The symmetry in this case involves translation in time, but no mirror-flip. Temporal symmetry provides organizational structure that is perceived as a Gestalt.

If a person makes a gesture with the left hand and then makes the same gesture with the right hand, then we could say the person exhibits *spatiotemporal symmetry*. Both spatial bilateral symmetry and temporal symmetry were involved in the production of these coordinated gestures. Similarly, if two people face each other and one makes a movement with the right arm and the other person then mimics that movement with her or his left arm, then this dyad exhibited spatiotemporal mirror symmetry. We will explore spatiotemporal mirror symmetry in dyads as one person mimics another while they stand face to face.

## 1.1   Symmetry formation in conversation

There are many features of conversational behavior that give rise to symmetry. The most striking of these, obvious to even the casual observer, is that people tend to mimic each other's posture during conversation (Lafrance 1985). This mirror symmetry in conversants' posture and gesture is not complete. In fact, if one conversant mimics her or his conversational partner too closely this will almost always be immediately noticed by the partner and can become a source of discomfort and embarrassment for the partner being mimicked.

What communicative purpose does this mirror symmetry in posture and gesture serve during conversation? One possible explanation is that by creating mirror symmetry a conversant is attempting to physically access the somatosensory inner state of his or her interlocutor. To the extent that cognitive or emotional states are correlated with physical postures, gestures and facial expressions, one may be able to trigger the experience of an internal state by placing one's body in a particular position, making a particular gesture or expression. Thus, one might expect that the formation of symmetry in conversation might signify common internal states and thus facilitate communication (Bavelas, Black, Chovil, Lemery, & Mullett 1988).

A second possible explanation is that by forming symmetry with an interlocutor a conversant is attempting to express agreement or empathy. This explanation is fully congruent with the first explanation since it is predicated on the idea that the conversant has information leading him or her to believe that creating symmetry will be understood by the interlocutor as conveying that the conversant is experiencing or has knowledge of a similar inner state as is being experienced by the interlocutor.

The second explanation does not account for the phenomenon of embarrassment caused by excessive symmetry. However, if symmetry formation is a communicative strategy for understanding, when a conversant forms too much symmetry with her or his interlocutor, it might signal too much access to the interlocutor's inner state causing an embarrassing sense of loss of privacy.

## 1.2   Symmetry breaking in conversation

As symmetry is formed between two conversants, the ability of a third party observer to predict the actions of one conversant based on the actions of the other increases. In this way we can consider there to be an increased redundancy in the movements of two conversants as the symmetry between them increases. Formally, the opposite of redundancy is information as proposed by Shannon and others (Redlich 1993; Shannon & Weaver 1949). Thus, when symmetry is broken, the third party observer of a conversation's postures and gestures would be surprised. The observer's previously good predictions would now be much less accurate. In this way we can consider the breaking of symmetry as a method for increasing the nonverbal information communicated in a conversation.

We consider the interplay between symmetry formation and symmetry breaking in posture and gesture to be integral to the process of communication. The spatiotemporal structure of the formation and breaking of symmetry is likely to be diagnostic of a variety of social and cognitive aspects of the dyadic relationship. This diagnostic should be relatively insensitive to the semantic content of the conversation, but instead will express the large scale interpersonal prosodic nature of the dyadic interaction.

## 1.3  Perception–action loops and mirror systems

Mirror systems as studied by Rizzolati and colleagues (Rizzolatti & Arbib 1998) are hypothesized to provide a link between the integration of the perception of an individual's self-movement and the perception of another individual's mirrored movement. In macaques, it has been found that the same neuron will fire when a particular movement is made by the monkey or when the monkey observes the same movement made by a researcher. Since there exists a class of these mirror neurons that (a) fires only in the presence of a movement that includes an actor, an action and an object acted upon, and (b) is located in the macaque's brain in an area that is likely to correspond to an area in the human brain implicated in syntax (Broca's area), there has been speculation that such a mirror system may be an evolutionary step on the road to the development of language (Gallese & Goldman 1998).

Given these assumptions, we wonder: How well might such a mirror system perception–action loop perform in ideal circumstances? Is there a difference between the situation in which both individuals attempt to synchronize with each other as compared with when one person provides a leading stimulus and another follows? What are the time lags that separate leaders and followers when both can hear an external synchronizing stimulus?

## 2. An experiment in symmetry and dance

In order to answer the previous questions and to further explore the nature of dyadic perception–action loops, we designed an experiment in which individuals listened to a repeating rhythm and were asked to either lead or follow their partner in a free-form spontaneous dance. During the dance, both participants' movements were tracked and recorded to a computer. Thus, the temporal structure of the symmetry formation and symmetry breaking between the two participants in a dyadic perception–action loop involving the mirror system could be examined.

### 2.1 Participants

Six dyads of young adults were recruited from undergraduate psychology classes at the University of Notre Dame. Each dyad consisted of one male and one female participant. Some participants were previously acquainted and others had not previously met. Data from one dyad was not used since one of the participants in the dyad refused to follow instructions.

### 2.2 Apparatus

An Ascension Technologies MotionStar 16 sensor magnetic motion tracking system was used to track the motions of participants. Eight sensors were placed on each individual: one on the back of a baseball cap worn tightly on the head, one strapped just below each elbow using a neoprene and velcro around-the-limb strap, one held to the back of the back of each hand with an elastic weightlifting glove, one held to the sternum with a neoprene and velcro vest, and one strapped just below each knee with a neoprene and velcro around-the-limb strap. Each sensor is connected to the MotionStar system computer with a long cable. Thus each individual had a bundle of 8 cables that were gathered and positioned behind them in order to provide the minimum of interference with movement.

### 2.3 Methods

Participants were strapped into the sensors and led into the measurement room where they were instructed to stay within a 1 m × 1 m square marked with tape on the floor of the room. The two regions were 1.5 m apart at their nearest edge. Headphones were worn by each participant and they were then instructed that during each trial they would hear a repeating rhythm lasting approximately 40 seconds during which time they were to dance in synchrony with rhythm without touching each other. Prior to each trial they would be instructed over the headphones whether to lead or follow their partner during that trial.

Each trial stimulus consisted of a repeating rhythm synthesized by computer to have a beat interval of 200 ms and either 7 or 8 beats per repeating measure. Eight stimuli were chosen from a set of stimuli with known properties of perceived segmentation (Boker & Kubovy 1998). At the beginning of each trial each person was either instructed to lead or follow. Thus there were four instruction conditions: (1) person A leads and person B follows, (2) person A follows and person B leads, (3) both person A and B lead, and (4) both person A and B follow.

## 3.   Results

In order to simplify the analysis of the movements, data from the eight sensors attached to each dancer were combined in order to calculate an overall velocity at each sample interval in the following manner. First the sternum sensor position was subtracted from each of the other sensors so that all limb positions and the head position were relative to the position of the trunk. Then the velocity along each axis for each sensor was calculated as

$$v_{ij}(t) = \frac{x_{ij}(t+3) - x_{ij}(t-3)}{6(1/80)} \tag{1}$$

where $v_{ij}(t)$ is the velocity at sample $t$ for sensor $i$ along axis $j$, and $x_{ij}(t-3)$ is the position in centimeters relative to the trunk at sample $t-3$ for sensor $i$ along axis $j$. The difference in the two sampled positions is divided by $6(1/80)$ since there are six intervals between the two samples and each interval is 1/80th of a second. Thus the estimated velocity at each sample time $t$ is effectively low pass filtered to remove high frequency noise and is expressed in units of cm/sec.

Finally the results of all velocity calculations at each time $t$ were combined as a root mean square to give the overall movement of an individual as

$$\bar{v}(t) = \left( \frac{\sum_{i=1}^{8} \sum_{j=1}^{3} v_{ij}(t)^2}{24} \right)^{\frac{1}{2}}. \tag{2}$$

Thus the root mean square velocity $\bar{v}(t)$ gives an estimate of the total activity for a dancer at time $t$. While this overall estimate of velocity does not give any estimate of the accuracy with which spatial symmetry is formed, it does give an estimate of the overall amount of temporal symmetry when it is analyzed using cross-correlational analysis.

The root mean square velocity was calculated for each trial and was then predicted using a mixed effects model grouping by subject within dyad. Predictor variables in the model were the sex of the subject, the length of the repeating rhythm and the instruction category. There was no significant effect of sex on the overall RMS velocity during the dance. There was an effect of length of the rhythm (p < 0.01) such that rhythms with 8 beats per measure produced higher velocities than rhythms of length 7.

## 3.1   Cross-correlation of two dancers' velocities

In order to gain a summary estimate of the symmetry between dancers' overall velocities, we calculated the lagged cross-correlation of the two dancer's RMS velocities during short intervals of time. For two second windows of time, 160 samples of velocity, we calculated the correlation between the two dancers where the onset of the windows was lagged by values on the interval $-2\,\text{sec} \leq \tau \leq +2\,\text{sec}$. Thus a trial resulted in an $T \times I$ matrix of correlations where, for a target time $t$ during the trial, a vector of correlations $r_{ti}$ was calculated as

$$r_{ti} = \begin{cases} r((x_t, \ldots, x_{t+w}), (y_{t-j}, \ldots, y_{t-j+w})) & \text{if } j \geq 0 \\ r((x_{t+j}, \ldots, x_{t+j+w}), (y_t, \ldots, y_{t+w})) & \text{if } j < 0 \end{cases} \tag{3}$$

where the index $i$ is on the integer interval $1 \leq i \leq I$, $I$ is an odd number, $r()$ is the Pearson product moment correlation function, $j$ is a lag index calculated as $j = (i-1)-((I-1)/2)$, and $w$ is the total number of samples within $a$ window. In the present case, we chose $w = 160$ and $I = 321$ so that given an 80 Hz sampling rate, we correlated 2 second windows lagged by as much as 2 seconds from each other.

A detailed examination of example trials gives an idea of the time course of how the symmetry forms between individuals and how it can be broken. In Figure 1 are plotted matrices of cross-correlations with grays indicating the value of the correlation in each cell. Each row of these graphs plots one target time during 20 seconds of the trial with time since beginning of the trial on the vertical axis. Each column of the graph plots one lag value. A lag of zero is in the column at the center of the horizontal axis, person A is leading for columns to the left of center, and person B is leading for columns to the right of center.

When both participants were instructed to follow, the cross-correlations were likely to appear as in Figure 1a. A short, five second period of settling of the two participants occurred at the beginning of the trial during which time they each followed the others' movements. Then as they became still and stood and stared at each other, there was no temporal structure to their velocities, which were near zero.

When both participants were instructed to lead, very often temporally symmetric motions would become evident. Even though each was instructed to lead, participants would mutually entrain into spatiotemporal symmetry. An example of this type of behavior can be seen in Figure 1b which plots 20 seconds from the middle of a trial in which both participants were instructed to lead. The strong vertical dark gray bands, which peaks are separated by approximately 1400 ms and occur during the interval 15 seconds to about 27 seconds from the beginning of the trial, indicate the presence of strong spatiotemporal symmetry in the dyad's motions. However, at around 27 seconds there is a breaking of symmetry. It may be that one or the other of the participants became aware that they were following the other and decided to break the symmetry in order to take the lead.

Figure 1c and 1d plot example trials in which one dancer was leading and one dancer was following. In Figure 1c is plotted an example of the quick formation of strong, stable symmetry in a trial as exhibited by the vertical purple bands beginning around 8 seconds into the trial. As a contrast, Figure 1d plots an example trial at the other end of the spectrum, in which only weak symmetry of motion was established.

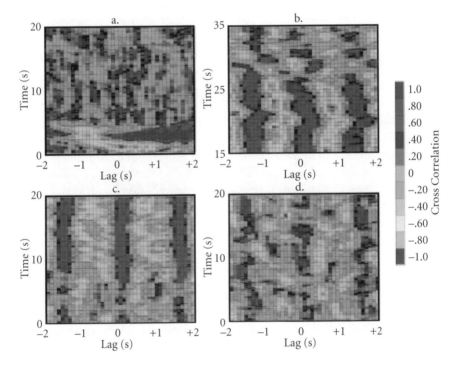

**Figure 1.** Cross-correlation matrices plotted for twenty seconds from four example trials. The scale on the right denotes the color assigned to each value of cross correlation. (a) The beginning of a trial in which both subjects were instructed to follow. (b) Twenty seconds from the middle of an example trial in which both subjects were instructed to lead. (c–d) The beginning of two example trials in which one subject was instructed to lead and the other to follow.

## 4.  Implications for mirror systems

The results of the current experiment demonstrate that when individuals coordinate their movements in a lead-follow or lead-lead dance to a repeating rhythm, the overall velocity of each individual entrains reliably with the length of the perceptual segmentation from the auditory stimulus. Thus, cyclic movement is created with a frequency that matches the frequency of the repeating rhythm and stable, reliable, and near zero between-individual phase lags are present. While it may be unremarkable that this situation occurs when one individual is instructed to lead and the other to follow, it is somewhat unexpected that this small phase lag between-individual entrainment occurs when both individuals are instructed to lead. Thus strong spatiotemporal symmetry was observed even when both individuals were instructed to lead: that is to ignore any symmetry that occurred between them.

The strong lead-lead correlation between individuals' velocities, while reliably smaller than that of the lead-follow condition, is so strong that it suggests that entrainment of cyclic movements between individuals may be especially easy. One possible mechanism for this entrainment may be a spatiotemporal component to mirror systems. If this type of entrainment is easier than other forms of mimicry, then this would provide a possible explanation for the near universal use of a cyclic movement of the head to nonverbally indicate agreement or disagreement during conversation. The primary information that needs to be communicated nonverbally by a listener during conversation, understanding or misunderstanding, would be likely communicated by a method that would allow the quickest recognition of entrainment between individuals. If this argument holds, we expect a special bias toward cyclic movement in mirror systems: a bias towards spatial mirror symmetry and temporal translational symmetry.

## Acknowledgements

This work was supported in part by NIA grant R29 AG14983-01 and a grant from the University of Notre Dame. We wish to thank Elizabeth Marsh who assisted in gathering these data. Corresponding author: Steven M. Boker, Department of Psychology, The University of Notre Dame, Notre Dame Indiana 46556, USA; sboker@nd.edu; http://www.nd.edu/~sboker.

# References

Bavelas, J. B., Black, A., Chovil, N., Lemery, C. R., & Mullett, J. (1988). Form and function in motor mimicry: Topographic evidence that the primary function is communicative. *Human Communication Research, 14*(3), 275–299.

Bertamini, M., Friedenberg, J. D., & Kubovy, M. (1997). Detection of symmetry and perceptual organization: The way a lock-and-key process works. *Acta Psychologica, 95*(2), 119–140.

Boker, S. M., & Kubovy, M. (1998). The perception of segmentation in sequences: Local information provides the building blocks for global structure. In D. A. Rosenbaum & C. E. Collyer (Eds.), *Timing of behavior: Neural, computational, and psychological perspectives* (pp. 109–123). Cambridge, MA: MIT Press.

Gallese, V., & Goldman, A. (1998). Mirror neurons and the simulation theory of mind-reading. *Trends in Cognitive Sciences, 2*(13), 493–501.

Kubovy, M. (1994). The perceptual organization of dot lattices. *Psychonomic Bulletin & Review, 1*(2), 182–190.

Lafrance, M. (1985). Postural mirroring and intergroup relations. *Personality and Social Psychology Bulletin, 11*(2), 207–217.

Lindenmayer, A., & Prusinkiewicz, P. (1990). *The algorithmic beauty of plants.* New York: Springer Verlag.

Palmer, S. E., & Hemenway, K. (1978). Orientation and symmetry: Effects of multiple, rotational, and near symmetries. *Journal of Experimental Psychology: Human Perception and Performance, 4*(4), 691–702.

Redlich, N. A. (1993). Redundancy reduction as a strategy for unsupervised learning. *Neural Computation, 5,* 289–304.

Rizzolatti, G., & Arbib, M. A. (1998). Language within our grasp. *Trends in Neuroscience, 21*(5), 188–194.

Senechal, M. (1990). *Crystalline symmetries: An informal mathematical introduction.* Philadelphia: Adam Hilger.

Shannon, C. E., & Weaver, W. (1949). *The mathematical theory of communication.* Urbana: The University of Illinois Press.

# Part III

# Mirror neurons system and the evolution of brain, communication, and language

# On the evolutionary origin of language*

Charles N. Li and Jean-Marie Hombert
University of California, Santa Barbara, USA / Laboratoire Dynamique
du Langage, Lyon, France

## 1. Origin vs. evolution of language

In recent years there has been a flurry of scholarly activities on the origin of language. New scholarly societies have been formed; conferences have been organized;
books, edited and written. In all of these activities, a common theme prevails as it
appears in the titles of conferences, articles and books. This common theme is,
"the evolution of language". It implies that language was the communicative behavior of hominids. Yet no one would assume that all our phylogenetic ancestors
had language, if language designates the casual, spoken language of anatomically
modern humans.[1] Even if we confine ourselves to the family of hominidae, no
one would assume that the earliest hominids such as the Ardipethecus ramidus
and the Austrolopithecines,[2] had language. It was about 6–7 million years ago that
the first hominids began to evolve away from the quadrapedal, knuckle-walking
great apes by embarking upon the evolutionary pathway of developing bipedal lo-
comotion.[3] At the beginning, they were on average a little more than 1 meter in
height and 30 kilograms in weight with a cranial capacity at approximately 400 c.c.
Among contemporary primates, they would be much more ape-like than human-
like. Given their anatomical difference from humans, it seems sensible to consider
their communicative behavior distinct from human casual, spoken language. It follows, then, the evolution of their communicative behavior is not the evolution of
language. Nevertheless, their communicative behavior evolved, as did the communicative behavior of all early hominids evolve toward the emergence of language.
The investigation of the origin of language is, therefore, an enterprise concerned
with the evolution of the communicative behavior of our hominid ancestors, NOT
the evolution of language. The evolution of language concerns linguistic change.
It is diachronic linguistics. The origin of language is not diachronic linguistics.
Chronologically the study of the evolution of language starts from the time when

language emerged, whereas the study of the origin of language ends at the point in time when language emerged. This distinction does not belittle the significance of the research effort probing into the older and older layers of human language. Nor does it dismiss the importance of the proto-human language if and when its features can be inferred. The distinction must be made for one important reason, and that is the ways casual, spoken language changes and the ways hominid communicative behavior changes are fundamentally different.

## 2.  The evolutionary change of communication vs. linguistic change

The fundamental difference lies in the fact that early hominid communicative behavior, not human language, is subject to the constraints of Darwinian evolution. In other words, the evolution of our hominid ancestors' communicative behavior involved natural selection and genetic mutation. A change of their communicative behavior in the direction toward language was adaptive in the sense that it enhanced their life expectancy and their reproductive success. Those hominids who made the change achieved a higher level of fitness than those hominids who failed to make the change. The reason is that a change moving the hominids' communicative behavior one step closer to human language would imply greater communicative efficiency. Greater communicative efficiency would entail greater ease with which valuable experience and knowledge could be passed from one individual to another and from one generation to another. Rapid and efficient transmission of knowledge conferred an immense competitive advantage to the hominids for securing resources and possibly vanquishing others, including other species of hominids whose communicative behavior was less developed in the direction toward language. Given that hominids within the genus of Homo and possibly most gracile species of the Autrolopithecine genus are generalists who did not specialize in any specific ecological niche, the competitive advantage conferred by a more effective communicative behavior may explain why modern humans are the only surviving species within the taxonomic family of hominids. In the animal kingdom, the only other case of a single surviving species in a family is the ant-eating African aardvark! Typically different species of a family specialize in different ecological niches. Consider, for example, the felines and the Darwinian finches of the Galapagos. When two hominid species happened to co-exist as generalists and the communicative behavior of one species were more effective than that of the other species, there would be a good possibility that the communicatively more advanced species would eliminate the other through competition!

The change of hominid communicative behavior toward human language began with symbolic communication. By symbolic communication, we mean the use of symbols each of which represents directly, consistently and exclusively an entity in the world. The emergence of the first symbolic communicative signal among hominids is not only an important evolutionary landmark but also represents a quantum leap from non-human primate communication. Prior to this landmark development, the communicative signals of hominids should not be qualitatively different from non-human primate communicative signals. Non-human primate communicative signals are not symbolic. They have functions, not meanings. Consider, for example, the well-known warning calls of the African vervet monkeys. One indicates the warning uttered by the signaler when it notices the presence of a reptilian predator. Even though such a warning call differs from the other warning calls connected to the presence of a mammalian predator, an avian predator, a Masai herdsman, or some other potentially dangerous animals, it is not a symbol that represents a reptilian predator. It merely indicates the function of the vocalization in the presence of a reptilian predator. Seyfarth and Cheney (1999) note that vocal production, i.e. delivery of acoustically defined calls, among apes and monkeys appears fully formed shortly after birth, suggesting that vocal production may be largely innate. In addition, Aitken (1981) and Pandya et al. (1988) conducted experiments on monkeys showing that their vocal production was mediated primarily by the central (periaqueductal) gray area of the mid-brain, a phylogenetically very old set of neurons responsible for arousal and motivational states in all vertebrates. Although Seyfarth and Cheney (1999) point out that the development of vocal usage (vis-à-vis production) as well as the development of responses to the calls of others do require some learning at least for vervet monkeys, their study does not alter the fact that (1) non-human primate communicative signals are not symbolic,[4] and (2) the production of non-human primate communication is mediated primarily by the central gray area of the mid-brain.[5] In the case of vervet monkey's warning calls, the only role of the neocortex involves associating a particular involuntary vocalization with a specific situation. The vocalization is involuntary because it is probably associated with fear aroused by the situation.[6] Hence an infant vervet possesses the adult repertoire of vocalization. The learning during ontological development involves the correct coupling of one involuntary vocalization with one specific dangerous situation. Each expression can be graded according to intensity. But it is only the coupling process that is mediated by the neocortex, and this coupling process, according to Seyfarth and Cheney, requires learning. The neural mechanism we have just sketched for non-human primate communication contrasts sharply with the neural mechanism of the production of causal, spoken language. The production of casual, spoken language is primarily mediated by the neocortex. The emotional/motivational state of the speaker can be viewed as a coterminous but neurologically separate dimension of speech expressed primarily in prosody. It is, therefore, not surprising that participants in casual spoken language can talk about things that are remote in time and space from the location of the conversation. This is the "displacement" feature of human language that Charles Hockett (1960) pointed out. It does not exist in non-human primate communication because a non-human primate communicative signal tends to be associated with the emotional or motivational reaction to a particular situation including the animal's own internal hormonal state.

Even though the onset and expansion of symbolic communication in hominid evolution represent a break from non-human primate communication, the process of change before the origin of language remains an evolutionary change. Such a change typically involves a slow and gradual Darwinian process that requires hundreds and thousands of generations.[7] It is adaptive in the sense that it improves the fitness of the hominids. Linguistic change, the change of language after its origin, however, is by and large tied to society and culture. It has nothing to do with genetic mutation, natural selection, life expectancy or reproductive success. Language changes constantly. Our pronunciation changes, our vocabulary changes, our ways of speaking change, and our grammar changes. Confusing the evolution of language with the origin of language may mislead researchers into attributing features of language to the communicative behavior of my evolutionary ancestors before the emergence of language.

### 3.  The emergence of language vs. the emergence of anatomically modern human

Many paleoanthropologists believe that language emerged together with anatomically modern humans. That is, casual spoken language coincided with the emergence of anatomically modern humans in Africa some 150–130 thousand years ago (Walker & Shipman 1996). However, there is a confluence of evidence from paleo-demography, molecular genetics, and a variety of archaeological discoveries, which suggest that the crystallization of language may not have coincided with the emergence of anatomically modern humans.[8] This confluence of evidence has led us to postulate that language emerged around 80–60 thousand years ago, several tens of thousand years after the appearance of anatomically modern humans. We will briefly summarize the evidence:

1.  Around 60,000–40,000 years ago, the size of human population began its first explosive increase. According to F. A. Hassan's study of demographic archaeology, the dramatic increase in human population started at the end of the Middle Paleolithic period at about 40,000 years ago (Hassan 1981). Figure 1 is modeled after Hassan (1981: 196).

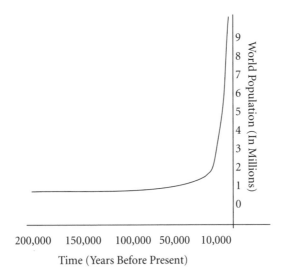

**Figure 1.** Estimates of human population

The French paleo-demographer, Jean-Noel Biraben, independently arrived at a similar conclusion in his "Essai sur l'evolution du nombre des Hommes", (1979). According to Biraben's estimate, the world population increased by 500% around 40,000 years ago.

The population explosion during the period of 60,000–40,000 years ago is also confirmed by the study of mitochondria DNA (m-DNA) phylogeny (Sherry et al. 1994) on the basis of polymorphism and average mutation rate.[9]

The second major population increase in human history occurred at the beginning of the Neolithic period, around 10,000 years ago. The driving force behind this second population explosion is well known: the development of agriculture.

Question: What caused the first explosion of human population between 60,000 to 40,000 years ago?

Whatever the cause may be, it must have the potential of facilitating all aspects of human activity and social interaction and consequently enhancing human life expectancy and survival rate.

2.  At around 40,000 year before present, a "Big Bang" of art occurred. The oldest preserved rock paintings discovered to date are the red ochre figures of half- human and half-beast found in the Fumane Cave northwest of Verona at 36,500–32,000 years old and the Grotte Chauvet paintings of animals at approximately 32,000 years old (Balter 1999). The artistic sophistication of the Grotte Chauvet paintings includes such refined techniques as shading and perspective, suggesting a long period of the development of artistic concepts and

skills before the creation of the Grotte Chauvet painting. Personal ornamentation is another facet of the Big Bang of art. The oldest ornaments in the form of beads and pedants carved out of ivory are 35,000 years old (White 1986). These ornaments are conceptually, symbolically and technically complex, suggesting the work of a modern human mind.

Question: Is the Big Bang of art a consequence of the emergence of language, which facilitates my intellectual capability?

3.  At around 50,000–40,000 years ago, the beginning of the Upper Paleolithic period, tools, like art, in stark contrast to all other earlier tool kits, began an unprecedented acceleration of diversification and specialization. This development in tool variety and complexity was worldwide. If we plot the trajectory of change in stone-tool technologies in terms of number of distinct tool types against time, the curve obtained strongly resembles that of the population change. It shows a long stasis characterized by a relative flat line until the end of the Middle Paleolithic and the beginning of Upper Paleolithic when the curve begins to shoot up vertically. Figure 2 is modeled after Lewin (1993:33).[10]

The Upper Paleolithic tools include hafted blades that are at least twice as long as they are wide and numerous types of hafted small geometrically shaped tools such as chisels and files for carving and making bone instruments. They indicate a level of sophistication involving design and symbolism previously unattained in hominid history.

Question: What is the reason behind this explosive development of tools?

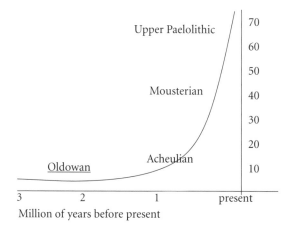

Figure 2.  Number of distinct stone tools in hominid history

4.  The colonization of Australia occurred approximately 60,000 years ago. At the time, because of glaciation, Australia, Papua New Guinea and Tasmania formed one continuous land mass, while many of the present day islands of the Indonesia archipelago were connected with the Malaysia peninsula of Asia. Reaching Australia from Asia entailed the crossing of deep, fast-moving ocean water of approximately 100 kilometers. Such sea-crossing required social organization, collaborative effort, sophisticated planning, some skills, equipment and knowledge of navigation.

Question: What enabled humans to cross deep, fast-moving ocean water at that time but not before?

To sum up, these four pieces of evidence collectively point to a new cognitive capacity for sophisticated culture emerging during the period of 80,000–60,000 years ago. We cannot attribute this new cognitive capability to a larger brain, because human cranial capacity, if anything, has decreased since the dawn of anatomically modern human at around 150,000–120,000 years ago. In fact, the significant time gap between the first occurrence of anatomically modern humans and the first indication of a capacity for modern culture prompted Donald Johanson and Blake Edgar to pose the following question in their 1996 book, "From Lucy to Language",

> This is one of the key unanswered questions in paleoanthropology today. Is it possible that the brains of early Homo sapiens were simply not yet wired for sophisticated culture? The modern capacity for culture seems to have emerged around 50,000 year ago, and with it, behaviorally modern humans who were capable of populating the globe.                (Johanson & Edgar 1996:43)

Interestingly, the noted paleoanthropologist, Richard Klein, made a similar observation. Klein suggested that a *hidden evolution* of the brain, unrelated to its size and shape took place some 50,000 years ago, and that hidden evolution accounted for human's modern capacity for sophisticated culture and cognition (Klein 1989).

We submit that Klein's notion of a hidden evolution of the brain is exactly the same as the answer to the question posed by Johanson and Edgar, and the answer to Johanson and Edgar's question is also the answer to the four questions we have posed in my discussion of the confluent evidence. In our opinion, Klein's "hidden evolution of the brain" is a new deployment of cognitive ability brought about by the emergence of language. In other words, the crystallization of hominid communicative behavior into language is the underlying reason for all of the three pieces of evidence: the first and sudden surge of human population, the Great Bang of art, the explosive development of tools, and the crossing of deep, fast-moving ocean water separating Asia from Australia.

If language emerged after the arrival of anatomically modern humans, how and when it emerged? What are the mechanisms underlying the evolution of ho-

minid communicative behavior? According to our research, there are three mechanisms and four processes of evolution that are especially important. We will briefly
sketch these three mechanisms and four processes.

## 4.    Four evolutionary processes leading to the emergence of language

### 4.1    Reduction of the gastrointestinal tract

The reduction of the gastrointestinal tract is a necessary concomitant development
of the increase in encephalization in hominid evolution. The reason is that an enlarged brain consumes an enormous amount of energy that has to come at the
expense of some part of a homeostatic system of the hominid anatomy. The brain
of a newborn infant, for instance, consumes 60% of the energy it takes in. Leslie
Aiello and Peter Wheeler provide detailed analysis of this evolutionary process in
a series of papers from 1995 to 1998. What enables the G.I. tract to decrease in
hominid evolution is the change of diet. The change of diet in hominid history is
inferred from archaeological evidence, the size of the fossilized jaw and the detailed
properties of the fossilized teeth: their size, shape, striation, surface structure and
the thickness of the enamel. The change is in the direction of greater nutritional
value. Increased nutrition of ingested food facilitated the evolutionary process of
decreasing the G.I. tract. Meat and sea food, of course, are the most nutrient-rich
food. They became part of the diet of hominids in the genus Homo. Cooked food
also facilitates digestion and makes it possible for the shrinking of the G.I. tract.
Cooking can also enhance the sugar content of a variety of tubers. However, the
earliest uncontroversial date for hearth is 400,000–300,000 year before present. Lee
Berger claims that hearth existed in one of the hominid site in South Africa at approximately 900,000 years before present (Berger 2000). If Berger is correct about
the South African hearth, cooked food might very well have played a role in the
reduction of the G.I. tract in hominid evolution. A diet of 60% cooked tubers,
about the proportion used in modern native African diet according to Wrangham
et al. (1999), will increase coloric intake by approximately 43%. Wrangham and
his colleague estimates that every square kilometer in Tanzania's savanna wood-
land, similar to the habitat of most early hominids, contains 40,000 kilograms of
tubers today. They argue that cooked tubers, more so than meat, made possible the
evolution of large brain, smaller teeth, shorter arms and longer legs, and even male-
female bonding. The hypothesis put forth by Wrangham et al. is supported by the
thesis that women, especially grandmothers, played a critical role in the evolution
of Homo erectus by being the food gatherers (O'Connell et al. 1999).

## 4.2  Enlargement of the vertebral canal

Ann MacLarnon and Gwen Hewitt (1999), provides detailed analysis and convincing arguments to demonstrate that the thoracic vertebral canal enlarged in hominid evolution during the period of 1.6 million years ago to 150,000 years ago for the purpose of enhancing thoracic innervation of the intercostal muscles controlling breathing during speech. Their analysis demonstrates that all other possible reasons for the enlargement of the vertebral canal were invalid. In other words, the anatomical evolutionary process of enlarging the vertebral canal in hominid history is an adaptation to enhance the vocalization capability.

## 4.3  Descent of the larynx

The descent of the larynx is another gradual evolutionary process that occurred among the species of the genus Homo. It resulted in the gradual formation of an L-shaped vocal tract which serves to facilitate articulation. The fossil evidence of this process is poor because the key to the descent of the larynx is a specially shaped hyoid bone. Even in modern humans, the hyoid bone is miniscule. Fossil remain of the hyoid bone is, therefore, predictably scarce. But we do know from the paper written by Arensburg in 1990, that the Kebara Neanderthal in Israel possessed the specific hyoid bone required by a descended larynx.

## 4.4  Increase in encephalization

The large size of the human brain in comparison to other primates is prominently manifested in the neocortex.[11] The neocortex, the newest outer "skin" or "bark" in evolutionary terms, plays a critical role in all human cognitive behaviors. Because of its enormous size, it endows human beings with a prodigious cognitive memory and other capabilities.[12] A prodigious cognitive memory is a pre-requisite for language because, beyond vocabulary and grammar, every language has an enormous set of idiosyncratic ways of saying things. Let us explain.

Any person who is fluent in two or more unrelated languages or more is likely to have noticed that being fluent in a language requires much more than internalizing the grammar and acquiring a good vocabulary of that language. One needs to know how to say things in a language, and how to say things in a language requires an enormous amount of knowledge beyond syntax and morphology. A person can master all of the grammatical principles of a language, possess a large vocabulary in that language, but if that person has not learned the myriad ways of saying things in that language, s/he will not speak like a native speaker. In other words, such a person's utterance is likely to be unidiomatic or not in agreement with the ways native speaker say things. For example, in most Romance languages, the way to say,

"I am hungry", is literally "I have hunger." If a Spaniard says in English, "I have hunger" to mean "I am hungry", the Spaniard has not made a grammatical mistake in English. His utterance is simply unidiomatic, i.e. not in accord with the way native speakers say it!

Most polyglots have witnessed interesting and amusing examples of unidiomatic utterances by non-native speakers. The important point is that the ways of saying things tend to be unique to a language or a group of closely related languages, and they are not confined to a few special expressions.

The New Zealand linguist, Andrew Pawley, has written eloquently about this aspect of language. So has his teacher, the American linguist, George Grace (1987). I will quote from an article by Pawley (1991),

> A language can be viewed as being (among other things) a code for saying things. There are a number of conventions that constrain how things should be said in a language generally or in particular contexts. Here I will mention only the general maxim: *be idiomatic*. This means, roughly, that the speaker (author, translator, etc.) should express the idea in terms that native speakers are accustomed to. For example, if you ask me the time and my watch shows the little hand pointing just past the 5 and the big hand pointing to the 2, an idiomatic answer would be 'It's ten past five', or 'It's five ten'. A reply such as 'It's five o'clock and one sixth' or 'It's five-sixth of an hour to six' or 'It's six less than fifty' would not count as idiomatic. To break the idiomaticity convention is to speak unnaturally.                                        (Pawley 1991:433)

The implication of this important characteristic of language is that linguistic behavior requires a prodigious memory. The neocortex of our brain must be able to store a vast amount of knowledge acquired through learning: the vocabulary, the grammar, and the myriad ways of saying things. We wish to emphasize that this knowledge is acquired through learning. We may be genetically predisposed toward acquiring language ontologically. Since language is our species-specific communicative behavior, there is nothing unusual for humans to be genetically predisposed toward acquiring language. Every species in the animal kingdom is either genetically programmed or predisposed to develop its species-specific communicative behavior. Earlier discussion points out that vocal production in non-human primates is largely innate, although vocal usage and proper communicative response to conspecifics require some learning. Acquisition of the first casual spoken language by children, however, requires a great deal of learning and a long, arduous process in comparison with the ontological development of the communicative behavior among non-human primates.[13] The human predisposition toward acquiring a casual spoken language does not imply an innate template of language-specific principles and parameters as Chomsky (1986) and Pinker (1994) claim.[14] What is innate, in our opinion, is the architectural and chronotopic development of the human brain in ontogeny, which channels the human infant's attention to the linguistic and social interaction of his/her environment and enables the human infant to learn a complex symbolic behavior requiring, among other things, a prodigious memory. The acquisition of language is, then, a complex interplay between this innate predisposition and the language environment (Elman, Bates et al. 1996; Elman 1999). From his/her linguistic environment, a child learns the vocabulary, the grammar and the myriad ways of saying things in a language. Having a large neocortex for our physical size constitutes an important aspect of the genetic predisposition toward acquiring a language. But human beings are not innately endowed with any knowledge of how to say things in any language. The numerous ways of saying things in a language require a long process of learning and tremendous amount of memorizing beyond the vocabulary and the grammar.[15]

Since language requires a large cognitive memory because of the vocabulary and the myriad ways of saying things, the expansion of the neocortex in hominid evolution must be co-related with the origin of language.

We should point out that in spite of the fact that increase in encephalization has received the most attention in the research on the origin of language, the other three evolutionary processes, namely, the decrease of the gastrointestinal tract, the increase of the vertebral canal and the descent of the larynx, are equally important.

We have summarized the four evolutionary processes that accompanied the development of hominid communicative behavior. We will now briefly delve into the three underlying evolutionary mechanisms.

## 5.  Three evolutionary mechanisms underlying the emergence of language

### 5.1  Duplication of Hometic genes

The mechanism underlying the sudden origin of phenotypic characteristics whether anatomical, physiological or behavioral is the duplication of the master regulatory genes, the so-called Homeotic genes.

Sudden origin of phenotypic features complements the classical Darwinian evolutionary change, which tends to be gradual and incremental. But sudden origin is also Darwinian. What is unusual about it is the nature of underlying genetic change, namely the duplication of the master regulatory genes, the homeotic genes.

Homeotic genes specify the synthesis of Transcription Factors, which turn on or off structural genes in a developing embryo. Turning on or off structural genes will determine the synthesis of certain enzymes and the growth or the absence of growth of specific physical structures. A minor illustration of a change in these master regulatory genes in human beings is the growth of six fingers on one hand.

A major illustration would be the development of an otherwise normal embryo without a head, due to the deletion of one such master regulatory gene, called LIM-1.

Most of the human homeotic genes turn out to be products of gene duplications at different times in evolution. Gene duplication as an evolutionary innovation has two distinct advantages. First, gene duplication can accomplish in one swoop what may have taken eons of time to create through the cumulative effect of gradual and piecemeal evolutionary changes in each of the original genes. Secondly, when a master regulatory gene is duplicated, the duplicated gene may undergo mutations, and therefore, perform new functions, because the original gene continues to perform its old functions that are necessary for the survival of the organism.[16] If the new functions are favored by natural selection, we obtain a sudden origin of new phenotypic manifestation and possibly a new species. The consequence of duplicating the genes regulating structural development is that the resulting structures should also show signs of duplication. However, the duplicated structures will be modified if the duplicated genes have undergone mutation. An obvious example of repeated structures in humans are the vertebrate column. The brain also contains many repeated structures, for example, the radial units of the embryo in its early stage of development, which are ultimately responsible for the size and architectonic pattern of the neocortex (Rakic 1988). These repeated structures could have arisen phylogenetically from the duplication of regulatory genes.

Recently, it has been discovered that genes which regulate the formation of the neocortex of the mammalian brain, known as Emx-1 and Emx-2, are duplicated and mutated copies of the genes that control head and brain formation in fruit flies. So the most advanced portion of my brain goes back to a very humble origin (Allman 1999)!

We wish to make clear that the sudden origin of phenotypic characteristics does not imply any suggestion of the sudden origin of language. On the contrary, by our reckoning, it took approximately 1.5–2 million years for hominid communicative behavior to evolve into casual spoken language. The relevance of the mechanism for the sudden origin of phenotypic characteristics to the origin of language lies in the development of the brain in hominid history. Aside from our earlier speculation of the evolutionary increase of radial units based on Rakic's research, Allman (1999) also suggests that many areas of the neocortex could have arisen in evolution from duplications of pre-existing areas as a result of genetic mutation. This evolutionary mechanism de-mystifies the dramatic expansion of the hominid brain during the past two and half million years, a relatively short duration on the evolutionary scale of time.

## 5.2  Change of the developmental clock

The second evolutionary mechanism concerns the change of the developmental clock. Developmental clock designates the length of ontological development of an animal or an organ of an animal.

The molecular mechanism determining the length of the molecular clock, however, is complex, involving regulatory genes as well as a feedback system consisting of inter-cellular communication. This evolutionary mechanism is important to the origin of language. Human language, as we all know, is inextricably connected with our large brain. One reason for the proportional large brain in Homo sapiens vis-a-vis the great apes, for example, is that the developmental clock for the human brain is lengthened considerably. The brain size of a human infant is not very much larger than that of an infant chimpanzee. But the human infant brain continues its developmental path for nearly twenty years. A chimp brain stops expansion three months after birth.[17]

While the lengthening of the developmental clock for the brain is partly responsible for the increase in encephalization among hominids, slowing down the developmental clock of the body also plays a role in creating a large human brain in proportion to body size. Slowing down the developmental clock for the body means terminating the developmental process long before the human body can reach a stage and size commensurate with the brain. This result is known as the *decrease in somatization*. An example of the slowing down of the human developmental clock is the late eruption of human teeth. In apes, for example, the deciduous teeth come out soon after birth. In human infants, the deciduous teeth continue to erupt well into the second year. In apes, the molars erupt immediately after deciduous teeth come out. In humans, the third molars, the so-called wisdom teeth, do not erupt until either late teens or early twenties.

Even though molecular biologists have not yet elucidated the full picture of how developmental clock is determined, we know that a change in developmental clock does not necessarily require long term, cumulative genetic mutations. In other words, the change of developmental clock for the body and the brain in hominids could occur suddenly. These changes in part explain the relatively large number of hominid species during the five million years before the emergence of anatomically modern humans.

## 5.3  The causal role of behavior in evolution

The third evolutionary mechanism is the causal role of behavior in evolution. This is an evolutionary mechanism that tends to be overlooked in contemporary, genetically based theory of evolution. James Mark Baldwin (1896) was the first evolutionary theorist to suggest that the behavior of animals can influence the course and the direction of the evolution of their own species. By now we know that among vertebrates, the penetration of a new habitat, for example, is typically initiated by behavior rather than genetic mutation. The best known cases are the Darwinian finches of the Galapagos islands. A new habitat will unleash a new set of forces of natural selection operating on the animal and move the animal into a new direction of evolution.

Ernst Mayr (1963) took up Baldwin's theory and wrote, "a shift into a new niche or adaptive zone is, almost without exception, initiated by a change in behavior. ... the importance of behavior in initiating new evolutionary events is self-evident" (p. 604). More recently, Plotkin (1988) provides a collection of insightful articles on the role of behavior in evolution. In particular, Plotkin observed that social learning directly impinged upon the biological evolution of hominids and hominoids. Social learning, of course, is the process through which innovative behavior can pass from the innovator to its social cohorts and onto the next generation. Social learning is also critical in first language acquisition. Decades ago, the emphasis in research on first language acquisition was the creative aspect of children's language acquisitional process. Empirical studies in the last twenty years have demonstrated that while there is definitely a creative aspect in children's acquisition of language, social learning and imitation are extremely important. In the evolutionary development of hominids, the speed and capacity of social learning and imitation expanded dramatically. The expansion has a neurological base in the increase in encephalization. More specifically, the increase of mirror neurons in the neocortex plays a significant role in the expansion of hominids' capacity of learning and imitation.[18] Mirror neurons are found in many areas of the neocortex including the Broca's area. Since they are responsible for both production and perception, any increase in the number of mirror neurons in the association neocortex will enhance the ability of learning and imitation. We assume that the increase in the number of mirror neurons in hominid evolution is proportional to the increase in encephalization.

## 6.   A humble beginning of symbolic communication

A variety of evidence from linguistics, psychology and child language suggests that representing some concrete object with a communicative signal is the least cognitively demanding of all symbolic communicative behavior. For this reason, we submit that the first step in the co-evolution of hominid brain and hominid communicative behavior is the naming of a concrete object.

Symbolic behavior, as we all know, has tremendous adaptive value. Just being able to name even one concrete object, such as a predator or a prey or a food item, would confer a significant competitive edge to a group of hominids. Suppose an early hominid, for example, a Homo erectus, in a flash of creative innovation, first invented one communicative signal symbolizing some concrete object, his or her social group would be able to learn such symbolic communicative signal from the innovator. As a consequence, this group of Homo erectus would have a competitive edge for survival and reproduction over other hominids. We assume that before the occurrence of the first symbolic communicative signal, the communicative behavior of hominids was not significantly different from the design of contemporary non-human primate communication. In particular, such an assumption implies that like non-human primates, early hominids communicative signals have functions only but not meaning. The genius of inventing the first symbolic signal lies in switching from the mid-brain to the neocortex as the primary neural substrate for signal production. This is why the invention of the first symbolic communicative signal is such an important innovation. At the time of the innovation, the part of the neocortex being hijacked for mediating the production of the first few symbolic communicative signals could be the Broca-Wernicke region of the association neocortex, which probably directed certain motor behavior in response to auditory input. This first innovation then sets the stage for the co-evolution of communicative behavior, brain, culture, size of social group and other anatomical innovations. The advantage of postulating behavior rather than genetic mutation initiating the co-evolutionary process should be obvious. New communicative behavior can be passed on to other members of a social group and to future generations through learning and imitation. If the initial symbolic communicative behavior had to be engendered by a genetic mutation, then those who did not undergo such a genetic mutation would not and could not have the behavior. It is highly improbable that an entire social group of hominids all underwent the same genetic mutation at the same time. If only one hominid underwent such a mutation, this hominid would stand out as a freak among its peers since no one else could produce or understand its new communicative behavior. In such a case, even without taking into consideration the normal effect of genetic drift, it would be highly improbable that such a genetic trait resulting from mutation could spread and thrive. A social freak among a group of hominids or any other animals would have a slim chance to survive. Ostracization would be its immediate fate.

Given the scenario in which a hominid of the genus Homo, in a flash of creative innovation, invented a linguistic sign, some questions immediately jump to mind: How realistic is such a scenario? Is it just wishful thinking or is there some evidence for it? Wouldn't symbolic communicative behavior, even at the most elementary level of having one or two symbols for some concrete objects, require a qualitatively different brain? The questions are interrelated. The fundamental issue hinges on whether or not simple symbolic communicative behavior requires a brain with a language module. We believe that it does not. Instead of postulating a language module in my brain, we would like to introduce the concept of *cognitive reserve*.

## 7.   Cognitive reserve

By cognitive reserve, we mean cognitive capability that is not fully utilized or manifested in the normal repertoire of behavior of a mammal. The various projects training great apes to manipulate linguistic symbols are evidence for the apes' cognitive reserve specifically in the domain of symbolic communication. Regardless of the controversy surrounding the degree of success of these projects involving the chimps Sarah and Washo, the bonobo Kanzi, the gorilla Koko, there is no doubt that these great apes are able to acquire and use some linguistic signs after extensive and intensive training. It is true that in their natural environment, apes' communication is strictly non-symbolic and they give no indication of developing symbolic communication. It is only through human intervention that they succeed in acquiring some linguistic symbols. The important point is that they have the cognitive reserve for acquiring and using some linguistic symbols even if the process of acquisition is highly unnatural. We know many mammals can be trained by humans to perform a great variety of impressive cognitive feats that are not included in their natural behavioral repertoire. Much of the trained behavior is evidence in support of what we call the mammal's cognitive reserve. It exists in all mammals endowed with a neocortex: the ability to perform some novel behavior that is not expected in its normal repertoire. It probably exists in other animals to a lesser extent.

From an evolutionary perspective, the existence of cognitive reserve is not at all surprising. In fact, it is expected. Evolution does not create a central nervous system without any reserve for unexpected demands or unexpected change of environmental or ecological conditions. Mammals without such reserve capacity are unlikely to survive very long in a changing world, and the world is always changing and never short of the unexpected.

The evidence for cognitive reserve, however, extends beyond the success of human effort to train great apes to acquire linguistic symbols. There are cases of mammalian behavior in their natural environment that suggest a level of cognitive capability which far exceeds what is manifested in their usual behavioral repertoire. The Japanese macaque on Kojima island which acquired the methods of cleaning sand-covered potatoes and effectively sorting grain from sand is one of the best known examples. The first discovery of using a stone/wood hammer and a suitable base as an anvil for cracking nuts by the chimps in Tai forest or the first discovery of fashioning a twig/straw into a probe for fishing out termites by the chimps of Gombe are also strokes of genius that attest to the existence of cognitive reserve.

The importance of the concept of cognitive reserve and the earlier discussion of behavior initiating a new direction of evolution is that they provide the theoretical underpinning for postulating that the dawn of symbolic communication was initiated behaviorally by a hominid in the genus Homo. In particular, this behavior is the creation of a communicative signal refering to a concrete object. This signal, because of its adaptive value, was transmitted through social learning to the social cohorts of that hominid and then to the next generation. Thus began a new direction of evolutionary development of the hominids: the co-evolution of brain, symbolic communicative behavior, decrease of the gastro-intestinal tract, increase of the vertebral canal, descent of the larynx and the enhancement of material culture. The various components of this co-evolutionary process are mutually reinforcing, like an arm race, one egging on the others.

## 8.    Spoken vs. written language

It took approximately 1.5–2 million years for hominid communication to evolve into full-fledged language as I know it nowadays. 'Full-fledged language' designates casual spoken language. It is not written language, which differs *significantly* from casual spoken language in terms of vocabulary, grammar as well as coherence and organization. Contrary to the common belief, written language is not just spoken language written down. Written language is a recent cultural invention with approximately 5000 years of history, representing a crowning cultural achievement and a critical cultural instrument of great importance. But it is an inappropriate base for inferring the structure and properties of language at its point of origin. When hominid communicative behavior evolved into language, it is a spoken form of communicative vehicle for social interaction involving more than one participant. It is not a written language. If we infer the properties of language at its point of origin from contemporary casual spoken language, we will be free from the burden of figuring out how hominids gradually evolved a communicative behavior characterized by the logicity, an extremely high level of coherence and a tightly structured organization in written language. In other words, the evolution of hominid communicative behavior into casual spoken language is a stage of evolution that is completely distinct from the evolution of casual spoken language into written language in terms of chronology, process and content. The first stage is a biological evolution within the Darwinian framework. The second stage is a cultural development that has nothing to do with the Darwinian notions of natural selection and random mutation. Even though linguists are perfectly aware of the difference between casual spoken language and written language, few take the trouble of extracting data from carefully transcribed casual conversation out of the academic setting. For many, linguistic data is obtained through introspection of how they think they utter a sentence in their own native language. Such a sentence is, at best, a token of the formal written language rather than casual spoken language for many reasons of which the most important one is that such a sentence is independent of any communicative context. The entire communicative context, linguistic or non-linguistic, visual, auditory or tactile, of any casual conversation serves as such a rich source of information to the interlocutors that renders the grammar, the diction and the organization of casual speech significantly different from those of written language.

## 9.   Toward the crystallization of language

The first stage of the evolutionary development toward casual spoken language is the increase of communicative symbols for concrete objects, e.g. food, predator, objects for landmark, different animals, different plants. During this stage, the creation of each new symbol represents a stroke of genius by a hominid, and the establishment of each newly created symbol in the repertoire of the communicative signals of the social group to which the creator belongs, requires social and cultural transmission. The social group most likely consists of close kin in the beginning before it extends to a more distantly related clan. It is important to realize that the entire process is an evolutionary event. It did not happen every day. It did not happen every year, and it probably did not happen in every generation. We must avoid unconsciously projecting our frame of mind onto the evolutionary scene involving our hominid ancestors. They had neither the cognitive capability nor the cultural environment we have. They were at the beginning of a long evolutionary path that ultimately led to the emergence of language. They did not have language yet. We believe that the onset of symbolic communication began with Homo erectus. There are several reasons for our belief: (a) The Homo erectus brain at 800–950 cc is considerably larger than the brain of all earlier hominids, including Homo habilis, the first species of the Homo genus. (b) The Homo erectus brain shows an increase in cerebral asymmetries. (c) They are the first hominids which migrated out of Africa and reached as far as Asia and Indonesia evidenced by the famous fossils of Peking man and Java man. The migration suggests an expanding population, which in turn, suggests a higher level of fitness, probably caused by improved communicative capability. (d) As Holloway (1995) points out, since the time of the Homo erectus, the evolution of the hominid brain showed a gradual increase in volume, refinement and asymmetries that could not be allometrically related. In other words, the Homo erectus brain evolved exclusively for the purpose of greater cognitive capacity. We believe that this evolutionary process of the brain is correlated with the gradual evolution of the symbolic communicative behavior.

Having a few communicative symbols for concrete objects, however, *is not tantamount* to being aware of the abstract principle of associating symbolic communicative behavior with concrete objects, even though the symbol itself is a token of this principle. In other words, there is a significant difference between using a communicative symbol for a concrete object and being aware of the principle underlying that act of creation. The various projects training apes to manipulate human linguistic symbols illustrate this difference. Sarah, Washoe, Kanzi and Koko may be able to master a good number of linguistic symbols. But there is no indication that they are aware of the underlying principle of association between a sign and what it signifies. Thus, the appearance of communicative signals that signify concrete objects 1.5–2 million years ago did not imply the dawn of language. As we have stated earlier, the addition of each new communicative signal that symbolizes another concrete object is a significant step along the evolutionary pathway toward the emergence of language. Each evolutionary step occurs on the evolutionary scale of time. There isn't a rapid cascade of new linguistic symbols following the initial appearance of a linguistic symbol in the communicative repertoire of some Homo erectus. Furthermore the use of each linguistic symbol was transmitted socially. That transmission process also took time. The case of the Japanese macaques on Kojima island provides some hint on the speed of transmission during the early phase of the evolution of hominid communicative behavior toward language. After the female genius macaque innovated the behavior of washing sand-covered potatoes in sea water, it took four years for the behavior to spread among eight members of the troop, all of whom happened to be the immediate kin of the female innovator. The slow pace of cultural transmission is also observed in chimpanzees learning of nut-cracking in the wild. It takes a young chimp five to six years to fully master the art of cracking nuts, sometimes with the help of its mother (Gibson & Ingold 1993). Close social tie obviously facilitated the learning of a new behavior. Learning, nevertheless, was far from being instantaneous. In contrast, anatomically modern humans learn simple skills and acquire new behaviors with nearly lightening speed. As we mentioned earlier, our speed of learning is probably facilitated by the larger number of mirror neurons we have.

Regarding the creation of communicative symbols for concrete objects, each act of creation typically involved serendipity in a highly motivating and possibly stressful situation. Besides the act of creative innovation, the expansion of linguistic symbols co-evolved with the increase in encephalization, enhancement of culture, growth in the size of social group and population, and at various points in time significant anatomical innovations. Increase in encephalization was necessary because of the demand of greater cognitive capacity and memory for handling communicative symbols, and because the increase of mirror neurons improved the speed and capacity for learning.[19] Enhancement of culture was necessary because it facilitated the spread of newly created linguistic symbol. Growth of the size of social group and population was necessary because the more hominids acquired linguistic communicative symbols, the more likely a new genius would emerge to create an additional linguistic symbol for another concrete objects. The anatomical innovations as we have already pointed out, also involve the decrease of the G.I. tract, the expansion of the thoracic nerves, the decrease in somatization and the descent of the larynx. These changes emerged through a co-evolutionary process. They did not occur in a few generations. On the one hand, the development of hominid communication toward language needed this multifaceted co-evolutionary process; on the other hand, the development of hominid communication in the direction of language, because of its adaptive value, pushed our hominid ancestors down the evolutionary path which led to these multifaceted innovations. The end product of this complex co-evolution that went on for approximately 1.5–2 million years is language. However, the pace of the development was not constant. For most of the two million years, the development was characterized by stasis. The increase of the number of linguistic symbols moved at a snail's pace. Toward the end of the two million years, i.e. around the time of the emergence of anatomically modern humans, the rate of development began to accelerate. If we plot the 1.5–2 million years of evolutionary development as a curve with the vertical axis representing the rate of change of hominid communicative behavior toward language and the horizontal axis representing time, the shape of the curve will be very similar to the curves showing the increase of hominid population and the development of stone tools. The first segment of the curve is a line with a very gentle slope characterizing primarily stasis for most of the 1.2–2 million years. The second segment of the curve is a sharp turn into a steep climb characterizing a dramatically fast approach toward language during the final 100–150 thousand years. The sharp turn signals that a critical number of linguistic symbols has been reached, and the symbols began to expand from designating concrete objects to actions, activities, events, experience, thought. At this juncture, the concatenation of linguistic symbols became a naturally emerging phenomenon. For instance, when a hominid's vocabulary was large enough to include items denoting action or activity, it would follow that the hominid understood the relation between an actor and an action or an agent and an activity. The concatenation of an actor with an action would emerge naturally.

Expressing an actor or agent with an activity suggests the incipience of grammar in the sense that there is a concatenation of words to form a larger communicative signal. When there is concatenation, there is, at the minimum, the syntactic phenomenon of word order. But syntax in the sense of word order does not require any quantum cognitive leap. As we have already pointed out, the notion of activity or action implies the existence of an agent or an actor. If a hominid had a word for an action, we can assume that the hominid already understood that an action required an actor to execute it. As for stabilizing a word order, it is a social convention, negotiated consciously or unconsciously by the members of a community.

What about all of the other grammatical structures beyond word order found in contemporary spoken languages?

We have by now historical linguistic data that account for the emergence of nearly all grammatical conventions, be it inflection, derivation, subordination, conjunction, interrogative, imperative or subjunctive. Linguists have been able to elucidate the precise processes and mechanisms by which such grammatical constructs may emerge in a language. Grammaticalization is one of the most important mechanisms in the emergence of grammar (Traugott & Heine 1991). Grammaticalization began to occur as the hominids started to link symbolic signals into larger communicative units.

What about the notion of generating sentences that is the foundation of generative grammar?

In our discussion of the defining characteristics of language at its crystallization, we did not mention recursive function or generativity. Yet ever since Chomsky's famous publication of "Syntactic Structures", many scholars including most linguists consider recursive function the unique defining feature of human language, e.g. Pinker (1995). Indeed, if one surveys the literature on language, one cannot fail to notice the omnipresence of the concept of recursive function or generativity. It depicts the speaker's ability to generate an indefinitely large number of sentences from a finite vocabulary with a finite set of syntactic rules. Let us briefly examine recursive function and generativity.

In elementary formal logic, one of the concerns is the device needed for producing strings of symbols. The simplest device can be expressed in what is called 're-writing rule'. A re-writing rule is a rule which re-writes one symbol into a sequence of symbols. A trivial example of re-writing rule has the following form:

$$S \rightarrow aaa$$

This rule states that the symbol on the left of the arrow, 'S', is to be 're-written' as the sequence on the right of the arrow, '*aaa*'. In an artificial language for computers, I can specify that the symbol 'S' designates a sentence, and according to this re-writing rule, a sentence in this formal language is represented by a string of three '*a*'s.

Now suppose in an artificial language we have two re-writing rules which can be applied repeatedly:

S → a
S → aS

If we apply the first rule, we obtain a sentence consisting of one 'a'. If we apply the second rule and then apply the first rule to the output of the second rule, which is 'aS', we obtain a string of two 'a's, namely, 'aa'. If we apply the second rule twice, the first round we get 'aS', the second round we obtain 'aaS' (the output of applying the second rule to the 'S' of 'aS'). Take the result 'aaS' and now apply the first rule to the 'S', we have 'aaa', a string of three 'a's. It should be obvious now that we can obtain as long a string of 'a's as we wish simply by applying the second rule a sufficient number of rounds.

Let's assume that in our artificial language, there is only one vocabulary item, namely, the letter 'a', and let's further assume that the sentences in this artificial language are composed of a string of 'a's. With these two rules, we can generate an infinite set of sentences in this artifical language, each of which consists of a different number of the 'a's:

The property that the second rule, S → aS, has is called 'resursiveness'. Such a rule is called a 'recursive rule' because the symbol on the left of the re-writing rule recurs on the right. In natural languages, embedding and conjunction are grammatical devices that have this recursive property if one wishes to express grammatical rules in the form of re-writing rules, e.g.

S → S and S

This rule states that a sentence in English, 'S', can be re-written as two conjoined sentences with the grammatical word, 'and', performing the role of conjunction. Theoretically one can keep on conjoining sentences, or keep on embedding sentences so that the final product can be indefinitely long. The notion of 'generating sentences' is based on the concept of re-writing rule in logic.

If you can have indefinitely long sentences, you will have an indefinitely large number of sentences. The key notion is infinity conveyed by the expression 'indefinitely long' and 'indefinitely large'. Because there is an infinite or indefinitely large number of integers, 1, 2, 3, 4, 5, …, no one can claim to have the largest integer. We can talk about the set of all integers. But it is an infinite or indefinitely large set![20]

Theoretically the number of possible sentences in English is indefinitely large because theoretically 'the longest English sentence' does not exist. If one chooses to describe English syntax or certain aspect of English syntax in terms of re-writing rules, one can claim that a recursive function is needed. However, one never conjoins or embeds an indefinitely large number of sentences in either spoken or written language. "Indefinitely large number of sentences" or "infinitely long sentences" are *theoretical* possibilities. In order to understand whether or not recursive property is a unique defining feature of human language, we must find out if there is a *theoretical* possibility of describing animal communicative behavior with recursive function.

Consider the songs of the humpback whales (Payne 1995). A male humpback whale song is composed of units impressionistically described as grunt, moan and squeak, which combine to form 'phrases'. Phrases are in turn combined into 'themes'. A song is made of a sequence of themes. We do not know if a phrase or a theme serves as a functional unit conveying some message. We do know that the song as a whole has a definite communicative function. It advertises to the females an individual male's presence and physical fitness for mating. In the study of animal communication, the song is called a courtship signal. The song may also serve to fend off competing males and convey territoriality during the mating season. For our purpose, the most important aspect of the humpback whale song is that it is usually sung in repetition, sometimes exceeding half an hour of time. The repetition indicates a gradation of the intensity of the signaler's emotional state. The more the repetition, the greater the desire of the male to attract the female and the more it demonstrates the male's physical fitness. Hence, repetition is not communicatively redundant. It has communicative significance.

We will describe the whale song in terms of re-writing rules. Let 'S' be the symbol for the courtship signal. Let '$a$' be the symbol for one song. In order to account for the entirety of the courtship signals of the humpback whale, we need the following two re-writing rules one of which is recursive:

$$S \rightarrow a$$
$$S \rightarrow aS$$

Just like our earlier example of an artificial language composed of strings of '$a$', we obtain an infinite set of possible humpback whale courtship signals, each of which represents a point along the continuum of the male's emotional state and his physical fitness. This infinite set is represented by the set of strings of '$a$'s. Each string denotes one bout of singing which may contain any number of repetitions of the song: '$a$', '$aa$', '$aaa$', '$aaaa$', ...

Of course, no whale sings indefinitely. Such a fact, however, is no more or no less meritorious than the fact that no human being conjoins or embeds sentences indefinitely! The issue here is theoretical capability. In real life, humpback whales often sing continuously for half an hour or more. It is difficult to think of a human being taking half an hour to utter one sentence.

One can also argue that the singing of some song birds have the recursive property. Most song birds repeat their songs as the humpback whale does. One only needs a sleepless night in the spring time and listen to the mocking bird which often sings continuously for several hours. The stronger and healthier the bird, the more repetitions it sings during the mating season, and each repetition signals a higher notch of intensity and therefore, the physical fitness and the motivation of the male singer.

In conclusion, the generative hypothesis of human language does not seem to serve any purpose. It has diverted the attention of scholars toward artificial problems that have little or no bearing on the nature of language. One could argue that there is a potential to represent some aspects of a language with a recursive function. But as we have pointed out, there is also a potential to represent some animal communicative signals with a recursive function. In short, recursive function is not a property uniquely attributable to human language. The important characteristics of language related to recursiveness is the creativity demonstrated by native speakers. Creativity in language use, however, is primarily based on the principle of analogy. There is no need to use the recursive function to account for linguistic ingenuity. Analogy provides the mechanism for creating new expressions and utterances. Metaphors and the ways we say things are two of the most important bases for the creation of new utterances through analogy. Both metaphors and the ways we say things are two of the most critical components of language.

Since the 1960's, most linguists have accepted Chomsky's idea that linguistic data for syntactic-semantic research is best obtained by the linguist through introspection of his/her own native language. However, academics spend a lifetime reading and learning to write and speak in the academic style. When they use themselves as a source of data, naturally the data that wells up in their mind are isolated sentences from their academic, written language.

When Chomsky and generative linguists talk about the ease with which a child acquires fluency in a native language within two years in early childhood, they are talking about casual, spontaneous spoken language. But when they describe the syntactic structures of language on the basis of introspected data, they are describing tokens of academic, written language. We have already noted that the difference between casual, spoken language and academic written language is enormous. There is an even greater difference between the acquisition of the first spoken language and the acquisition of literacy. Learning to write has none of the spontaneity and ease that characterizes the acquisition of the first spoken language. All human infants, with some rare exceptions due to deformity, are destined to acquire a spoken language. Not all humans are destined to acquire literacy. Learning to write requires education, assistance, guidance and years, if not decades, of practice. There is as much gradation in the quality of written language as there are educated people in a society. In short, written language is a codified cultural artifact for the purpose of creating records without the same kind of situational and contextual information and without prosody, facial expression, body posture and physical contact, all of which are available in speech interaction. Written language is not the result of biological evolution. As we have pointed out, it is a cultural product.

Finally we wish to point out that much of what generative linguists consider as canonical grammatical constructions are formalized or conventionalized in written language. Since most linguists speak or try to speak in the style they write, the canonical grammatical constructions are transported into their spoken language. They may believe that such grammatical constructions are the mental prototypes of the language, and the data of casual spoken language represent what Chomsky (1965) calls the "degenerate" data that are fragments of the full forms. This belief is further reinforced by the fact that formal written language carries greater social prestige than casual spoken language because of the written genre's associations with literacy, education and social status. Thus, the perception that the canonical grammatical constructions of formal, written language are the mental prototypes of language is based on a social prescription enforced through education, literacy and value system. From the perspective of evolution, language is first and foremost a human communicative behavior. If we are to study the human communicative behavior, we must base our study on casual, spoken language transcribed with the utmost fidelity and not viewed as fragments of stylistic conventions. Casual spoken language does not have the grammar of formal written language. In a forthcoming article, Paul Hopper, after describing and analyzing data on several syntactic constructions from a spoken English corpus, concluded,

> Corpus studies suggest that the "degenerate" data are the true substance of natural spoken language, and that what my grammars give me then are normativized assemblies of these fragments that tend to impress themselves on me as mental prototypes because of their greater social prestige.
>
> (Hopper, forthcoming)

We find it ironic that in the empirical investigation of language, which is a human behavior with an evolutionary history, it is necessary to defend the importance of authentic, unedited behavioral data collected from casual, spoken language.[21] Obviously for many linguists, such data are fragmented and unimportant. Such an attitude has impeded the investigation of the evolutionary origin of language.

In conclusion, this paper represents a condensation of 6 million years of hominid evolution and a sketch of a diverse array of information from many disciplines that are relevant to the evolutionary origin of language. Many important topics are left out and many others receive only a brief cursory presentation. As a consequence, the paper reminds me of an old Chinese saying: "Flower appreciation on a galloping horse." For blurred images and obscure landscape, we apologize. However, it is important to note that during the past decade major contributions toward an understanding of the origin of language have come primarily from the neurosciences and paleoanthropology. We hope that we are successful in demonstrating that linguistics can also contribute toward an understanding of the origin of language, once we move beyond the mist created by the generative paradigm.

## Notes

* The authors are grateful to CNRS for supporting their research on the origin of language. We are also grateful to Joan Bybee, Noam Chomsky, Paul Hopper, Frits Kortlandt, Guido Martinotti, Alain Peyraube, Maxim Stamenov, Edda Weigand and Bruce Wilcox for their invaluable comments and suggestions.

1. It will be clear later in this paper that the definition of language as the casual spoken language of human beings is of extreme importance. I use the term 'anatomically modern humans' to circumvent the confusion caused by a proliferation of taxonomic terms such as Early Homo sapiens, Archaic Homo sapiens, Homo sapien sapiens, etc. Compared to the hominids of the past 250,000 years, anatomically modern humans have a gracile skeleton characterized by long bone shape, a specific depth and extent of muscle insertion, a thin cranial wall and mandibular body, a high, domed cranium, a reduced jaw, and the absence of a prominent browbridge over the eyebrow, i.e. no supraorbital torus.

2. For a succinct and comprehensive analysis of the current hominin taxonomy, see Wood and Collard (1999). A new discovery, however, poses additional challenge to the already controversial hominin taxonomy. On December 4, 2000, French and Kenyan paleoanthropologists announced the discovery of "Millennium Ancestor" (Orrorin tugeensis) in the Tugen hills of Kenya's Baringo district in the Great Rift Valley. The fossil remains include various body parts belonging to five individuals. The fossils have not been dated yet. But the strata where the fossils lay buried show an age of 6 million years. If the dating proves correct, these fossils would be approximately 1.5–2 million years older than the Ardipithecus. They would yield exciting information of the earliest evolutionary development of hominids. Preliminary report suggests that the Millennium Ancestor was about the size of a modern chimpanzee and capable of walking upright as well as tree-climbing. The discoverers of the Millennium Ancestor, Brigitte Senut and Martin Pickford, hypothesize that all Australopithecines belong to a side branch of the hominid family tree, and the Millennium Ancestor, not Lucy, the Austrolopithecus afarensis, is the direct ancestor of modern humans. They base their hypothesis on three key factors: (1) the age of the fossils at 6 million year. (2) The Orrorin's femurs which point to some level of bipedalism. (3) The molars of the Orrorin which are small, squared and thickly enameled. These features of the molars remain with anatomically modern humans.

3. The evolution of bipedalism took several million years to complete. It involved the change of the skeleton from the skull to the toe, the redesign of the nervous system and the change of muscular structure from the neck down. Even though Lucy, the famous Austrolopithicus afarensis who lived more than 3 million years ago, was a fully functional bipedal hominid, the changes involved in bipedalism did not complete until the emergence of the Homo erectus.

4. Non-human primate communicative signals are typically multi-modal, involving visual as well as auditory, and sometimes tactile channels. But their visual (such as facial expression, body postures) and tactile communicative signals are even more transparent as manifestations of their emotional and motivational states.

5. The deceptive use of communicative signals among primates, which has been observed among several species, would involve cognition beyond the involuntary vocalization stimulated by an external circumstance. Deception, however, is not frequently observed among primates, even though it suggests that the use of a communicative signal for deception is subverted by the neo-cortex.

6. There is an amusing incidence involving a chimp discovering a cache of delectable food at Jane Goodall's camp in Gombe. It immediately went behind a tree and covered its mouth so that its involuntary food call cannot be heard and its facial expression of excitement cannot be seen by its companions. This episode is significant because it demonstrates that (1) the chimp is aware of its own emotional reaction to the sudden discovery of delectable food, and (2) through its neocortex, it is trying to conceal its emotional states expressed by its communicative signals. Similar incidences involving other primates have been reported by ethologists, for example, Cheney and Seyfarth's (1990) account of deception by vervet monkeys.

7. The change of some animal communicative signals may be culturally transmitted, e.g. the courtship songs of the white-crowned sparrow are known to have dialectal differences. In such cases of the change of animal communicative behavior, the classical Darwinian evolutionary process does not apply and the time of change may be very short.

8. If language crystallized several tens of thousands of years after the emergence of anatomically modern humans, the polygenesis of language would be possible. The issue of monogenesis vs. polygenesis of language is briefly discussed in Li (2002, 2003).

9. The m-DNA contains only 37 genes and 16569 base pairs. The small number of genes and base pairs make it easy to examine the variability of m-DNA in different individuals. Most important of all, mitochondrial genes are maternally transmitted, although recent investigations show that rare leakage of paternal m-DNA into a fertilized ovum is possible. If the source of m-DNA is exclusively maternal, then variation of the m-DNA can only be caused by mutation. Thus a molecular clock based on an average mutation rate in the m-DNA tends to be reliable. For an informative discussion of the mitochondrial DNA and its relevance to human evolution, see Cann (1995).

10. Before Tim White unearthed the fossils of Austrolopithecus Ghari in Ethiopia in 1997, the Oldowan in Kenya is the oldest known stone tool technology. The Oldowan tools date from 2.5 to 1.7 million years ago, and they are associated with the emergence of the genus Homo. However, Austrolopithecus Ghari, which is dated 2.5 million years ago, used stone tools which were carried from a site more than 50 miles away from the location of the Ghari fossils. This discovery nullified the long-standing belief that stone tools were a Homo invention. The Acheulian technology emerged with the Homo erectus. The major difference between the Oldowan and the Acheulian is the addition of the hand ax, the cleaver and the pick in the Acheulian technology. The Mousterian technology contained a larger range of tool types than the Acheulian. However, the Mousterian technology, associated with the Neanderthals, did not exhibit much technological improvement over the Acheulian.

11. For an informative discussion of the evolution of the human brain and a comparison of the human brain with animal brains, see Falk (1991) and Roth (2000).

12. Some animals have a better memory for certain sensory experience than humans. For example, dogs and cats are better than humans in remembering olfactory experience. This fact, however, does not imply that dogs and cats have a larger capacity for cognitive memory. Their olfactory perception is much more acute than that of human beings. Their greater ability to perceive and differentiate odors is connected to their better olfactory memory.

13. For a succinct summary of children's acquisition of grammar, see Bates and Goodman (1999).

14. We take note of the fact that Chomsky's current theoretical stance is considerably different from his 1986 pronouncements. In his new Minimalist Program (Chomsky 1995), grammar is largely derived from the lexicon. If we are correct in assuming that what is considered innate by Chomsky and his followers is the newest version of the so-called "Universal Grammar", which is austere and minimal, the issue of representational innateness for language behavior is practically moot.

15. In some grammars, one finds some sporadic discussion of some particular ways of saying things. Typically such a grammar concerns a language unrelated to the Indo-European language family. The authors are motivated to discuss some ways of saying things in those languages because many of these ways of saying are bizarre from the Indo-European perspective. For example, in most grammars of Sub-Sahara African and East Asian languages, 'serial verb construction' is usually presented because it is a construction that does not occur in Indo-European languages.

16. A gene carries the information for coding a particular protein, and each protein plays an important role in the anatomy and physiology of an animal. Hence the mutation of a gene, by and large, is deleterious because the mutation may impair the synthesis of a particular protein which is essential for survival.

17. Although a chimp brain stops expanding three months after birth, its development including myelination and neuronal connections is not complete until approximately five years of age.

18. Mirror neurons were discovered in the laboratory of Giacomo Rizzolatti. For more information on language and mirror neurons, see Rizzolatti and Arbib (1998).

19. There is no hard evidence for the increase of mirror neurons in the evolutionary development of hominids. I do know from the work of Rizzolatti et al. (1988) that mirror neurons are phylogenetically old. In fact, the discovery of mirror neurons first occurred in experiments involving monkeys (Rizzolatti et al. 1996). My claim of the increase of mirror neurons in hominid evolution is based on the inference that as the hominid brain increased in size, the number of mirror neurons increased correspondingly.

20. Another trick connected with the notion of infinity is the correct, but seemingly counter-intuitive, fact that there are as many even integers (2, 4, 6, 8, 10 ...) as there are integers (1, 2, 3, 4, 5, ...). To prove this claim, I perform the simple operation of multiplying each integer by the number 2. This operation does not affect the number of integers. But after the operation, the set of integers becomes the set of even integers without any change in the total number of elements in the set. The reason is that 'the total number of elements

in the set' can lead to confusion. This 'total number' is no longer an integer. It is an infinity, or more precisely, a countable infinity. There are other kinds of infinities in mathematics. None of them is an integer.

21. There are by now many corpora of carefully transcribed data of casual conversations. One corpus used by me is the Santa Barbara Corpus of Spoken American English created by John Du Bois (2000).

## References

Aiello, L., & Wheeler, P. (1995). The expensive tissue hypothesis. *Current Anthropology, 36,* 199–221.

Aitken, P. G. (1981). Cortical control of conditioned and spontaneous vocal behavior in rhesus monkeys. *Brain and Language, 13,* 636–642.

Allman, J. M. (1999). *Evolving Brains.* New York: Scientific American Library.

Arensburg, B. (1989). A middle Paleolithic hyoid bone. *Nature, 338,* 758–760.

Baldwin, J. M. (1902). *Development and Evolution.* New York: Macmillan.

Balter, M. (1999). New light on the oldest art. *Science, 283,* 920–922.

Bates, E., & Goodman, J. C. (1999). On the emergence of grammar from the lexicon. In B. MacWhinney (Ed.), *The Emergence of Language* (pp. 29–80). Mahwah, New Jersey: Lawrence Erlbaum Publishers.

Berger, L. (2000). *In the Footsteps of Eve.* Washington, DC: National Geographic Adventure Press.

Biraben, J.-N. (1979). Essai sur l'evolution du nombre des Hommes. *Population, 1,* 13–25.

Cann, R. L. (1995). Mitochondrial DNA and human evolution. In J.-P. Changeux & J. Chavillon (Eds.), *Origins of the Human Brain* (pp. 127–135). Oxford: Clarendon Press.

Cheney, D. L., & Seyfarth, R. M. (1990). *How monkeys see the world: Inside the mind of another species.* Chicago: The University of Chicago Press.

Chomsky, N. (1957). *Syntactic Structures.* The Hague: Mouton Publisher.

Chomsky, N. (1965). *Aspects of a Theory of Syntax.* Cambridge, MA: MIT Press.

Chomsky, N. (1986). *Knowledge of Language: Its nature, origin, and use.* New York: Praeger.

Chomsky, N. (1995). *The Minimalist Program.* Cambridge, MA: MIT Press.

Du Bois, J. (2000). *Santa Barbara Corpus of Spoken American English.* CD-ROM. Philadelphia: Linguistic Data Consortium.

Elman, J. (1999). The emergence of language: A conspiracy theory. In B. MacWhinney (Ed.), *The Emergence of Language* (pp. 1–28). Mahwah, New Jersey: Lawrence Erlbaum Publishers.

Elman, J., Bates, E., Johnson, M., Karmiloff-Smith, A., Parisi, D., & Plunkett, K. (1996). *Rethinking Innateness: A connectionist perspective on development.* Cambridge: MIT Press.

Falk, D. (1991). 3.5 million years of hominid brain evolution. *The Neurosciences, 3,* 409–416.

Gibson, K., & Ingold, T. (Eds.). (1993). *Tools, Language and Cognition in Human Evolution.* Cambridge: Cambridge University Press.

Grace, G. (1987). *The Linguistic Construction of Reality*. New York: Croom Helm.

Hassan, F. (1981). *Demographic Archaeology*. New York: Academic Press.

Hockett, C. (1960). The origin of speech. *Scientific American, 203*, 88–96.

Holloway, R. (1995). Toward a synthetic theory of human brain evolution. In J.-P. Changeux & J. Chavaillon (Eds.), *Origins of the Human Brain* (pp. 42–54). Oxford: Clarendon Press.

Hopper, P. (forthcoming). Grammatical constructions and their discourse origins: Prototype or family resemblance? In *Proceedings of the LAUD 2000 Conference on Cognitive Linguistics*. Duisburg, Germany: LAUD.

Johanson, D., & Edgar, B. (1996). *From Lucy to Language*. New York: Simon & Schuster.

Klein, R. (1989). *The Human Career: Human biological and cultural origins*. Chicago: University of Chicago Press.

Lewin, R. (1993). *The Origin of Modern Human*. New York: Scientific American Library.

Li, C. N. (2002). Some issues in the origin of language. In J. Bybee (Ed.), *Papers in Honor of Sandra Thompson*. Amsterdam: John Benjamins, in press.

Li, C. N. (2003). *The Evolutionary Origin of Language*. New York: Regan Books, Harper-Collins Publishers.

MacLarnon, A., & Hewitt, G. (1999). The evolution of human speech: The role of enhanced breathing control. *American Journal of Physical Anthropology, 109*, 341–363.

Mayr, E. (1963). *Animal Species and Evolution*. Cambridge: Harvard University Press.

O'Connell, J. F., Hawkes, K., & Blurton Jones, N. G. (1999). Grandmothering and the evolution of Homo erectus. *Journal of Human Evolution, 36*(5), 461–485.

Pandya, D. P., Seltzer, B., & Barbas, H. (1988). Input-output organization of the primate cerebral cortex. In H. Steklis & J. Erwin (Eds.), *Comparative Primate Biology: Neurosciences* (pp. 39–80). New York: Liss.

Pawley, A. (1991). Saying things in Kalam: Reflections on language and translation. In A. Pawley (Ed.), *Man and a Half: Essays in Pacific anthropology and ethnobiology in honor of Ralph Bulmer*. Auckland: The Polynesian Society.

Payne, R. (1995). *Among Whales*. New York: Bantam Doubleday Dell Publishing Group.

Pinker, S. (1994). *The Language Instinct: How the mind creates language*. New York: William Morrow.

Pinker, S. (1995). Facts about human language relevant to its evolution. In J.-P. Changeux & J. Chavaillon (Eds.), *Origin of the Brian* (pp. 262–283). Oxford: Oxford University Press.

Plotkin, H. C. (1998). *The Role of Behavior in Evolution*. Cambridge: MIT Press.

Rakic, P. (1988). Specification of cerebral cortical areas. *Science, 241*, 170–76.

Rizzolatti, G., & Arbib, M. (1998). Language within my grasp. *Trends in Neuroscience, 21*, 188–194.

Rizzolatti, G., Camarda, R., Fogassi, L., Gentilucci, M., Luppino, G., & Matelli, M. (1988). Functional organization of inferrior area 6 in the macaque monkey: II. Area F5 and the control of distal movements. *Experimental Brain Research, 71*, 491–507.

Rizzolatti, G., Fadiga, L., Gallese, V., & Fogassi, L. (1996). Premotor cortex and the recognition of motor actions. *Cognitive Brain Research, 3*, 131–141.

Roth, G. (2000). The evolution of consciousness. In G. Roth, & M. Wullimann (Eds.), *The Evolution of Brain and Cognition*. New York: Wiley.

Seyfarth, R. M., & Cheney, D. L. (1999). Production, usage, and response in nonhuman primate vocal development. In M. Hauser & M. Konishi (Eds.), *The Design of Animal Communication* (pp. 391–417). Cambridge, MA: MIT Press.

Sherry, S. T., Rogers, A. R., Harpending, H., Soodyall, H., Jenkins, T., & Stoneking, M. (1994). Mismatch distributions of m-DNA reveal recent human population expansions. *Human Biology, 66*, 761–775.

Traugott, E., & Heine, B. (Eds.). (1991). *Approaches to Grammaticalization*, Vols. I & II. Amsterdam: John Benjamins.

Walker, A., & Shipman, P. (1996). *The Wisdom of Bones: In search of human origins.* New York: Knopf.

White, R. (1986). *Dark Caves, Bright Visions.* New York: The American Museum of Natural History.

Wood, B., & Collard, M. (1999). The human genus. *Science, 284*, 65–71.

Wrangham, R. W., Jones, J. H., Laden, G., Pilbeam, D., & Conklin-Brittain, N. (1999). The raw and the stolen: Cooking and the ecology of human origins. *Current Anthropology, 40*(5), 567–594.

# Mirror neurons, vocal imitation, and the evolution of particulate speech

Michael Studdert-Kennedy

Haskins Laboratories, New Haven, CT, USA

In all communication, whether linguistic or not, sender and receiver must be bound by a common understanding about what counts: what counts for the sender must count for the receiver, else communication does not occur... the processes of production and perception must somehow be linked; their representation must, at some point, be the same.

(Liberman 1996: 31)

## 1. Introduction

Language has the peculiar property of developing in the child, and presumably evolving in the species, under "conspicuously intersubjective circumstances" (Quine 1960: 1). Unlike other modes of action, such as walking, swimming, manipulating objects, eating, which engage humans with their purely physical environments, language engages them with one another. The physical environment certainly mediates their mutual engagement, through the sounds that speakers make and listeners hear, but the forms and functions of those sounds are specific to, uniquely determined by, and uniquely accessible to members of the human species. In other words, as the epigraph to this paper implies, language as a mode of communicative action only exists by virtue of the matching subjectivities of its users.

Neural support for this intersubjective process may come from "mirror neurons" of the type discovered by Rizzolatti, Fadiga, Gallese and Fogassi (1996) in the macaque monkey – neurons that discharge both when an animal engages in an action and when it observes another animal engage in the same action. Indeed, the quotation from Alvin Liberman above was also the epigraph to Giacomo Rizzolatti's and Michael Arbib's paper "Language within our grasp" (1998), of which the closing words were another quotation from Liberman: "... one sees a distinctly linguistic way of doing things down among the nuts and bolts of action and perception, for it is there, not in the remote recesses of the cognitive machinery, that the specifically linguistic constituents make their first appearance" (1996:31). What follows is a brief account of what "the nuts and bolts of [phonetic] action and perception" seem to consist of, and a speculative account of where and how mirror neurons might fit "among the nuts and bolts".

Like Rizzolatti and Arbib (1998), I shall argue that, evolutionarily, an orofacial mode of expression and communication, perhaps supplemented by brachiomanual gesture, is the likely immediate antecedent of speech. What I want to emphasize, however, is the puzzle posed by a critical discontinuity in the path from gesticulation to speech, namely, the shift from an iconic, analog mode of mimetic representation (Donald 1991, 1998), in which meaning is intrinsic to and, in some sense, isomorphic with representational form, to the arbitrary, digital mode of speech and language, in which meaning is extrinsic and socially ordained.

My hypotheses are as follows. First, mirror neurons exist in humans as components of mirror systems that support recognition and replication (i.e. imitation) of the actions of a conspecific. Second, facial imitation, a capacity unique to humans, is supported by a specialized mirror system. Third, vocal imitation, also unique among primates to humans, coopted and perhaps coevolved with the facial mirror system. Finally, I assume that the organization of a mirror system is both somatotopic and functional; that is to say, specific parts of the motor anatomy (e.g. hand, finger, lips, tongue) engaging in specific modes of action (e.g. grasping, twisting, smacking, protruding) activate a mirror system. Accordingly, if we are to extend the concept of a mirror system to the "nuts and bolts" of speech, we must take into account both the components of the vocal apparatus and how those components combine to effect phonetic action.

We begin with a summary account of the role of speech in language.


## 2.   The particulate structure of language

Language is unique among systems of animal communication in its unbounded semantic scope. Other animal vocal systems are limited, so far as we know, to a few dozen fixed signals bearing on the immediate situation with respect to predators, food, sex and various social contingencies. Humans, by contrast, can speak about whatever they choose, present or absent, past or future, concrete or abstract, real or imaginary, in a flexible, adaptively appropriate manner. Language derives its unique generative power and scope not simply from human cognitive reach (which may indeed be as much a consequence of language as a cause), but from its dual hierarchical structure of phonology and syntax. At the lower level, phonology, a few dozen meaningless sounds (consonants and vowels) are repeatedly sampled, permuted and combined to form an unlimited vocabulary of meaningful units (morphemes, words); at the upper level, syntax, the meaningful units are repeatedly sampled, permuted and combined to form an unlimited variety of phrases and sentences.

This dual pattern (Hockett 1958) is a special case of a general principle common to all systems that "make infinite use of finite means" (Chomsky 1972; von Humboldt 1836/1964), including physics, chemistry, genetics and language. Abler (1989) who first recognized the commonalty across these domains called it "the particulate principle of self diversifying systems" (see also Studdert-Kennedy 1998, 2000). According to this principle, a small number of elementary units or particles (atoms, chemical bases, phonemes) are repeatedly combined to form larger units (molecules, genes, words) with different structures and broader functions than their components. To fulfil their combinatorial function the elementary units must be discrete (categorical), invariant (context-free) and meaningless (devoid of intrinsic function). They must be discrete, so that when they combine with other units, they do not blend into an average, but retain their integrity to form new integral units with novel structures and functions that cannot be predicted from the properties of their constituents; they must be invariant, so that they retain their identity from one context to another (i.e. their commutability); they must be meaningless, so that the same unit can be repeatedly combined with other units to form different units of meaning, or function.

For the evolutionary biology of language, a central concern is the nature and origin of the elementary units. Uncertainty concerning the units arises for several reasons. First, the intuitively given units of linguistic description, consonants and vowels, have no status outside language. Phonetically, they are defined by their function in the formation of the consonant-vowel syllable, a fundamental unit of every language; linguistically, they are defined by their contrastive function in distinguishing words. They are therefore part of what an evolutionary account, undertaking to derive language from its non-linguistic precursors, must explain. Second, consonants and vowels are not primitive units, but compounds, analogs of the molecule, not the atom. According to the standard structuralist formulation, they are "bundles of features" (e.g. Jakobson & Halle 1956:8). Features, however, are purely static, descriptive properties of a segment, unsuited to the dynamic properties of speech either as a motor act or as an acoustic signal.

What we evidently need is a unit of articulatory action, with which we can trace development from the prelinguistic mouthings of an infant to the purposive phonetic acts of a competent speaker. For this we turn to the only explicit model of speech as a mode of motoric action ever proposed, the articulatory, or gestural, phonology being developed by Browman, Goldstein, Saltzman and their colleagues at Haskins Laboratories (Browman & Goldstein 1986, 1992; Fowler & Saltzman 1993; Saltzman 1986; Saltzman & Munhall 1989).

### 3.   Articulatory phonology: Gestures as units of phonetic function

The term "gesture" is often used informally to refer to an intentional movement of the speech articulators. In the framework of articulatory phonology "gesture" has a precise, technical definition as the process of forming and releasing a constriction at a certain point in the vocal tract. The phonetic function of a gesture is to set a value on one or more vocal tract variables that contribute to shaping a vocal tract configuration by which the flow of air, driven through the tract by the lungs, is controlled, so as to produce a characteristic pattern of sound.

Figure 1 displays the tract variables and the effective articulators of a computational model for the production of speech at its current stage of development (Browman & Goldstein 1992). Inputs to the model are parameters of equations of motion for forming and releasing constrictions. Constrictions can be formed within the oral, velic, or laryngeal subsystems. Within the oral subsys-

| | tract variable | coordinative structure |
|---|---|---|
| LP | lip protrusion | upper & lower lips, jaw |
| LA | lip aperture | upper & lower lips, jaw |
| TTCL | tongue tip constrict location | tongue tip, tongue body, jaw |
| TTCD | tongue tip constrict degree | tongue tip, tongue body, jaw |
| TBCL | tongue body constrict location | tongue body, jaw |
| TBCD | tongue body constrict degree | tongue body, jaw |
| VEL | velic aperture | velum |
| GLO | glottal aperture | glottis |

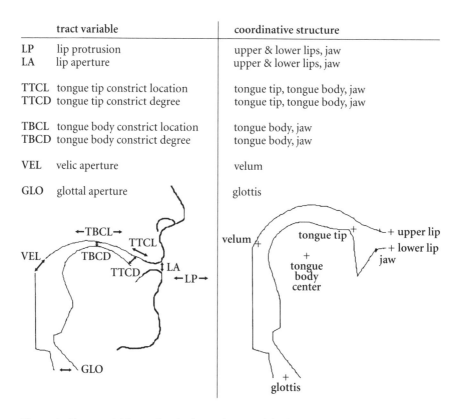

**Figure 1.** Tract variables and articulators in a model of speech production.

tem the parameters specify the primary end-effector of the constriction (lips, tongue tip, tongue body), one of nine discrete locations from lips to pharynx, one of five degrees of constriction (closed for stop consonants, "critical" for fricative consonants, narrow, mid, or wide for vowels), and a "stiffness" value specifying rate of movement. A velic gesture opens the nasal port for nasal consonants. The larynx is assumed to be set for voicing (glottal adduction), unless a devoicing gesture (glottal abduction) is activated. Thus, a phonetic segment may comprise a single gesture (e.g. lip closure for /b/), two gestures (e.g. lip closure and velic lowering for /m/), or more gestures (e.g. lip protrusion with critical postalveolar constriction and glottal abduction for /ʃ/, the first sound of the word *shop*).

The gestures for a given utterance are organized into a larger coordinated structure represented by a gestural score. The score specifies the values of the dynamic parameters for each gesture, and the period over which the gesture is active. Figure 2 (top) schematizes the score for the word, *nut* (['nʌt]), as a sequence of partially overlapping (i.e. coarticulated) activation intervals. Each gesture has an intrinsic duration that varies with rate and stress. Correct execution of an utterance therefore requires accurate timing of each gesture itself and accurate phasing of gestures with respect to one other. Timing, we shall see shortly, is a constant source of difficulty for a child learning to speak.

A distinction, critical to the present discussion, must be drawn between a concrete instance of a gesture and the gesture as an underlying abstract control structure that coordinates the movements of articulators. As an abstract coordinative structure (Fowler, Rubin, Remez, & Turvey 1980; Turvey 1977), a gesture is defined by its goal, or function, not by the actions of particular muscles, or the movements of particular articulators. For example, as indicated in Figure 1, lip aperture is determined by the coordinated movements of upper lip, lower lip and jaw; but the contributions of the three articulators to any particular instance of lip closure vary with context.

Experimental evidence for coordinative structures comes from studies in which one articulator is perturbed during cooperative execution of a gesture by several articulators. For example, Kelso, Tuller, Vatikiotis-Bateson and Fowler (1984) unpredictably and transiently checked the upward movement of the mandible into the syllable-final stop of a speaker uttering the syllable /bæb/: compensatory upward movement of the mandible and increased downward movement of the upper lip to effect bilabial closure occurred within 20–30 msec of the perturbation, too fast for central reprogramming to have occurred. When the same perturbation was applied to the mandible during upward movement for the syllable-final fricative of /bæz/, compensation included increased upward movement of the tongue tip, but not downward movement of the upper lip. Thus, perturba-

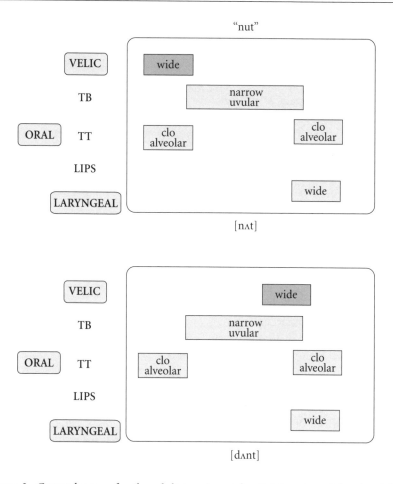

**Figure 2.** Gestural scores for the adult target, *nut* [ˈnʌt] (above), and for the child's erroneous attempt, [ˈdʌnt] (below)

tion of the same articulator elicited instant response from a different cooperating articulator, as a function of gestural goal.

Abstractly, then, a gesture is a coordinative structure controlling an equivalence class of articulator movements, where equivalence is defined by function or goal, the achievement of a certain vocal tract configuration. Thus, we arrive at the first requirement of any hypothesis concerning the role of mirror neurons in language development and evolution, namely, a unit of phonetic (and so ultimately linguistic) function.

## 4. Behavioral evidence for gestures from speech errors

### 4.1 Adults: Random speech errors in the laboratory

Evidence for discrete units in speech production comes from fluent, random errors, or "slips of the tongue" in which whole units are omitted, transposed or exchanged (e.g. Fromkin 1971; Shattuck-Hufnagel 1987). In adult speech the most frequent units of exchange are whole words (e.g. "We have a laboratory in our computer") or whole segments (e.g. "shunsine" for "sunshine"). Single gesture errors (e.g. "tebestrian" for "pedestrian", where labial and alveolar closures are exchanged) are rare. Single gesture errors seem to be rare partly because we often cannot distinguish them from whole segment errors (e.g. "packback" for "backpack"), partly because errors can be partial and can go undetected (Mowrey & MacKay 1990), and partly because adults have acquired higher order coordinative structures that routinise control of the gestural combinations that repeatedly recur in segments.

Nonetheless, single gesture errors can be induced experimentally by "tongue twisters", (e.g. Mowrey & MacKay 1990; Pouplier, Chen, Goldstein, & Byrd 1999). The latter group of experimenters asked a subject to repeat phrases with alternating initial consonants (e.g. "sop shop") at rapid metronome-controlled rates for 10 seconds; they recorded tongue and lip movements with a magnetometer. Note that in "sop shop", /s/ is produced with one oral gesture, raising the tongue tip, while / ʃ / is produced with two, raising the front part of the tongue body and protruding the lips. By far the more frequent error was / ʃ / for /s/, not the reverse. On some of these errors, both tongue body raising and lip protrusion for / ʃ / occurred, on others only tongue body raising without lip protrusion. Dissociation of two gestures that normally cohere to produce an integral segment, / ʃ /, shows that the two gestures are independently controlled.

The strongest behavioral evidence for gestures as segmental components comes, however, from children's first attempts to imitate adult words, during the narrow window of development between the first imitative attempts and the emergence of segments as units of motor control.

### 4.2 Children: Systematic speech errors in natural imitations of early words

The earliest vocal unit of meaning in a young child is probably the prosodic contour (Menn 1983), the earliest segmental unit of meaning the holistic word (Ferguson & Farwell 1975). Early words are said to be holistic because, although they are formed by combining gestures, gestures have not yet been differentiated as context-free, commutable units that can be independently combined to produce new words; nor, *a fortiori*, have recurrent gestural combinations yet been integrated into the cohesive units of segmental control that seem to emerge with vocabulary growth during the 2nd–3rd year of life (Jaeger 1992a, b). What we often see, then, in a child's early words are unsuccessful attempts to achieve the amplitude, duration, and relative phasing of gestures for which the child has successfully recognized both the end-effectors (lips, tongue body, tongue tip, velum, larynx) and, for oral gestures, their rough locus of constriction.

We draw our examples from a study of a North American child, "Emma", conducted over a 4 month period around her second birthday (Studdert-Kennedy & Goodell 1995). None of the words discussed below immediately followed an adult model utterance; all words were therefore spontaneous delayed imitations drawn from the child's long term memory. Emma had already developed a favorite gestural routine, a /labial-consonant/vowel/alveolar-consonant/vowel/ sequence, that seemed to serve as an armature for phonetic construction of words and as an articulatory filter on word selection (Vihman 1991; Vihman & De Paolis 2000): she tended to choose for imitation words to which the routine could be appropriately applied. This bias on word selection is evident in Table 1.[1] The first word, *raisin*, illustrates three common types of gestural error in early child speech: (1) omission of one gesture from a two-gesture combination: the child omits tongue tip retroflexion for [r], but retains lip protrusion, yielding [w]; (2) error of constriction amplitude, widely attested in the acquisition literature: full closure instead of the critical fricative constriction for [z], yielding [d]; (3) vowel-consonant harmony: the unwanted final vowel following the release of [d] is assimilated to the point of consonant closure, yielding [i].

Notice that neither the semi-vowel, [w], nor the consonant, [d], appears in the target. Intrusion of unwanted segments, a commonplace of child word-learning at this stage of development, is difficult to explain if we assume standard, context-free segmental or featural primitives organized over the syllable (e.g. MacNeilage & Davis 1900). Intrusions follow naturally, however, if we assume gestural primitives organized over the domain of the word. For example, errors often arise because gestures "slide" along the time line (Browman & Goldstein 1987) into misalignment with other gestures, yielding segments not present in the target. Thus, in the second word of Table 1, *berry*, lip protrusion for [r] slides onto the preceding vowel to yield rounded [u]; an error of amplitude and tongue shape on the remaining tongue tip retroflexion for [r] then yields the unwanted [d]. Notice, incidentally, that while initial [r] in *raisin* becomes [w], medial [r] in *berry* becomes [d], a result that a featural account would not predict because a given segment carries the same featural predicates regardless of context, and so should be subject to the same motoric errors.

On the third word, *tomato*, Emma evidently omits both the first syllable and glottal abduction for medial [t]; she then allows velic opening for [m] to spread, or slide, into alignment with alveolar closure for [t], yielding [n]. Figure 2 (bottom) illustrates gestural sliding in one of Emma's attempts to say *donut*, where the second

**Table 1.** Some target words and a two-year old child's attempts to say them, illustrating paradigmatic and syntagmatic gestural errors (from Studdert-Kennedy & Goodell 1995). Parentheses around letters in target words indicate syllables evidently omitted in child's attempts.

| Target word | Consonant gestures in adult form | Child's attempts |
|---|---|---|
| raisin<br>['rezn] | [r]:tongue tip postalveolar retroflection + lip protrusion; z: critical alveolar constriction; [n]: alveolar closure + velic lowering. | ['weːˈni]<br>['weːnˈdi],<br>['weːˈdi] |
| berry<br>['berɪ] | [b]:bilabial closure;<br>[r]: as above. | ['buːˈdi] |
| (to)mato<br>[tʰəˈmetʰəʊ] | [t]: alveolar closure + glottal abduction;<br>[m]: bilabial closure + velic lowering. | ['meːˈnə] |
| (ele)phant<br>['ɛlɪf(ə)nt] | [f]: critical labiodental constriction + glottal abduction; [n]: as above; [t]: as above. | ['amˈbin],<br>['aˈmin],<br>['aˈfin],<br>['aˈpin],<br>['aˈbin] |
| (hippo)potamus<br>['hɪpəˈpatəməs] | [p]: bilabial closure + glottal abduction; [t]: as above; [m]: as above; [s] critical alveolar constriction + glottal abduction. | ['apɪnz] |
| apricot<br>['æprɪˈkat] | [p]: as above; [r] as above; [k] velar closure + glottal abduction; [t]: as above | (1) ['aɪbəʷaʰ aː],<br>(2) ['a pəˈgʌ]<br>(3) ['əˈfuˈkaː]<br>(4) [ʰʌfəˈtsaː]<br>(5) ['gelˈgʌˈpaː]<br>(6) ['ŋəᵗʰapʷəˈtʰ aː] |

syllable emerged as [dʌnt]. The switch from [nʌt] to [dʌnt] evidently follows from a simple error of timing: velic opening slides away from syllable initial [n] (so that alveolar closure and release now yield [d]) into alignment with alveolar closure (but not release) for final [t], yielding [nt].

The next three words are more complicated. The perhaps unexpected words, *elephant* and *hippopotamus*, were names for pictures in a book. For both words the child adopted a tactic that she favored for words or phrases of three or more syllables: she lowered her jaw and substituted the wide pharyngeal gesture of vocalic [a] as a sort of place-holder for the initial syllable or syllables. Here, five attempts at *-phant* ([fənt]) all include a labial gesture, but only ['aˈfin] achieves the correct labiodental [f], only ['aˈbin] and ['aˈpin] the required accompanying glottal abduction. Other attempts are voiced throughout and include an intrusive high front vowel, [i], harmonizing with alveolar closure for [n]; in two attempts, ['am'bin] and ['a'min], velic lowering for [n] slides into alignment with labial closure for [f], yielding [m].

For *hippopotamus*, Emma substituted her idiosyncratic [a] for the first two syllables. She then compressed the remaining three syllables, *-potamus*, into one, [pɪnz], built around her labial-alveolar routine. Figure 3 illustrates the surprisingly simple process by which she may have accomplished the transformation. She correctly executed labial closure and glottal abduction for [p] and alveolar closure for [t], thus linking the lip and tongue tip gestures of her favored routine, as indicated by the arrows in the figure. She omitted labial closure for [m] and glottal abduction for [t] and [s]; she omitted the low back tongue body gestures for the vowels, roughly harmonizing the vocalic nucleus to the following alveolar closure; and she allowed velic lowering for omitted [m] to slide into alignment with alveolar closure

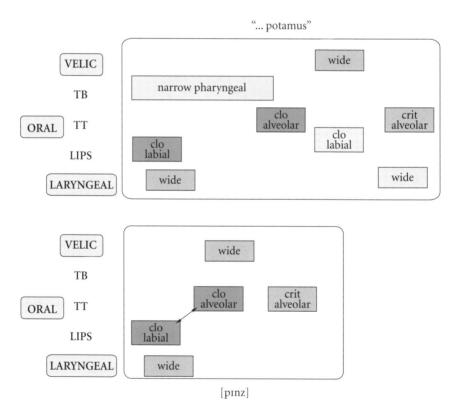

**Figure 3.** Gestural scores for the final three syllables of the adult target *hippopotamus*, ['potəmʌs] (above) and for the child's attempt at the word, ['apɪnz] (below)

for [t], yielding [n]. The outcome of these maneuvers was [pɪnz], a syllable composed of four segments, three of which did not appear in the target. Once again, we have an attempt at an adult target that is segmentally bizarre, but readily intelligible, if we take the primitives of phonetic action to be gestures rather than segments.

Finally, let us briefly consider Emma's attempts at *apricot* ([ˈæprɪˈkat]). The word is obviously difficult, with its pattern of alternating glottal abduction for voiceless stops and glottal adduction for vowels, its zigzag front-back-middle pattern of consonant location, and its sweep (low to high, front to back) around the vowel space, making a total of a dozen intricately interleaved gestures to be executed in less than half a second. Not surprisingly, Emma's attempts are highly variable. Yet she captures several properties of the word accurately in different tokens: the alternating pattern of glottal abduction and adduction (tokens 3, 4, 6): the constriction location and amplitude of the final vowel (1, 3–6); the rough location of at least two out of three consonantal gestures (2–6); and, setting aside the initial velar intrusions of 5 and 6, the labial–lingual sequence of these gestures. At the same time, every token includes at least one segment not present in the model: [b], [s], [f], [ts] or [ŋ]. With the exception of [f] (an error in the exact location and amplitude of the word's labial constriction) and of the intrusive velic lowering for [ŋ] all these errors arise from a failure of gestural timing or coordination: for the affricate, [ts], a relatively slow release of [t], yielding the intrusive frication of [s]; for [b], [g] and [ŋ] a failure of glottal abduction during oral closure. Yet glottal abduction was not always omitted: on the contrary, repeated attempts, as called for by [p], [k], [t], are evident not only in the correct execution of at least one of the set in every token except the first, but also in several erroneous intrusions: the brief aspiration (superscript [h]) inserted in tokens 1, 4, and 6, and the whispered initial vowel of token 2. Evidently, the child had recognized the repeated glottal abductions of the adult model, but could not always phase them correctly relative to the rapid sequence of oral gestures: glottal abduction, like the velic lowering and lip protrusion of earlier examples, repeatedly slid out of alignment with its target consonant closures.

## 5. The somatotopic organization of phonetic gestures

Two aspects of the data we have reviewed are of particular interest for the present discussion. First, the child executes consonants with their precise, categorical loci of constriction more accurately than the less precise, continuously variable vowels. Second, despite the often egregious discrepancy between target and copy, the child activates the end-effectors of the gestures composing an utterance with surprising

Table 2. Initial consonants in early words of 4 English-learning children in three half-hour sessions during transition from babbling to word use (13–16 months approximately): Tabulation of data from Stanford Child Phonology Project in Appendix C of Vihman (1996).

| Target | Number correct | Number of single feature errors | | | Number attempted |
|---|---|---|---|---|---|
| | | place | voicing | other | |
| Bilabial [p, b, m, w] | 88 | 3 | 46 | 3 | 140 |
| Alveolar [t, d, n, r, l, s, z] | 46 | 14 | 25 | 1 | 86 |
| Velar [k, g] | 27 | 4 | 9 | 0 | 40 |
| Total | 161 | 21 | 80 | 4 | 266 |
| % of errors | | 20 | 76 | 4 | |

accuracy. Errors in glottal and velic gestures tend to be errors of phasing rather than omission; and while errors of precise location do occur for oral gestures (e.g. /s/ for /ʃ/, /p/ for /f/), the end-effector (lips, tongue tip, tongue body) and so the rough locus of constriction (i.e. place of articulation) tends to be correct. In other words, gestural errors on consonants tend to be errors of amplitude or phasing rather than of omission or end-effector. Table 2 lends further support to this claim with data from 4 children in the Stanford Child Phonology Project (Vihman 1996, Appendix C): 80% of single gesture errors on voicing (glottal action) or manner (primarily, gestural amplitude), 20% on place of articulation.[2]

Whether biases in types of gestural error arise from difficulties in perception or in production, we cannot easily tell. We do know that children tend to apply an "articulatory filter" (Vihman 1991; Vihman & De Paolis 2000), choosing words to say that match their available articulatory routines. We also know that a child often repeatedly corrects itself, without adult guidance, until it reaches an acceptable approximation to the target adult form: for example, Emma's repeated attempts at *elephant* and *apricot*. Both selection and avoidance of words and the unguided self-correction of words suggest that a child has an adequate perceptual representation, but cannot easily coordinate gestures to achieve an acceptable match in her own speech. There may, of course, be no general answer to the question: similar errors may arise from different sources in different utterances. In any event, what is important for the present discussion is that, whatever the source of error, the relative accuracy with which end-effectors are activated in children's early words argues that gestures are somatotopically represented. Thus, we arrive at the second requirement for any hypothesis concerning the role of mirror neurons in language development and evolution, namely, somatotopic representation of function.

How might somatotopic representation of the vocal tract have arisen evolutionarily? As a way into this question, let us consider what we have learned in recent years about how infants imitate facial configurations.

## 6.  Correspondences between facial and vocal imitation in infants

Infants are capable of manual, vocal, and facial imitation. Facial imitation is unique among the three because the infant can neither see nor hear the consequences of its own facial movements, nor can it feel the muscle activities of the faces it imitates. The connection between the facial patterns it sees and the kinesthetic sensations of its own actions is therefore intermodal, or cross-modal. Much of what we know about this cross-modal process comes from the sustained research program of Meltzoff and Moore (1997, and the many papers cited therein).

Meltzoff and Moore (1997, Table 1, p. 181) list ten characteristics of infant facial imitation as follows. (1) Infants imitate a range of acts, including mouth opening, lip protrusion, tongue protrusion, eye blinking and cheek motion. (2) Imitation is specific (e.g. tongue protrusion leads to tongue protrusion, not lip protrusion, and vice versa). (3) Literal newborns imitate. (4) Infants quickly activate the appropriate body part. (5) Infants correct their imitative errors. Infants imitate (6) novel acts, (7) absent targets, and (8) static gestures. (9) Infants recognize being imitated. (10) There is developmental change in imitation.

Infants begin to imitate speech several months later than they begin to imitate faces – in the laboratory, around the 4th–5th month of life, outside the laboratory, in the second half of the first year (Vihman 1996). Otherwise, most of the characteristics that Meltzoff and Moore (1997) attribute to infant facial imitation appear also, *mutatis mutandis*, in infant imitation of speech. To account for these characteristics, Meltzoff & Moore (1997) propose a model for facial imitation that they term Active Intermodal Matching (AIM); the model can readily be extended to vocal imitation. The authors argue that facial imitation is: (i) representationally mediated, (ii) goal directed, and (iii) generative and specific.

Each of these properties of the AIM model accords with properties of child speech production briefly described in the previous section. First, a child's early words are not necessarily immediate imitations of an adult model. In fact, all Emma's words cited above were spontaneous utterances prompted by the sight of their referent and drawn from long term memory, that is, from a stored representation of an adult utterance and/or of her own previous attempts at the word. Second, children's repeated attempts at self-correction, unguided by an adult, demonstrate that imitative utterances are goal-directed attempts to match a target. Finally, the flexibility of children's responses, the repeated use of specific gestures or gestural routines to imitate new words, demonstrates the generative capacity necessary for building a large vocabulary from a small repertoire of gestures.

Three theoretical concepts that Meltzoff and Moore (1997) propose to account for facial imitation also conform nicely with what we evidently need to account for the imitation of speech: (i) body part ("organ") identification; (ii) mapping of movements to goals; (iii) a cross-modal metric of equivalence between the acts of self and other. The first step in both modes of imitation is to identify the body part, the facial "organ" or articulator, to be moved: head, brow, jaw, cheek, lips, tongue for the face; lips, tongue tip, tongue body, velum, larynx for speech. The capacity to identify corresponding body parts renders self and other commensurate, establishing organs or articulators as units of cross-modal analysis. Whether such correspondences are innate or emerge through experience of movement (perhaps beginning *in utero*) is a matter of great interest, but need not concern us here.

The gradual mapping of articulator movements to phonetic goals has long been assumed to be the function of infant vocal babble (e.g. Fry 1966). Meltzoff and Moore (1997) propose an analogous function for "body babbling" by which the infant learns the mapping between muscle activations and facial organ relation end-states (facial configurations). Many muscles contribute to moving a facial organ into a certain relation with another (e.g. tongue between lips, tongue protruding beyond lips, and so on), just as many muscles contribute to a given speech gesture or configuration of the vocal tract; these muscles, as we have seen constitute a coordinative structure specific to a particular goal (Fowler, et al. 1980; Turvey 1977). Body babbling, like vocal babbling, serves then to build muscular activations into coordinative structures and to map these structures, onto the end-states, or goals, of facial organ relations.

Thus, organ relations provide the framework, the cross-modal metric of equivalence between the faces of self and other that "... render[s] commensurate the seen but unfelt act of the adult and the felt but unseen facial act of the infant" (Meltzoff & Moore 1997: 185). Extending this account to manual, vocal and other modes of imitation, we may hypothesize that relations among body parts serve as a general metric by which all imitative acts are compared with their models. Vocal tract configuration would then be the metric of equivalence between utterances of child and adult; and the motor theory of speech perception (Liberman 1991; Liberman & Mattingly 1985), although originally formulated quite without reference to the facts of vocal imitation, would prove to be a special case of a general principle.

Where now might we expect mirror neurons to fit into the AIM model of facial imitation and into its possible extended version in vocal imitation? If the function of mirror neurons is, as Rizzolatti and Arbib (1998) have proposed, to represent the actions of others in the process of recognizing, understanding, and (perhaps) imitating them, we must suppose that these neurons are components of a system that mediates between a perceived movement, or gesture, and activation of the coordinative structure that controls execution of that movement or gesture. If we further assume, with Rizzolatti and Arbib (1998), that a "closed system" of expressive facial communication was an evolutionary precursor of "the open [i.e. particulate and combinatorial] vocalization system we know as speech" (p. 192), how might the transition from closed analog facial configuration to open digital vocal gesture have come about?

## 7.   From analog face and cry to particulate speech

Both human and non-human primates express and communicate their emotions by both face and voice (Darwin 1872/1965; Hauser 1996). Indeed, the two modes are so intricately related that we often cannot tell "… whether the sounds… produced under various states of the mind determine the shape of the mouth, or whether its shape is not determined by independent causes, and the sound thus modified" (Darwin 1872/1965:91). Nonetheless, we do know that functional (meaningful) contrasts among primate calls and cries are conveyed by continuous variations in fundamental frequency, amplitude, and rhythm, effected by actions of the larynx and jaw; humans have indeed carried such variations over into the prosody of their speech. Non-human primates may also effect spectral contrasts in vocal tract resonances (formant pattern) by continuously variable actions of lips and jaw (Hauser 1996:180–186); but we have no evidence that they execute communicatively effective spectral contrasts by the discrete actions of supralaryngeal articulators (lips, velum, tongue), characteristic of human speech. Nor do we have evidence for non-human primate use of the voice for arbitrary symbolic reference.

We are thus confronted by three apparent discontinuities between human and other primates in use of their facial and vocal apparatus. Humans alone (i) imitate facial and vocal action, (ii) modify vocal spectral structure by discrete, categorically distinct actions of the articulators, (iii) use vocal sounds for arbitrary symbolic references. Here, I will argue that the key to all three discontinuities may lie in the gradual evolution of the mirror neuron system for representing and effecting manual action, already present in lower primates, into a mirror system for brachio-manual imitation in the apes and early hominids. Incipient capacities for brachio-manual imitation in modern chimpanzees are evident both in the wild (termite fishing, nest building) and in captivity (sign language). Under the increasing social and cultural pressures of early hominid groups, the mirror system for brachio-manual imitation was perhaps genetically duplicated and gradually adapted first to facial, then to vocal imitation.

How, when, or why the leap into facial, and later vocal imitation occurred, we may never know. Donald (1991, 1998), however, offers a lead into the questions. He has built a powerful argument for the necessity of a culture intermediate between apes and *Homo sapiens* and of some prelinguistic mode of communication as a force for social stability and cohesion in that culture. The fossil evidence points to *Homo erectus* as the transitional species. The species was relatively stable for over a million years and spread out over the entire Eurasian land mass, leaving in its tools and in its traces of butchery and fire, evidence of a complex social organization well beyond the reach of apes.

Viewing the modern human mind as a hybrid of its past embodiments, Donald finds evidence for an ancient prelinguistic mode of communication in the brachio-manual gestures, facial expressions, pantomimes, and inarticulate vocalizations to which modern humans may still have recourse when deprived of speech. "Mimesis" is Donald's term for this analog, largely iconic, mode of communication. Mimesis is more than mere mimicry, because it includes a representational dimension, a capacity for conscious, intentional control of emotionally expressive behaviors, including facial expression and vocalization, of which no non-human primate is capable. Donald proposes, indeed, that mimesis "... establishe[d] the fundamentals of intentional expression in hominids" (1998:60). Certainly, such a development must have occurred at some point to bridge the vast gap in communicative competence between apes and humans; and a mimetic culture, as conceived and richly elaborated by Donald, seems to be fully consistent with the evolutionary speculations of others (e.g. Bickerton 1991; Deacon 1997; Hurford et al. 1998; Knight et al. 2000).

Donald goes on to propose that mimesis also "... put in place... the fundamentals of articulatory gesture from which all languages are built" (1998:65). Again, some such development must surely have taken place, but here we face difficulties. However long its reach or broad its scope, mimesis was still an analog mode of iconic gesticulation, facial expression, and inarticulate grunts and cries. Even if we accept the postulated mimetic culture, we are still left with two key questions: How did communication shed its iconicity? How did vocal communication go digital, or particulate?

What I wish to argue here is that solutions to both problems may be implicit in the act of imitation itself. Conceptually, imitation of a novel act entails three steps: (i) analysis of the target act into its components; (ii) storage of the analyzed structure for a shorter or longer period, depending on the interval between model and copy; (iii) reassembly of the components in correct spatiotemporal order. Repeated analytic acts of imitation gradually induce particulation of the imitative machinery; particulation thus fragments the holistic function, or iconic image, into its non-iconic components, opening a path into arbitrary meaning.

Consider, first, the neuroanatomical differentiation and somatotopic representation of the human hand and arm (Fadiga, Fogassi, Pavesi, & Rizzolatti 1995; Fadiga, Buccino, Craighero, Fogassi, Gallese, & Pavesi 1999). In its earliest forms, as indeed perhaps often still today, brachio- manual imitation might have been mediated through a mirror system for representing a holistic function, such as throwing a stone, grasping a branch, seizing food. Such acts would have been motivated by recognizing the goals of conspecific acts. The form of the imitated acts would then have been shaped automatically by the actor's morphology and by the demands of the physical context. As fingers, thumbs, wrists came to be engaged in more and more diverse physical acts, they would have become increasingly differentiated and capable of independent action, free to engage in arbitrary non-iconic conventional gestures, common in many cultures today (e.g. Morris, Collett, Marsh, & O'Shaughnessy 1979).

Similarly, we may suppose, facial imitation might initially have been (and may often still be) mediated by observation and empathetic experience of the emotion expressed by the face of a conspecific; the holistic configuration of the face would then have been automatically shaped by the emotions and physiology common to model and imitator. As with the hand, increasing frequency and diversity of facial expression would have encouraged differentiation and independence of parts of the face – as in raising the eyebrows, winking, pursing or pouting the lips, flaring the nostrils, protruding the tongue. Individual "organs" of the face would thus become free for arbitrary imitative acts, as in the studies of Meltzoff and Moore (1997).

Finally, a similar process of differentiation must also have occurred for the vocal tract, but its origins are obscure. Unlike configurations of hands or face, and unlike the expressive actions of larynx and jaw in primate cries and calls, holistic configurations of the supralaryngeal articulators neither lend themselves to iconic representation, nor express emotions through some intrinsic physiological function. How then did such configurations come to take on meaning?

Two paths into holistic meaning seem possible, neither of them compellingly persuasive. One path is through phonetic symbolism, as proposed by Rizzolatti and Arbib (1998:193). The obvious iconic poverty of the vocal tract, just remarked, does not encourage one to see this as a route to any but the most minimal vocabulary. A second path might be through the adventitious effects of meaningful facial expressions on concurrent vocal sounds, as remarked by Darwin (1872/1965:91) in the citation above. Again, the range of possible effects is not encouraging.

A third, perhaps viable, way out of the impasse is to follow Carstairs-McCarthy's (1998, 1999) radical lead and turn the problem on its head. For many subtle, deeply argued reasons that I have neither space nor competence to summarize here, Carstairs-McCarthy proposes that language-independent changes in the vocal tract (induced by bipedalism, although the precise mechanism is not, in my view, essential to his argument) endowed early hominids with a much increased range of syllables with which concepts of objects and events could be associated. The habit of vocal play (Halle 1975; Knight 2000) may have increased still further the stock of syllabic patterns, much as we observe it today in infant babble

(MacNeilage & Davis 2000). Early hominids thus found themselves with a surfeit of syllables in search of meanings rather than of meanings in search of an expressive mode. Differentiation of the syllable into its component gestures would have followed under pressure for speed and economy, again much as we observe the process in a child's early words, described above.

Be all this as it may, whatever the origin of the shift from iconic to arbitrary symbolic reference, prior evolution of particulated mirror systems for hand and face would have facilitated evolution of vocal imitation, and so of particulate speech, by offering a neural mechanism that the vocal tract could coopt.

## 8.  Conclusion

The proposed evolutionary path from manual to facial to vocal imitation rests on the hypothesis that all three modes of imitation are mediated by mirror neuron systems of representation. The hypothesis is consistent with Alvin Liberman's (1996:31) recognition that a link between perception and production is necessary for successful linguistic (or any other form of) communication. The hypothesis may also be amenable to experimental test by brain imaging or other techniques, perhaps along lines suggested by the elegant research of Luciano Fadiga and his colleagues (1995, 1999).

## Acknowledgements

My thanks to Louis Goldstein for advice, discussion, and the figures. Preparation of the paper was supported in part by Haskins Laboratories.

## Notes

1.  Transcriptions of child utterances as strings of phonetic symbols does not imply that the child has developed independent control over segments. A phonetic symbol is simply a convenient shorthand for segments, or combinations of gestures, heard by an adult transcriber.

2.  If we score errors by precise locus of constriction, as is typically done in studies of random errors, rather than by end-effector, as I have done here, we increase the opportunities for error; place errors are then proportionately more common than other types in both adults (van der Broecke & Goldstein 1980) and children (Jaeger 1992a, b).

# References

Abler, W. (1989). On the particulate principle of self-diversifying systems. *Journal of Social and Biological Structures, 12*, 1–13.

Bickerton, D. (1990). *Language and Species*. Chicago: Chicago University Press.

Browman, C. P., & Goldstein, L. (1986). Towards an articulatory phonology. *Phonology Yearbook, 3*, 219–252.

Browman, C. P., & Goldstein, L. (1987). Tiers in articulatory phonology, with some implications for casual speech. In J. Kingston & M. E. Beckman (Eds.), *Papers in Laboratory Phonology, I* (pp. 341–376). New York: Cambridge University Press.

Browman, C. P., & Goldstein, L. (1992). Articulatory phonology: An overview. *Phonetica, 49*, 155–180.

Carstairs-McCarthy, A. (1998). Synonymy avoidance, phonology and the origin of syntax. In J. R. Hurford, M. Studdert-Kennedy, & C. Knight (Eds.), *Approaches to the Evolution of Language* (pp. 279–296). Cambridge: Cambridge University Press.

Carstairs-McCarthy, A. (1999). *The Origins of Complex Language*. Oxford: Oxford University Press.

Chomsky, N. (1972). *Language and Mind*. New York: Harcourt, Brace and World (revised edition).

Darwin, C. (1872/1965). *The Expression of the Emotions in Man and Animals*. Chicago: University of Chicago Press.

Deacon, T. W. (1997). *The Symbolic Species*. New York: Norton.

Donald, M. (1991). *Origins of the Modern Mind*. Cambridge, MA: Harvard University Press.

Donald, M. (1998). Mimesis and the executive suite: Missing links in language evolution. In J. R. Hurford, M. Studdert-Kennedy, & C. Knight (Eds.), *Approaches to the Evolution of Language* (pp. 44–67). Cambridge: Cambridge University Press.

Fadiga, L., Fogassi, L., Pavesi, G., & Rizzolatti, G. (1995). Motor facilitation during action observation: A magnetic stimulation study. *Journal of Neurophysiology, 73*, 2608–2611.

Fadiga, L., Buccino, G., Craighero, L., Fogassi, L., Gallese, V., & Pavesi, G. (1999). Corticospinal excitability is specifically modulated by motor imagery: A magnetic stimulation study. *Neuropsychologia, 37*, 147–158.

Ferguson, C. A., & Farwell, L. (1975). Words and sounds in early language acquisition. *Language, 51*, 419–439.

Fowler, C. A., & Saltzman, E. (1993). Coordination and coarticulation in speech production. *Language and Speech, 36*, 171–195.

Fowler, C. A., Rubin, P., Remez, R., & Turvey, M. (1980). Implications for speech production of a general theory of action. In B. Butterworth (Ed.), *Language Production, I: Speech and talk* (pp. 373–420). London: Academic Press.

Fromkin, V. (1971). The nonanomalous nature of anomalous utterances. *Language, 47*, 27–52.

Fry, D. B. (1966). The development of the phonological system in the normal and the deaf child. In F. Smith & G. A. Miller (Eds.), *The Genesis of Language: A Psycholinguistic Approach* (pp. 187–206). Cambridge, MA: MIT Press.

Halle, M. (1975). Confessio grammatici. *Language, 51*, 525–535.

Hauser, M. D. (1996). *The Evolution of Communication*. Cambridge, MA: MIT Press.

Hockett, C. F. (1958). *A Course in Modern Linguistics.* New York: MacMillan.

Humboldt, W. von. (1836/1972). *Linguistic Variability and Intellectual Development.* Translated by G. C. Buck, & F. A. Raven. Philadelphia: University of Pennsylvania Press.

Hurford, J. R., Studdert-Kennedy, M., & Knight, C. (Eds.). (1998). *Approaches to the Evolution of Language.* Cambridge: Cambridge University Press.

Jaeger, J. J. (1992a). 'Not by the chair of my hinny hin hin': Some general properties of slips of the tongue in children. *Journal of Child Language, 19,* 335–336.

Jaeger, J. J. (1992b). Phonetic features in young children's slips of the tongue. *Language and Speech, 35,* 189–205.

Jakobson, R., & Halle, M. (1956). *Fundamentals of Language.* The Hague: Mouton.

Kelso, J. A. S., Tuller, B., Vatikiotis-Bateson, E., & Fowler, C. A. (1984). Functionally specific articulatory cooperation following jaw perturbations during speech: Evidence for coordinative structures. *Journal of Experimental Psychology: Human Perception and Performance, 10,* 812–832.

Knight, C. (2000). Play as precursor of phonology and syntax. In C. Knight, M. Studdert-Kennedy, & J. R. Hurford (Eds.), *The Evolutionary Emergence of Language* (pp. 99–119). Cambridge: Cambridge University Press.

Knight, C., Studdert-Kennedy, M., & Hurford, J. R. (Eds.). (2000). *The Evolutionary Emergence of Language.* Cambridge: Cambridge University Press.

Liberman, A. M. (1996). *Speech: A special code.* Cambridge, MA: MIT Press.

Liberman, A. M., & Mattingly, I. G. (1985). The motor theory of speech perception revised. *Cognition, 21,* 1–36.

MacNeilage, P. F., & Davis, B. (2000). Evolution of speech: The relation between ontogeny and phylogeny. In C. Knight, M. Studdert-Kennedy, & J. R. Hurford (Eds.), *Evolutionary Emergence of Language* (pp. 99–119). Cambridge: Cambridge University Press.

Meltzoff, M., & Moore, K. (1997). Explaining facial imitation: A theoretical model. *Early Development and Parenting, 6,* 179–192.

Menn, L. (1983). Development of articulatory, phonetic and phonological capabilities. In B. Butterworth (Ed.), *Language Production, II* (pp. 3–30). London: Academic Press.

Morris, D., Collett, P., Marsh, P., & O'Shaughnessy, M. (1979). *Gestures: Their origins and distribution.* New York: Stein & Day.

Mowrey, R. A., & MacKay, I. R. A. (1990). Phonological primitives: Electromyographic speech error evidence. *Journal of Acoustical Society of America, 88,* 1299–1312.

Pouplier, M., Chen, L., Goldstein, L., & Byrd, D. (1999). Kinematic evidence for the existence of gradient speech errors. *Journal of Acoustical Society of America, 106,* 2242(A).

Quine, W. V. O. (1960). *Word and Object.* Cambridge, MA: MIT Press.

Rizzolatti, G., & Arbib, M. A. (1998). Language within our grasp. *Trends in Neuroscience, 21,* 188–194.

Rizzolatti, G., Fadiga, L., Gallese, V., & Fogassi, L. (1996). Premotor cortex and the recognition of motor actions. *Cognitive Brain Research, 3,* 131–141.

Saltzman, E. (1986). Task-dynamic coordination of the speech articulators: A preliminary model. In H. Heuer & C. Fromm (Eds.), *Experimental Brain Research Series,* Vol. 15: *Generation and Modulation of Action Patterns* (pp. 129–144). New York: Springer.

Saltzman, E., & Munhall, K. (1989). A dynamical approach to gestural patterning in speech production. *Ecological Psychology, 1,* 333–383.

Shattuck-Hufnagel, S. (1987). The role of word-onset consonants in speech production planning: New evidence from speech error patterns. In E. Keller & M. Gopnik (Eds.), *Motor and Sensory Processes in Language* (pp. 17–51). Hillsdale, NJ: Lawrence Erlbaum Associates.

Studdert-Kennedy, M. (1998). The particulate origins of language generativity: From syllable to gesture. In J. R. Hurford, M. Studdert-Kennedy, & C. Knight (Eds.), *Approaches to the Evolution of Language* (pp. 202–221). Cambridge: Cambridge University Press.

Studdert-Kennedy, M. (2000). Evolutionary implications of the particulate principle: Imitation and the dissociation of phonetic form from semantic function. In C. Knight, M. Studdert-Kennedy, & J. R. Hurford (Eds.), *The Evolutionary Emergence of Language* (pp. 161–176). Cambridge: Cambridge University Press.

Studdert-Kennedy, M., & Goodell, E. W. (1995). Gestures, features and segments in early child speech. In B. de Gelder, & J. Morais (Eds.), *Speech and Reading* (pp. 65–88). Hove: Erlbaum (UK), Taylor & Francis.

Turvey, M. T. (1977). Preliminaries to a theory of action with reference to vision. In R. Shaw & J. Bransford (Eds.), *Perceiving, Acting and Knowing: Toward an ecological psychology* (pp. 211–265). Hillsdale, NJ: Lawrence Erlbaum Associates.

Van den Broecke, M. P. R., & Goldstein, L. (1980). Consonant features in speech errors. In V. Fromkin (Ed.), *Errors in Linguistic Performance: Slips of the tongue, ear, pen and hand* (pp. 47–65). New York: Academic Press.

Vihman, M. M. (1991). Ontogeny of phonetic gestures. In I. G. Mattingly & M. Studdert-Kennedy (Eds.), *Modularity and the Motor Theory of Speech Perception: Proceedings of a conference to honor Alvin M. Liberman* (pp. 69–84). Hillsdale, NJ: Lawrence Erlbaum Associates.

Vihman, M. M. (1996). *Phonological Development*. Cambridge, MA: Blackwell.

Vihman, M. M., & De Paolis, M. (2000). The role of mimesis in infant language development: Evidence for phylogeny? In C. Knight, M. Studdert-Kennedy, & J. R. Hurford (Eds.), *The Evolutionary Emergence of Language* (pp. 130–145). Cambridge: Cambridge University Press.

# Constitutive features of human dialogic interaction

## Mirror neurons and what they tell us about human abilities

Edda Weigand
Arbeitsbereich Sprachwissenschaft Westfälische Wilhelms-Universität
Münster, Germany

## 1. Mirror neurons and human abilities

The discovery and experimental proof of so-called mirror neurons seem to focus the interest of various sciences on the *central question of human abilities* (Rizzolatti & Arbib 1998). Human behaviour is the joint interest of all of us who, coming from different disciplines, are participating in this conference on mirror neurons and the evolution of brain and language. Human behaviour from a linguistic point of view can be considered to be human dialogic interaction. In 20th century linguistics there are mainly *two concepts of language*: the artificial concept of language as a sign system and the natural concept of language-in-use. If we accept the view that in language use we are dialogically interacting, the natural concept of language use can be identified as human dialogic interaction or as 'parole' or 'performance'. The dogma of language as a sign system has to be detected as a language myth based on the hypothesis that 'la langue' underlies 'la parole'. There is however no bridge interrelating the simple construct 'la langue' and the complex natural object 'la parole'.

Human dialogic interaction is the way in which language works, in which language presents itself as object-in-function. It is not an object on its own which could be separated from human beings and their abilities. Human beings are part of the world and part of the culture in which they live. Explaining human dialogic interaction therefore needs to explain it in the minimal dialogically autonomous unit which is the cultural *unit of the Action Game*.

In science, the challenge is the same for different disciplines. At the beginning of the new millennium, we are finally able to express this challenge more precisely: it is the *challenge of addressing the complex*. Every discipline selects a part of the complex which surrounds us as its object of study. Classical theorizing told us not to address the complex mix of order and disorder directly but to abstract from disorder and to establish a level below, in linguistics 'la langue' or, with some modification, 'competence'. A theory thus had to reduce complexity to simple rules. After two millennia of classical western thinking we have finally recognized the theory myth and feel able to tackle the problem of unstable systems.

Addressing the complex nevertheless means relating it to the simple in one way or another. *The simple and the complex* – that is the general scientific question. The simple may be sought either with respect to evolution, i.e. moving backwards in time, or with respect to the analysis of an already fully developed phenomenon without moving in time. Both perspectives should complement each other and in the end converge. *The evolutionary perspective* can be understood as attempt to trace a highly developed *complex phenomenon back to its simple origins*. I remember the well-known book by Gell-Mann (1994) on the quark and the jaguar which represent the simple and the complex from the point of view of physics. Compared with physics and the quark, the elementary particle, and the jaguar, the elegant wildcat, one might consider the *eloquent human being* as the *complex* in linguistics. *But what is to be considered the simple from the linguistic point of view?*

We might take the *mirror neuron as the simple in an evolutionary perspective.* However the experiments on mirror neurons, most excitingly, do not reveal them as simple in the sense of the quark. They are *complex units which integrate different dimensions*. The physiological object cannot be separated from its function. In order to recognize the mirror neuron, we have to recognize how this cell type functions. Material aspects are combined with cognitive and perceptual ones. The supposed simple mirror neuron presents itself as a complex integrated structure such as:

$$\begin{bmatrix} \text{biological–physiological structure} \\ \text{cognitive function} \\ \text{perceptual function} \end{bmatrix}$$

**Figure 1.** Mirror neuron

There is not a simple uniform unit *at the outset, but the integration of different dimensions,* different abilities, the material aspects of biology and the immaterial functions of intention and perception. Referring to the basic concept of consciousness, Maxim Stamenov (1997:278) problematizes the belief in a single and reasonably well-defined set of rules and calls a 'grammar of consciousness' a 'misleading metaphor'. *Recent research in neurosciences confirms* that different human abilities such as rationality, emotion, perception are interrelated and cannot be separated from biological-physiological structure (Damasio 1994; Schumann 1999). It is a *complex integrated whole*, the mirror neuron, from which the evolution of language has started. Matter and energy or function are perhaps one and the same phenomenon. The consequences for human abilities to be drawn from mirror neurons are far-reaching:

– First, *integration* of different abilities is a basic characteristic.
– Second, the mirror neurons characterize *human beings* at the very beginning *as directed* in various respects: *socially, dialogically, purposefully, intentionally, interactively, cognitively.*
– Third, *human beings are part of the world.* They can perceive the world only insofar as their abilities allow it. There is no reality as such, only reality as perceived by human beings through the filter of their abilities.

Integration in various respects characterizes our starting point. It also enters the *staircase model of the different disciplines* developed by Gell-Mann (1994: 111f.). At first sight the listing of disciplines might seem as an attempt to distinguish different levels but it should be read as a listing of different integrated dimensions, or as Gell-Mann puts it: "… while the various sciences do occupy different levels, they form part of a single connected structure."

↑psychology, linguistics, etc.

↑…………

↑biology

↑chemistry

↑physics

**Figure 2.** Staircase model

For linguistics, we can draw decisive conclusions from this basic fact of integration. It reveals *orthodox linguistic models and classical methodology as theory myth* insofar as they are based on discrete verbal items, patterns and rules. In trying to redefine linguistics, we have to give up the old concept of a theory which tells us that a *theory has to explain the complex by reducing it to rules.* We have to be finally prepared to *address the complex directly* which means addressing the problem of integration of order and disorder in unstable *open* systems. I will give a first sketch on how we might proceed in analysing complex dialogic interaction, thereby pointing to the simple basic components-in-function and trying to confirm them by the characteristics of the mirror neurons.

## 2.   The question of object and methodology

It is not only in the neurosciences that we have to say good-bye to some cherished traditional ideas, as Gerhard Roth (2000) emphasized in a recent article. For modern linguistics, the *distinction between performance and competence* is one of these ideas. Since antiquity, western science has predominantly tried to explain the complexity of performance by abstraction and reduction. At the beginning of the new millennium, it is time to leave this rooted orthodox thinking. Totally abstracting from features of performance and asserting that our object of study is rule-governed competence means confusing object and methodology and results in what Integrational Linguistics calls the language-and-communication myth (Harris 1981, 1998). Human competence has to be understood as *competence-in-performance*.

It is simply a myth to assume that our object, human dialogic interaction, could be represented by a closed rule-governed system of competence. Trying to describe the complex by simple rules cannot be the method we use to address the complex; it is the method of avoiding it. In the end, it turns out that the natural complex object has disappeared; rule-governed methodology has become a new artificial object. This basic methodological fallacy has been taken over from generative linguistics into pragmatics, too, by all those models which aim to describe communication as a rule-governed system, applying equally to both sides, the side of the speaker and the side of the interlocutor, among them the model of *dialogue grammar*, which I myself used ten years ago (e.g., Weigand 1989).

The main fallacy of rule-governed models lies in the belief that dialogic interaction functions like *chess*. In chess every move is totally rule-governed and both sides have equal rules except the rule for the first move. Dialogue however comprises more than doubling the speaker side. Dialogic competence has to be considered as the competence of different human beings, of different 'adaptive systems' behaving in ever-changing surroundings.

At the beginning of the new millennium, we should take the question of object and methodology seriously and recognize the turning point at which we have arrived in classical western thinking. The challenge is to find a way to address the complex directly without reducing it to rules. In modern physics this challenge has already been solved by quantum physics and its Principle of Uncertainty. In linguistics, too, we should finally accept that it is *time to open up the orthodox model*, to cancel its restrictions and to develop a *model of competence-in-performance*.

## 3.   Constitutive features of the object 'human dialogic interaction'

I take the title of my talk 'Constitutive features of human dialogic interaction' to be the analytical task of discovering the simple in the functioning of the complex. From this perspective, it will not be a surprise that the features I am going to list are quite different from the features indicated as 'defining characteristics of language', e.g. by Li (2003). Even if Li does not subscribe to the orthodox concept of language, he nevertheless works with a concept of language which does not really take account of dialogic interaction. He therefore can characterize language as a symbolic system. Considering however language as human dialogic interaction, we arrive at the following constitutive features:

-   One of the main characteristics of human behaviour results from the *integration* of different abilities. We interact by using different abilities together as communicative, i.e. dialogic means, mainly the abilities of speaking, of perceiving and of thinking. Even if we wanted, we cannot act by separating these abilities. There is no basic programme of language as a symbolic system to which an optional extra programme of communication is added. Human dialogic interaction is thus to be looked upon as complex human ability. Analogously, the mirror neuron, even if a minimal unit, represents a complex integrated whole.
-   Integration requires that the complex is addressed in a *holistic* way. The components have to be seen in their integral functioning. The mirror neuron can be characterized as an integral whole of matter and function, i.e. an *object-in-function*. Brain and mind cannot be separated.
-   Dialogic interaction is carried out by *different human beings*. It is an illusion to assume equal competence for them. Among the dialogical means on which interaction is based, there are cognitive means dependent on different world views, different cognitive backgrounds of individual human beings. *Meaning indeterminacy* thus results as a basic feature which has to be tackled by interactively negotiating meaning and understanding. The mirror neurons, too, function in the interrelation between different human beings. What the gestures mean is, to some extent, left to indeterminacy. Even if one might be tempted to judge different human beings at the beginning of evolution as quasi-the same, they are essentially different from the very outset having different intentions, different consciousness, different self-understandings. The subsequent divergence of their minds is due to different life stories in different surroundings.
-   Human beings are *slaves to their senses* (Harris 1981; Roth 2000). It is their senses, their abilities which filter their view of the world. The world, the language, the speaker must not be separated as was the case in the orthodox view.

They are interrelated in the minimal cultural unit of the Action Game within which human dialogic interaction is to be analysed. Already at the very outset, the mirror neurons characterize human beings as perceiving the world within the limits they allow.

- Human beings are *purposeful beings*. They are intentionally directed towards other human beings and towards the world. Assuming that mirror neurons signal the doing and the perceiving of gestures means that they confirm intentional behaviour, at least in the period of willed communication.
- Human beings are *social beings*. Their cognition is directed towards other social beings as is reflected in the discharging of the mirror neurons. We might therefore take dialogical purposes and needs as key concept to guide our analyses.
- Meaning to be negotiated in highly developed dialogic interaction is so complex that *not everything can be expressed explicitly*. Not everything needs to be expressed verbally, taking account of the integration of different means.
- Cognition plays a role as meaning and as means. There is however *no meaning as such*, only human beings *mean something*. Consequently meaning cannot be independently defined in advance but is dependent on the individual user and their abilities.
- As a consequence of these points, mainly of the fact that it is different human beings and that meaning is not defined, *open points* in human dialogic interaction have to be accepted. *Problems of understanding* are constitutive and not disturbing marginal factors. Meaning and understanding in principle are not identical on both sides of dialogic interaction. Interaction takes account of probability concepts such as preferences and habits of daily life. Its method therefore cannot be transference of a fixed invariable pattern but *negotiation of meaning and understanding*.
- Consequently, negotiation cannot be based on generalized rules only. We use *Principles of Probability* which help us to orientate ourselves in the range between order and disorder, between rules and chance. In this way, human beings behave like *complex adaptive systems* in using various techniques in order to adapt themselves to ever-changing particular conditions.

## 4. Exemplary analyses

Let me now illustrate some of these points by authentic examples:

### 4.1 Different claims are negotiated

Dialogic interaction does not mean doubling the speaker side but negotiating different positions of the interlocutors. The first example illustrates the whole complexity of the action game. Actions do not represent fixed patterns appropriate for specific conditions but depend on individual reasoning. In order to understand the sample dialogue, I must describe the situation because the text is not autonomous in itself but a component in the action game. In most cases, it is not sufficient to rely only on empirically registrable means as an observer as we do when analysing authentic texts from text corpora. We instead have to refer to the acting human beings, we have to understand their individual evaluation of action conditions and we have to perceive what is going on in the situation. These are exactly the prerequisites needed also for the mirror neuron-in-function.

The sample dialogue is a dialogue between a mother and her 18-year-old daughter (here in English translation). The daughter is preparing the meal in the kitchen and has a question about it.

(1)  Daughter   Come on!
     Mother     Please don't keep on disturbing me. I am working. When you
                were a small child it was alright for you to disturb me but
                now I can assume that you know: my mother is working and
                can decide when to ask me and can save your questions.
     Daughter   The food is for you, too, after all!

The daughter is expressing a speech act of the directive type claiming that it will be fulfilled: *Come on!* Her reasoning refers to the fact that she is preparing the meal for both. She does not accept her mother's reasoning according to which the daughter should not ask questions while her mother is working. In our example the different kinds of reasoning become verbally explicit, misunderstandings therefore are excluded. As complex adaptive systems we have to adapt ourselves in one way or another towards the position of the other partner and to negotiate our claims.

## 4.2  Misunderstandings and open points are constitutive

For the orthodox rule-governed view, misunderstandings are accidents which have to be avoided. As a method of self-defence, it is assumed they could be avoided by giving sufficient information. Authentic examples however prove the opposite: misunderstandings are inevitable. In dialogue, we start as individuals from different positions. Language as a form of life is inextricably interwoven with concepts of probability such as preferences which inevitably carry the risk of misunderstandings (Weigand 1999a).

The second example we are going to analyse is again a dialogue between the mother and her daughter. The mother enters the room while the daughter is playing the piano.

(2)  Mother      You are playing the piano again.
     Daughter    Shall I stop it?
     Mother      No, it doesn't matter. I'm going to work outside.

The first utterance of the mother points to the fact that we do not express our claims verbally by reference to rules only. *You are playing the piano again* can be understood as a representative speech act. But what does this mean? Does the mother appreciate that the daughter is playing the piano or is she angry? Intonation is not so clear as to decide this question. Everything is left to supposition. Even the daughter who knows her mother well is uncertain of what she meant in this individual situation. She refers to her mother's preference of working in silence without being disturbed by the piano and understands the utterance as a reproach. In this particular situation however this preference is not valid. With *Shall I stop it?* the daughter becomes the victim of a misunderstanding which however is immediately clarified by the next utterance of her mother.

This example demonstrates that there are inevitably open points in dialogue which cannot be avoided. Not everything is said explicitly; otherwise we could not start talking because we would have to reflect on the essential points to be expressed in order to avoid misunderstandings. How could a native speaker not trained in linguistics succeed? Where generalized rules come to their limits, *suppositions* have to account for individuality and chance.

## 4.3  Meaning is not defined

Orthodox linguistics means theorizing on the basis of fixed codes. Communication in this sense is thought of as the transference of fixed patterns, with meanings already defined in advance, equally valid for both sides. The following authentic examples demonstrate that such a view is an illusion.

(3)  If you are homeless, you will find a home in Hong Kong because there all are homeless.       (heard on German television, translated into English)

(4)  Change is the only constant in the life of a company.
                                   ('The Economist', March 25th–31st, 2000, p. 115)

If we really interacted on the basis of sign theory we would have to reject these examples as nonsense because they contain contradictions. In performance however we behave as complex adaptive systems and accept these examples. We negotiate their meaning in a way that *change* can be understood as *constant* or a person *being homeless* can nevertheless *have a home*. Examples of this type make very clear that we have to abandon the view of language as a sign system. In language use, there are no signs, no symbols having meaning independent of the speaker. As Wittgenstein told us, words in ordinary language use 'have meaning only in the stream of life'.

Dialogic interaction therefore is basically characterized by the *Principle of Meaning Uncertainty*. The so-called exact natural sciences have already introduced a Principle of Uncertainty in quantum physics. It is this point of meaning indeterminacy, or 'fuzzy' structures as Li (2003) calls it, which we have to accept in theory and address as a complex mix of order and disorder, of generalized rules and chance.

One of the major issues Deacon (1997:39–46) is concerned with in his book on 'the symbolic species' refers to the question of why there are no 'simple languages used by other species'. Such simple symbolic languages consisting of a 'very limited vocabulary and syntax' are artificial systems, each 'logically complete in itself', not adapted for communication. Why should they have evolved? Artificial systems might be constructed as symbolic systems. Human communication, however, is quite different from 'symbolic communication' in the sense of 'transmission of signs' (p. 70). We do not interact only by passing symbolic information. Consequently, the issue of a symbolic 'mode of thought' (p. 22) remains a hypothesis within the limits of the symbolic approach.

The structuralist and generativist view of 'how language works' (Pinker 1995:79–126) simply continues ignoring progress in pragmatics. How could we in the year 2000 still restrict our view to 'passing knowledge' or to 'conveying news', which is only one type of language action among others? Reducing the issue of 'how language works' to two 'tricks', the structuralist of 'the arbitrariness of the sign' and the generativist of the 'infinite use of finite media' (Pinker 1995:79–80), again is simply a 'trick', a 'myth' (Harris 1981) which should no longer need to be discussed. I take it up because it is presented as the truth of language, as 'the essence of the language instinct', 'intended for everyone who uses languages, and that means everyone!' (p. x).

## 4.4  Meaning is persuasion

If we accept that language is not an independent object but a complex ability of human beings, we have also to accept that it integrally combines important features dealt with separately in the orthodox view. Having based dialogic interaction on human beings and their needs, it becomes obvious that human beings will try to achieve their purposes more or less effectively. These efforts are precisely what has been dealt with separately as rhetoric in the orthodox view. In the model of the action game rhetoric becomes a constitutive integral principle which does not always result in specific verbal expressions but in cognitive strategies which are dependent on specific ideologies. Meaning is persuasion. Often it is not explicitly expressed.

Let us have a look at a short advertising text from 'The Economist' which demonstrates human beings as purposeful beings who keep their real purposes concealed in an attempt to get them across by suggestive means.

(5)   Wherever you are. Whatever you do. The Allianz Group is always on your side. For over 75 years we have successfully managed the assets of life insurance policy holders. This, together with the close cooperation of our global partners and the experience of our asset management team leads to improved long-term investment performance. It's no wonder then, that we were recently awarded the prestigious Standard and Poor's AAA rating. Maybe that's why we insure more Fortune 500 companies worldwide than anyone else. Allianz. The Power On Your Side.

('The Economist', March 25th–31st, 2000, p. 3)

Many points relevant to our discussion, which I can only briefly mention, become evident from this example:

– The verbal text is not an autonomous unit but only a component in the action game.
– It seems to be a monological text but nevertheless it is part of dialogic interaction with the reader.
– The action game is a cultural unit. You have to know many things in order to understand the publicity function of the text in the unit of the action game.
– The main message is not explicitly expressed: 'Join Allianz!' but left to suggestive means.
– Meaning is persuasion.
– Syntactic meaning can also be persuasive as can be seen from the heading *The Allianz Group is always on your side.* The indicative construction does not describe an existing event but only a potential or conditional fact insofar as it has to be complemented by something like *if you want.*
– Verbal and cognitive and also perceptual means (a picture is included) are integrated.
– Word meaning is on the one hand indeterminate, open to negotiation, for instance, in *to be on your side, successfully managed, power.* On the other hand word meaning is, at least in part, defined, due to a tendency of languages for specific purposes to name things unequivocally, for instance, in *life insurance, policy holders, long-term investment performance.*

These points demonstrate that it is simply absurd to assume pattern transference would be a useful method for describing texts. The complexity we have already found with the mirror neurons at the very outset confronts us here with high intensity. Human beings as complex adaptive systems negotiate meaning and understanding, using the text as verbal means in the action game and integrally using other means, cognitive and perceptual ones, to achieve their purposes effectively.

## 5.  Fundamentals of a theory of the dialogic action game

Let us now address the essential question of how we can cope with these features in a theory of human dialogic interaction. The challenge is to address the complex, which is the Dialogic Action Game as an unstable system in the range between order and disorder. The mirror neurons tell us that we are able to address the complex just at the very outset. In order to tackle this challenge in theory we have to be prepared to abandon classical western methodology as has been used for more than two millennia. We have to leave logical games and concentrate on human beings and their abilities.

The main point will be that we have to find a new way of theorizing which overcomes the limits of reduction to rules. We must not begin with the methodological fallacy: there have to be rules and we have to find them. We must first try to understand our object and then derive an adequate methodology from the object-in-function. Thus the theory will have two parts. I have already outlined the first part in listing constitutive features of the complex object, human dialogic interaction, and will now concentrate on the second part, an adequate methodology which describes how the object works.

As the complex is essentially a matter of probability, it requires a technique adapted to cases of probability, a technique which can be used by human beings. This technique has in my opinion to be made up of guidelines, principles or maxims which human beings can use in order to orientate themselves in the process of negotiation. When faced with the complex we first try to find regularities, rules or conventions. Where regularities come to an end we apply other techniques, i.e. mainly presumptions, associations, conclusions which may be based on generalizations but also on moment-to-moment decisions which take account of concepts of probability and chance. Even if there is support by rules, dialogic interaction is an event of performance, of actions performed by human beings who, in the end, are free to break any rule consciously or unconsciously. On the highest level of action, therefore, we orientate ourselves according to Principles of Probability and make individual decisions and use rules not as absolute techniques but as tools in our hands.

## 6.   Principles of probability: Basic and corollary ones

The Theory of the Dialogic Action Game is based on three basic principles and a set of corollary principles. I have dealt with these principles in detail in recent articles (e.g. Weigand 2000) and will now focus on how these principles are confirmed by the discharging of the mirror neurons.

The basic principles are the *Action Principle, the Dialogic Principle proper, and the Coherence Principle*. The **Action Principle** starts from the assumption that we are acting when speaking. Action, in general, is to be understood as the *correlation of purposes and means*. Referring to mirror neurons and the conclusion that the evolution of language and communication started with gestures, there must have been a dialogic purpose or claim being carried out by means of the gesture. It is dialogic action at the very outset which is signalled by the firing of the neurons. Communication as action in my view begins with intentional or willed communication. A possible involuntary precursor might be called communication, too, but cannot count as action. Gesture being carried out voluntarily may be traced back to movement produced involuntarily. If we however take integration seriously we might even consider the possibility that human beings are capable of intentional communication from the very beginning.

The action principle is a functional principle which is based on a dialogically orientated claim. The means at the outset were perceptual means, gestures, which include cognition. The concept of means already implies that they were used intentionally. Later on, verbal means were added in an integral way. Dialogic action can also be carried out by practical action itself. In the case of an initiative action, the claim is thus expressed by perceptual means; in the case of a reactive action, the claim is immediately fulfilled without a mediating utterance. The dialogic purpose or claim refers to a specific state of affairs which is functionally represented by reference and predication.

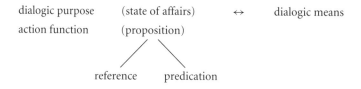

**Figure 3.** Dialogic action

Assuming that communication evolved from gestures, we can say it is the dialogic means of a gesture which signals a dialogic claim. The question arises what types of dialogic purposes or claims we have to distinguish in interaction. I dealt with this question in detail in other articles some time ago (Weigand 1989, 1991) and will restrict myself to briefly indicate the claims. In my opinion, our speech acts are predominantly based on two basic claims: a claim to truth and a claim to volition. Both claims correspond exactly to the basic mental states of belief and desire. The action function being based either on a claim to truth or a claim to volition refers to the propositional function, which consists of reference and predication.

Let me illustrate the structure of a speech act with the example of a directive speech act. A directive speech act is characterized by a claim to volition which is to be derived from the basic mental state of desire. This claim to volition refers to a specific state of affairs, for instance, that the interlocutor should help the speaker. The whole functional structure can be expressed by the means of a gesture which seems to be a perceptual means but integrally contains cognitive means as well.

Directive        [help (KP, Sp)]        ↔        gesture

claim to volition

**Figure 4.** Directive action (example)

In this way, from the very outset, the means are integrated: perceptual and cognitive means, later joined by verbal means. The thesis of the *symbolizing character of language* has to be *relativized in the light of integration of means*. Language has evolved not as a separate verbal ability of mastering symbols but as an integral part of a complex interactive ability. Single words or symbols are not the result of a monologic mental action of symbolizing as a 'mode of thought' but the result of a dialogic action of GIVING THINGS A NAME (cf. Weigand 2002a). In this sense, one-word utterances such as *apple* are actions. Their function depends on the framework of the action game: it can be a deictic-representative act 'this is an apple' or a directive act 'give me the apple', etc.

The *integration of means* is a result of the integration of human abilities which seems to be a necessary precondition for the evolution of communication. The integration of activities is a necessary condition of life. Human beings use different abilities in an integral way in order to communicate: speaking, thinking, and perceiving. These abilities have a *double, i.e. dialogic face*: speaking corresponds to understanding, thinking means, among other things, making assumptions and drawing inferences and assuming that the interlocutor will do the same, and perceiving refers not only to perception but also to producing means which can be perceived. The mirror neurons also confirm this double face insofar as they fire if a gesture is carried out or observed.

We are now in a position to transform Figure 3 to Figure 5:

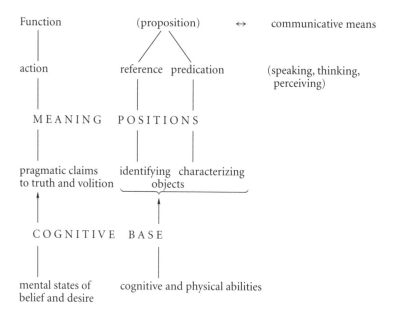

**Figure 5.** The speech act

Figure 5 starts from *the complex* phenomenological level of the dialogically ori-entated speech act, identifies *the simple* with regard to meaning positions, on the one hand, and indicates the integration of means, on the other. Meaning posi-tions are considered to be cognitive-social positions developed from a cognitive base. The mental states and abilities of the cognitive base can be understood as the brain-in-function.

A further question to be addressed concerning the Action Principle turns out to be the most important: it is the question of how to read the *arrow*, in Figure 5 the horizontal one which correlates functions with means. In orthodox models of fixed codes the arrow is read as conventional correlation. It is however simply not the case that utterances are related to speech act functions by a fixed gener-alized code in relation to specific situations nor is it the case that the set of so-called communicatively equivalent utterances is closed and could extensionally de-fine the speech act grammatically in a given language as I assumed some time ago (Weigand 1992). On the contrary, speech act functions and propositional mean-ings are dependent on the individual acting human beings who negotiate their positions in the unit of the action game. The arrow expresses the methodologi-cal technique of negotiation which in the end is the technique of using probability principles.

Let me illustrate the technique of negotiation by an example which we have already discussed as part of Example (2):

(6)   You are playing the piano again.
      You are playing the Game-boy again.
      (*Game-boy* is a trade name for a hand-held computer game)

Orthodox linguistics assumes fixed codes such as the code of symbolizing. Utterances like the Examples (6) cannot however be decoded by fixed codes. To a certain degree, they can be understood by reference to rules as declarative sentences which can be used for a representative speech act. However it remains open whether the utterance expresses joyful surprise or an angry reproach or just a statement. Presumptions come in which refer to the particular situation and individual psychological conditions of the speaker. In the end, the speaker alone knows what he/she meant by their utterance.

Very interestingly, the point of probability principles as a basic technique is also confirmed by mirror neurons. Initially, the complex dialogical claim has to be expressed by simple means. The means are integrated from the very outset, and the same means are used for different claims. Consequently, the correlating arrow cannot but indicate a probability relation.

Let me briefly point to Li's second characteristics of language which he calls 'the way we say things' (Li 2003, and in this volume). He is right in emphasizing that there are specific ways of using words in phrases or in multi-word expressions which are different from language to language (cf. Weigand 1998a). We do not refer and predicate with single signs free at our disposal. In my view, these multi-word expressions again make an argument against the thesis of symbolic communication. Thus we predicate in English by *to be hungry* whereas in the Romance languages we say *avere fame*; or we refer with the definite article in English but without an article in German in the phrase *to play the piano/Klavier spielen*. It is multi-word units or conventional phrases from which the utterance is constructed. Syntax plays its role on different levels – the level of the word, the phrase, and the utterance – as a combination of meaningful elements (Weigand 2002b). It is however the meaningful elements to start with, not abstract syntax as the orthodox view of the language myth makes us to believe. The fact that dialogic interaction is based primarily on more or less complex meaning combinations is confirmed by mirror neurons.

The Action Principle has to be seen in close connection with the **Dialogic Principle proper** which is the Principle of correlating action and reaction in the sequencing process of negotiation.

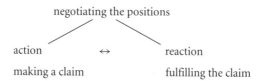

**Figure 6.** Complementary sequel of action and reaction

One of the characteristics of dialogic action is that the *single speech act cannot be considered communicatively autonomous*. Communicative actions are always dialogically orientated. Action and Reaction are not arbitrarily combined but interrelated by the same dialogic claim. The initiative action makes the claim, the reactive action fulfils it in a positive or negative sense or leaves the decision open. Aristotle's principle of the autonomy of the single act may relate to practical acts but not to communicative ones.

The mirror neurons confirm not only the Action Principle but also the Dialogic Principle proper. They signal either action or reaction by the same means of a gesture. The precise meaning and understanding has to be negotiated by cognitive and further specific perceptual means. Thus, to give one example, a directive speech act claiming help aims at a reactive speech act of consent. As it is primarily important to distinguish a positive reaction from a negative one, two simple different gestures have evolved: nodding and shaking heads.

DIRECTIVE [help (KP,Sp)]          ↔          CONSENT
claim made                                   claim fulfilled

gesture                                      gesture

**Figure 7.** Directive interaction (example)

From the general interactive function of negotiation or of trying to come to an understanding, specific speech act types and sub-types are to be derived by using functional criteria, for instance, criteria which differentiate the pragmatic claims or introduce propositional criteria (Weigand 1989, 1991). Dialogic interaction as a set of action games is primarily based on these complementary sequels of action and reaction which form the building blocks for the process of negotiation (Weigand 2000).

Let us now take up once again the crucial point of the *arrow*, which in the case of the Dialogic Principle proper correlates the initiative with the reactive speech act. Ten years ago, I understood the arrow as indicating conventional correlation. Assumptions, problems of understanding, open points were then excluded. Having in the meantime recognized our object of study as an open system, the arrow consequently is to be read as principle of probability. It is the initiative speech act itself which tells us via rational reasoning which reactive speech act is expected. Rationality, however, represents a limited concept in performance. Language-in-use, being part of our daily life, refers to life concepts, which are concepts of probability such as preferences and everyday habits. At the very outset, therefore, open points come in which are dealt with by suppositions and assumptions and might cause misunderstandings.

Referring to mirror neurons and an early evolutionary stage, it seems quite natural that the means are not assigned to meanings in a fixed way but are sometimes in danger of being misunderstood. They are only seemingly simple means such as gestures but in fact integrally combined with cognitive means, and they carry complex and from situation to situation changing meanings. *Conventions still have to be established.* Misunderstandings, a controversial point in linguistic theorizing, are quite natural events. They can be accepted because normally they are immediately clarified in dialogue.

Let me add a short remark to the third basic principle, the **Principle of Coherence**. On the basis of the integration of means which are not only verbal means, coherence can no longer be looked upon as relation between empirically registered points of a text. It has instead to be understood as a persistent attempt by the interlocutors to understand the means. Coherence therefore is established in the mind not in the text (Givón 1993). Such a conclusion might seem revolutionary for some linguists but it is revealed as an evident conclusion by reflecting on the way mirror neurons function.

The three basic principles are accompanied by **corollary principles**, among them Principles of Emotion, of Rhetoric, of Politeness, but also the Rational Principle or the Principle of Supposition (Weigand 1998b, 1999b). I cannot dwell on these individual principles but would like to make a few general comments. Starting from human beings and considering language as part of a complex human ability, it becomes quite obvious that this complex human ability is influenced by the conditions of human nature. Thus we are slaves of our senses, of our emotions, of our ability to reason. When interacting with other human beings and trying to negotiate our positions, we are guided by two opposing principles, one of effectively pushing our own purposes which is the *Principle of Rhetoric* and the other of respecting our partner which is the *Principle of Politeness*. Both principles are closely related and interact according to specific cultural conditions. Cultures differ in their view of social relationships and in the value they assign to the individual human being. Principles of Politeness therefore are a subtle indicator for the image a culture attributes to human beings, either with respect to the freedom of individuals or with respect to their role in the society.

Emphasizing principles of probability does not mean that we would not refer to rules, conventions, regularities. We try to structure the ever-changing complex according to regularities. Where regularities end we use suppositions and other means. The model which substitutes the orthodox view therefore is *an open model* which accepts misunderstandings and considers *human beings as complex adaptive systems* behaving in the range between order and disorder.

## 7. Mirror neurons and the view of human beings as complex adaptive systems

The *mirror neurons*, if we understand them correctly, *confirm precisely these fundamentals of dialogic interaction*. It is mainly the *integration* of different abilities which allows adaptive behaviour. Integration therefore represents the major challenge for science, especially as we have been used to clearly separate levels and categories in western thinking since antiquity.

Different disciplines must move closer together insofar as they are all related to human abilities. The main distinction is no longer to be seen between *natural and human disciplines* but between disciplines *related to human beings and technical, mechanistic ones*. This confronts us with the question *whether machines can simulate human dialogic interaction*. It is not only *consciousness* but also *integration of different abilities* which clearly distinguishes human beings and computers and makes their interaction different.

The question of mirror neurons and what they tell us about ourselves is however not restricted to integration in this sense. From the fact of integration, in my view, we can conclude that we cannot clearly separate the object and the way it functions. It is from the very beginning an *object-in-function*, combining biological, cognitive, social and other aspects. There are no separate entities, brain and mind, but one entity which is *matter as well as cognition* or matter and energy in integration. It is biology-in-function which relates the cell to the world and intentionally orientates human beings towards other human beings and the world. *Not only consciousness and integration but also intention, including dialogic intention,* distinguishes human interaction from machine interaction. Human beings behave from the very beginning as dialogically purposeful beings, manifesting themselves primarily as *the dialogic not the symbolic species*.

If these are conditions for every human being, one might pose the question why human beings behave as different human beings, as *individuals*. At the very outset, there are genetically different human beings who start to communicate, and there are different influences resulting from different life environments. Individuality will also result from the feature of intentionality. Why should individual human beings orientate themselves in the same way towards the world and other human beings if we do not accept that intentionality is a generalized, rule-governed feature?

Human beings are guided in their behaviour by intentionality and other perceptual and cognitive abilities which are not identical for every human being. They perceive that external events are determined or interrelated by different factors, by causation, by rational conclusions, by conventions, and by chance. They therefore adapt themselves towards different guidelines in their evaluation of the world and of themselves as being part of the world. In this process they refer to regularities and presumptions, generalizations and particularities, probabilities and chance. Moment-to-moment decisions include a certain leeway for innovation and creativity. Human beings thus do not behave according to a closed rule-governed system but according to an open unstable system which gives space to new perspectives and different views. Their interactive competence is a competence of addressing the complex performance, a *competence-in-performance* (cf. Weigand 2001).

Starting from this view, which dissolves the distinction between rule-governed competence and rule-breaking performance in the concept of competence-in-performance, we face the fact that at the beginning of the new millennium we have arrived at a turning point in science which directs us from the abstract simple to the natural complex. Leaving behind us classical Aristotelian methodology and Descartes' view of separate abilities, we have to tackle the problem of an integrational complex, a problem clearly outlined by Gell-Mann (1994), Damasio (1994), Prigogine (1998) and others in different disciplines. Consequently, in linguistics, too, and in related disciplines we should join our forces in order to understand and re-define our object, language, as an integral part of human dialogic interaction (Weigand 2002a and forthcoming). Language-in-function means language used by human beings for interactive purposes in integration with other dialogic means.

## References

Damasio, A. R. (1994). *Descartes' Error: Emotion, reason, and the human brain*. New York: Putnam.

Deacon, T. (1997). *The Symbolic Species: The co-evolution of language and the human brain*. London: Penguin Books.

Gell-Mann, M. (1994). *The Quark and the Jaguar: Adventures in the simple and the complex*. London: Abacus.

Givón, T. (1993). Coherence in text and in mind. *Pragmatics & Cognition, 1*, 171–227.

Harris, R. (1981). *The Language Myth*. London: Duckworth.

Harris, R. (1998). *Introduction to Integrational Linguistics*. Oxford: Pergamon.

Li, Ch. (2003). *The Evolutionary Origin of Language* [Regan Books]. New York: HarperCollins.

Pinker, S. (1995). *The Language Instinct*. London: Penguin Books.

Prigogine, I. (1998). *Die Gesetze des Chaos*. Frankfurt/M.: Insel.

Rizzolatti, G., & Arbib, M. A. (1998). Language within our grasp. *Trends in Neurosciences, 21*(5), 188–194.

Roth, G. (2000). Geist ohne Gehirn? Hirnforschung und das Selbstverständnis des Menschen. *Forschung & Lehre, 5,* 249–251.

Schumann, J. H. (1999). A neurobiological basis for decision making in language pragmatics. *Pragmatics & Cognition, 7,* 283–311.

Stamenov, M. I. (1997). Grammar, meaning and consciousness: What sentence structure can tell us about the structure of consciousness. In M. I. Stamenov (Ed.), *Language Structure, Discourse and the Access to Consciousness* (pp. 277–342) [Advances in Consciousness Research, 12]. Amsterdam & Philadelphia: John Benjamins.

Weigand, E. (1984). Lassen sich Sprechakte grammatisch definieren? In G. Stickel (Ed.), *Pragmatik in der Grammatik. Jahrbuch 1983 des Instituts für deutsche Sprache* (pp. 65–91) [Sprache der Gegenwart, 60]. Düsseldorf: Schwann.

Weigand, E. (1989). *Sprache als Dialog. Sprechakttaxonomie und kommunikative Grammatik* [Linguistische Arbeiten, 204]. Tübingen: Niemeyer.

Weigand, E. (1991). The dialogic principle revisited: Speech acts and mental states. In S. Stati, E. Weigand, & F. Hundsnurscher (Eds.), *Dialoganalyse III. Referate der 3. Arbeitstagung, Bologna 1990,* Bd. I (pp. 75–104) [Beiträge zur Dialogforschung, 1]. Tübingen: Niemeyer.

Weigand, E. (1998a). Contrastive lexical semantics. In E. Weigand (Ed.), *Contrastive Lexical Semantics* (pp. 25–44) [Current Issues in Linguistic Theory, 171]. Amsterdam & Philadelphia: John Benjamins.

Weigand, E. (1998b). Emotions in dialogue. In S. Čmejrková, J. Hoffmannová, O. Müllerová, & J. Světlá (Eds.), *Dialoganalyse VI. Proceedings of the 6th international congress on dialogue analysis, Prague 1998,* Bd. I (pp. 35–48) [Beiträge zur Dialogforschung, 16]. Tübingen: Niemeyer.

Weigand, E. (1999a). Misunderstanding – the standard case. *Journal of Pragmatics, 31,* 763–785.

Weigand, E. (1999b). Rhetoric and argumentation in a dialogic perspective. In E. Rigotti (Ed.), (in collaboration with S. Cigada), *Rhetoric and Argumentation* (pp. 53–69) [Beiträge zur Dialogforschung, 19]. Tübingen: Niemeyer.

Weigand, E. (2000). The dialogic action game. In M. Coulthard, J. Cotterill, & F. Rock (Eds.), *Dialogue Analysis VII: Working with dialogue. Selected papers from the 7th IADA conference, Birmingham 1999* (pp. 1–18) [Beiträge zur Dialogforschung, 22]. Tübingen: Niemeyer.

Weigand, E. (2001). Competenza interazionale plurilingue. In S. Cigada, S. Gilardoni, & M. Matthey (Eds.), *Comunicare in ambiente professionale plurilingue. Communicating in professional multilingual environments* (pp. 87–105). Lugano: USI.

Weigand, E. (2002a). The language myth and linguistics humanised. In Harris, R. (Ed.), *The Language Myth in Western Culture* (pp. 55–83). Richmond/Surrey: Curzon.

Weigand, E. (2002b). Lexical units and syntactic structures: words, phrases, and utterances. With particular reference to speech act verbal phrases. In C. Gruaz (Ed.), *Quand le mot fait signe.* Publications de l'Université de Rouen.

Weigand, E. (forthcoming). Dialogue analysis 2000: Towards a human linguistics. In M. Bondi & S. Stati (Eds.), *Dialogue Analysis 2000* [Beiträge zur Dialogforschung]. Tübingen: Niemeyer.

# Some features that make mirror neurons and human language faculty unique

Maxim I. Stamenov

Georg-August-Universität Göttingen, Germany /
Bulgarian Academy of Sciences, Sofia, Bulgaria

## 1. The puzzle of human language faculty

One can approach the problem about the uniqueness of human language faculty from different perspectives:

1. One can point out that humans have a referential potential of symbolic units (standardly identifiable as 'words') which outnumbers the capacities of the other biological species by several orders of magnitude. While other species have communicative calls (symbols) by dozens, humans have them in dozens of thousands. In psychological terms, human Long-Term Memory (LTM) must have a qualitatively different organization compared to LTMs of the other species;

2. One can point out the unique development of humans in terms of the neurophysiology supporting the performance of articulate speech with high speed and reliability. The complexity and speed in the motor performance required for the implementation of human speech amounts to a most complex motor performance every normal human individual can routinely execute. If we compare language capacity with musical capacity and performance, each normal human individual around the world without formal training whatsoever spontaneously performs on the level of a musical virtuoso from the point of view of the complexity of skills involved and speed of performance;

3. One can point out also the related feat of being capable to develop a phonological system for high-speed reliable categorization of the sounds of articulate speech in speech perception;

4. One can point out the unique development of formal structure of language, namely syntax. A frequently made claim is that the core of language as a separate capacity of human species (i.e., as different from other cognitive capaci-

ties) is identical to syntax. Syntax is a unique faculty of the mind, one without a second. This is the case, among others, because syntactic computation is referentially 'blind'. Becoming free from the snares of reference, it managed to optimize itself as close as possible to the level of formal perfection;

5.  One can point out that language faculty demanded for its development specific evolution in the capacity to understand conspecifics via the requirement of taking strategically the perspective of the other interlocutor (i.e., empathizing to her/him) in order to be able properly to interpret the content of her/his message. This capacity intentionally to 'transcend' one's own egocentric perspective apparently becomes possible only with the advent of language.

The list of these features, definitely, could be further developed and elaborated. In trying to trace the origin of language, attempts were made to identify the point of departure for the development of human-like language taking a single one from the set of the unique characteristics (i.e., 1–5) as the one initiating the movement toward fully blown language in a causally determined sequence (cf. Lyons 1991; Hauser 1996; Hurford, Studdert-Kennedy & Knight 1998; Trabant & Ward 2001; Li & Hombert, this volume, for a selection of alternative views on the problem). For example, a change in the vocal tract gave the possibility to some ancestors to start to vocalize in a better way. This change opened new horizons for expanding their repertoire of symbols. The broader repertoire of symbols gave ample opportunity to try to combine them. The latter development led to a combinatorial explosion and to the requirement to restrict the resulting unpredictability. This trouble was alleviated with the invention of syntax as a fast rule-governed processing of complex verbal stimuli, etc.

The scenario just sketched is formulated according to the logic of one-way cause-effect relationships developing one after another aspects of the future language faculty serially on the time span of innumerable millennia. The problem with scenarios like this is that language faculty, as we see it currently in action, involves unprecedented in its complexity set of computations distributed among several central systems in the mind, the latter being massively dependent on both feedforward and feedback processes. Different aspects of this faculty must have co-emerged in order to manage to be compatible on a wide scale. Is it possible to model co-emergence by means of serial cause-and-effect models? – this is a quite troublesome to face question.

One straightforward way to respond to the challenge is frankly to admit that we have currently no means to face it. We possess the language faculty that makes us unlike any other biological species on earth. The characteristics of this faculty look like nothing else in the universe, as we know it. This faculty was somehow implanted in us during the evolution. Evolution also took care to implement the language faculty as a deeply unconscious way of mental information processing, i.e., we have no access in principle to the way of representation and computation of language structure. This is essentially the position of Noam Chomsky (1993, 1994, 1995). It is both a radical and safe in its agnosticism position.

Correspondingly, for a linguist, the most consequent and the least troublesome way of dealing with Mirror Neurons System (MNS) and its potential purport for explaining the origin of language faculty is to dismiss it altogether. The logic behind such a dismissal would be a quite straightforward one: The monkeys have Mirror Neurons (MNs); the humans have MNs, too. The monkeys have no language; the humans have language. *Ergo*, MNS could not have been the causal agent initiating the 'crusade' toward the establishment of language. No technical elaborations and homologies in the anatomical structure of the brains of monkeys and men can help save the situation. If there was a significant breakthrough in the phylogenesis of humans leading to the formation of human language faculty, this was not the case *because* of MNS. To my mind, the simple logic supporting the thesis given above is unassailable if we are looking for a *single* causal factor triggering the development of the human language faculty.

The point of view of a sober linguist seems quite evidently to contrast with the 'linguistic turn' of the neuroscientists in interpreting the potential significance of the discovery of MNS. Fadiga and Gallese (1997) and Rizzolatti and Arbib (1998) explicitly interpreted this discovery as a way for reaching and deciphering the enigma of human language faculty. Their optimism was shared more recently by Ramachandran (2000). The proposals offered by Fadiga and Gallese (1997) and Rizzolatti and Arbib (1998) were founded, among others, on the following premises:

a.  Language skill has emerged through evolution by means of a process of preadaptation: specific behaviors and the nervous structures supporting them, originally selected for other purposes, acquire new functions that side and eventually supersede the previous one. The discovery of MNs may indeed provide a new neurobiological basis to account for the emergence of them, originally selected for other purposes, acquire new functions that eventually supersede the previous one;

b.  A continuity can be traced between language skill and pre-language brachio-manual behaviors, the primate premotor cortex being the common playground of this evolutionary continuity;

c.  The specialization for language of human Broca's region derives from an ancient mechanism, the MNS, originally devised for action understanding.

I think, however, that the *prima facie* purport for the fascination of the neuroscientists after discovering MNS specifically with language is due to a different reason: Only with the help of the comparison with language and the correlated with it potential for constructing mental representations as we have it today can we become aware of the real nature and specificity of MNS. In other words, before trying to construct scenarios of language origin and evolution based on MNS we must take care to analyse properly the nature of MNS itself. This can be achieved best by comparing it with the most advanced structurally and functionally mental representations of behavioral actions we are in possession of – the language-specific mental representations.

## 2.    What makes MNS peculiar from linguistic point of view

At the beginning of this section I will repeat once again the basic features of MNS. From some point of view, this may look like an unnecessary repetition in the present volume but I want to make sure that the description of the experimental results of MNS study is once again considered closely enough before entering the realm of elaborations and interpretations.

MNs respond both when a monkey performs a particular action and when the same action performed by another individual is observed. All MNs discharge during specific goal-related motor acts. Grasping, manipulating and holding objects are the most effective actions triggering their motor response. About half of them discharge during a specific type of prehension, precision grip (prehension of small objects by opposing the thumb and the index finger) being the most represented one. The most effective visual stimuli triggering MNs' visual responses are actions in which the experimenter or a second monkey interacts with objects with their hand or with their mouth. Neither the sight of the object alone or of the agent alone are effective in evoking the neuronal response. Similarly ineffective is imitating the action without a target object, or performing the action by using tools. In over ninety percent of MNs a clear correlation between the most effective observed action and their motor response is observed. In one third of them this correlation was strict both in terms of the general goal of the action (e.g., grasping) and in terms of the way in which it is executed (e.g., precision grip).

On the basis of its functional properties summarized very briefly above, MNS appears to form a cortical system in the brains of monkeys and primates that seems to 'match' observation and execution of motor actions. This 'matching', however, is of a quite specific type. Peculiar, even 'scandalous' about it from linguistic point of view, are several properties. Unlike other contributors to the present volume, I will claim that it has the following *differentia specifica* compared to language:

a.    the MNS is not interpersonal (intersubjective) in nature, i.e., it does not at all involve establishment of a relationship among two distinct selves – an agent and an observer;

b.    the MNS is not communicative in its function, i.e., it does not involve access to and sharing of experience (empathy, sympathy and/or imitative *Einfühlung*);

c.  the MNS does not support cognitive (language-like) mental re-presentations of propositional format. Correspondingly, it does not involve 'understanding' or 'interpretation' whatsoever of the echoed (internally mimicked) action. It just looks compatible in its structure to a behavioral situation describable by a transitive sentence structure in natural language. But the possibility to describe some behavior by some language-specific structure is by no means supposed to imply that the situation itself or the brain circuit 'representing' it possesses an isomorphic structure;

d.  the MNS is a low-level 'modular' (in the sense of Fodor 1983) mechanism, while the introduction of linguistic structure/meaning distinction requires the emergence and the development of at least two distinct 'central systems' (in the sense of Fodor 1983) processing structure and meaning and solving problems in the Working Memory (WM);

e.  the MNS is locked deictically to the immediate present, it is enacted in response to an actually observed here-and-now behavior, i.e. it apparently does not need the support from the LTM and/or of a general-purpose cognitive system like WM for the sake of performing off-line cognitive computations (cf. however Fogassi & Gallese, this volume, where an experiment is reported where a delay of up to 1,5 sec does not prevent a monkey from reacting with its MNS).

My conclusion is the following one: as much as MNS looks compatible in its performance to language-specific semantic structure (cf. Rizzolatti & Arbib 1998) in possessing a pattern that seems isomorphic to a transitive action initiated by an agent who manipulates an object (an inanimate patient), the really challenging aspects of this system emerge when one studies how it differs from language.

## 2.1  MNS is not intersubjective (interpersonal) but inter-embodied

Contrary to what has been written in interpreting the nature of MNS (cf. e.g., Goldman & Gallese 1998), I maintain that MNS in monkeys and primates is an *interpersonally (inter-subjectively) blind* tuning to what another individual of the same or similar in embodiment species is doing. It is a blind behavioral tuning because it does not serve a communicative purpose of intentionally sharing conscious experience between two selves (for the specificity of dialogic communicative interaction cf. Weigand 2000). Probably, a better way to deal with it would be to conceptualize it as an unconscious identification with the agent of the corresponding set of behavioral actions (either executed or observed). As will be seen below, however, it would be also difficult to interpret MNS action as due to an unconscious identification of the type taken into account, e.g., in psychoanalysis and psychology of personality.

Why MNS looks interpersonally blind from psychological and linguistic point of view? This seems to be the case because:

1. it is triggered automatically; the very visual availability of a third person manipulating an object would be enough to trigger the response;
2. it is enacted in an unconscious way (and cannot be potentially deautomatized and manipulated strategically);
3. it is due neither to a mapping nor to an identification of the observer with the agent of the action. This is the case because it does not require communicative (dialogical) first-person to second-person binding even in its minimal (root) form of shared attention (cf. however the latest report in Rizzolatti et al. 2001, for possible correction, in this respect). Furthermore, it is questionable to what degree it is due to matching of first-person observer to third-person agent. Rather, it is due to a resonance-based deictic (here-and-now) attunement of a quite peculiar sort;
4. it seems to function with a relatively high speed, i.e., faster than the off-line cognitive processes capable of strategic manipulation like thinking. *Ergo*, it must be noncognitive (nonpropositional) in the way of its implementation. Only the propositional format of thought, in my opinion, seems capable of supporting the distinction between two selves required for a theory-of-mind maintenance on-line between a nonspecular *ego* and a specular *alter ego* (for discussion, cf. Stamenov, forthcoming).

Due to the specific way of its implementation and functioning, MNS cannot support an intersubjective function based on 'theory of mind' (cf. Leslie 1987), as well as conscious perspective and role taking and/or empathy, as implemented by language-specific cognitive structure. (Behaviorists, if there are still around professionals sharing this belief system, should rejoice themselves most with the discovery of MNS. The latter system appears to display the shortest currently attested circuit between stimulus and response in *inter-embodied behavior*.)

The first two features mentioned above – about the automatic triggering and unconscious functioning – seem quite non-controversial to me from the demonstrations of the way MNS is activated in behaving animals. More controversial is (3), however. The appearance of intersubjectivity of MNS, to my mind, is an artefact of the conceptual differentiation in its functioning of two separate and different entities – of 'observer' and 'agent' – that are afterwards identified with (or mapped onto) each other. It is their mapping that makes the way MNS functions as if 'dialogically tuned' and potentially capable of supporting such high-level cognitive capacities like social learning and intersubjective sharing of experience (i.e., understanding). In the next section, I will make the point that the differentiation and afterwards the mapping of these two mental entities is a consequence of a logical analysis of the structure of observation and performance and has no evident correlates in the actual neural implementation of MNS itself.

I think that MNS, as a matter of fact, implements a 'direct', *resonance-based attunement* to a class of actions (the resonance-like functioning of MNS is suggested by Fadiga & Gallese 1997: 275, and re-iterated in Rizzolatti et al. 2001). It functions this way in order to get a component (part, member) of a single body into a direct attunement with a component of another single body of the same or similar species in respect to a particular class of actions.

The directness of rapport mentioned above guarantees the deictic function of MNS. This function aims at establishing a bond of agency here-and-now between observer's body member and agent's body member acting in the world. The binding between them is deictic (singular), i.e. valid for the current specious present moment of observation of an event in the external world. Due to its automaticity and unconscious activation it functions with a speed higher than the propositional off-line thought.

## 2.2 The case for the modularity of MNS

The MNS can serve as a striking illustration how a function like intersubjectivity, which seems crucially dependent on the central systems of the human mind constituting WM, can have as a 'forerunner' an 'encapsulated' local brain circuit. The point about modularity (including 'encapsulation') should be taken rather seriously. If we check the list of Fodor's (1983) distinctive features of mind's modules, the MNS fits the bill remarkably well:

1. input (to output) systems are domain specific. One of these systems according to Fodor himself is, e.g., the system for visual guidance of bodily motions (Fodor 1983: 47);
2. the operation of input (to output) systems is mandatory;
3. there is only limited central access (if at all) to the mental representations that the input (to output) systems compute;
4. input (to output) systems are fast (compared to slow off-line cognitive computing);
5. input (to output) systems are informationally encapsulated (they do not receive feedback from higher-level cognitive processes; in our case, the MNS has no access to facts that other systems 'know' about);
6. input (to output) analysers have 'shallow' output, e.g., in the case of vision it computes only the 'primal sketch' of the perceptual object that fits directly to how the object in question can be handled by the body (i.e., one of its members);
7. input (to output) systems are associated with fixed neural architecture;

8.  input (to output) systems exhibit characteristic and specific breakdown patterns (in our case these could be echolalia and echopraxia in humans, in the case of monkeys and primates the pathology of MNS remains to be investigated);

9.  the ontogeny of input (to output) systems exhibits a characteristic pace and sequencing (which in the case of MNS remains to be studied) (cf. Fodor 1983:47–101).

Thus, not only the peripheral faculties of perception but also some basic types of behavioral action important for the survival of the corresponding species may turn out to be inherited hardwired and perform as autonomous brain circuits. The fragmentariness (modularity) of brain's mind applies on a broader scale than it was previously envisaged. We may now claim that the mind is also modular in implementing specific action classes as specific sub-circuits of the body schema. The high specialization (and corresponding fragmentariness) of the multiple perceptual and action circuits can serve as a reason on its own for calling into existence of 'central systems' like language and consciousness. Their aim, from this point of view, would be to unite into a single gestalt the fragmentary body image.

## 2.3  The MNS Agent compared to the linguistic self

What we see in the demonstrations of a standard experiment in this paradigm is as follows. We see a monkey reaching, e.g., for a banana. The MNs fire. In the next condition of the experiment we see a monkey observing an experimenter reaching for a banana outside of the scope of her own reach. The MNs fire as if the monkey itself is reaching to grasp the fruit. Thus, we are led practically automatically to infer that we have a sort of a 'projection' of the observer-monkey to the agent-experimenter leading to an identification of the former with the latter. The structure of the situation, as seen and verbally described by a human mind, is projected as the mental and neural structure implemented in the monkey's mind and brain. In this way, we arrive at the scenario that presupposes the existence of two separate mental re-presentations with a transitive structure (subject plus direct object) that include an observer in the perceptual system and of an agent in the pre-motor system and their *matching* (mapping) as an outcome of MNS activation:

> The meaning of the observed action does not result from the emotion it evokes, but from a matching of the observed action with the motor activity which occurs when individual performs the same action. [...] What is important to stress here is that the proposed mechanism is based on a purely observation/execution matching system. The affective valence of the stimuli, even if possibly present, does not play a role in this 'understanding' system.
>
> (Rizzolatti et al. 1996:137)

What we have as hard evidence, however, is that it is the action itself that is tracked first both in perceiving and in preparing for action. We have no evidence whatsoever that they (agent and observer) are:

a.  first distinguished,
b.  in order afterwards to be identified with each other.

My point, in other words, is that this distinction is simply an artefact of the way the referred to experiments were re-presented and described. With this way of conceptualisation, the MNS automatically becomes a direct forerunner and prerequisite of all high-level social skills in monkeys, primates and humans.

What the available experimental studies actually show is the confluence of the perceptual and motor properties in the brain representation of the object (cf. Gallese 2000b; Fogassi & Gallese, this volume). One may object that this applies for the object of action but not for the subject of perception-cum-action continuum. However, a 'lonely' motor representation of how to handle an object seems to me highly implausible. This same specification must necessarily code the subject's capacities to manipulate the object. The very characteristics of the embodiment of the agent are already necessarily imprinted in the motor specification. This is, simply, the other side of the same coin of the 'instruction' how-to-handle-an-object. In this way, the latter specification codes the properties of both 'agent' and 'object'.

The differentiation between agent and observer is just an 'optical' illusion. There is no (unified) agent-self whatsoever involved in the way of action of MNS. There is no agent that presupposes and implies an executive control over the stages of performed action. There are no unified mental representations, etc. What we have at hand is a closed brain circuit plus inhibition to perform the serialized action in the case one is an 'observer'. The difference between 'observer' and 'agent' comes to the fore only after the activation of MNS as such in the garb of the general mechanism of inhibition (in the case one is supposed to become an observer) but not as an executive agency guiding strategically the realization of the intended behavior. The MNS prefigures the way the self functions while performing without self-agent whatsoever. It shows how one can function in an intelligent-looking way on-line without any sense of self whatsoever. And this applies to monkeys, primates, and humans alike.

The split between an implicit observer-self and an implicit or explicit agent-self can apparently become possible on a regular basis only with the advent of language. Its function is to differentiate (dissociate) a 'hidden observer' of a mental representation vs. an explicated self-actor in a mental representation, as in e.g., *I am eating a banana*. This is the split enabling self-consciousness as we know it from inner experience today. The linguistic hidden observer and even more so the explicable agent are not identical to the MNS agent. This is the case because both the hidden observer and the implicit central executive in charge of human behavior are functional components of the human WM and presuppose the availability of central systems while MNS 'agent' is modular and bound to a couple of action classes. Thus, what looks as 'the same' agent role from semantic point of view turns out to be implemented in humans in three different ways – as a MNS agent, as an implicit central executive and the explicit (self-conscious) self. If there is a link whatsoever between MNS and the self-agent it remains to be proven (cf. Rizzolatti et al. 2001:666, for an attempt to establish a link between MNS and the executive function of attention).

Because of its different genealogy and way of implementation, the linguistic self-actor has a constituency and a set of capacities that go far above the fixed relationship between an agent and a small object being manipulated (or eaten). It is capable of:

1.  identifying with or dissociating from the observed agent of a certain action;
2.  going off-line from the deictically present situation with its objects and participants;
3.  switching to a different mode of identification/projection (the I-thou communicative and I-s/he narrative mode) unlike the resonance (anonymous) mode of MNS;
4.  identifying the observer not only with an agent but also with an experiencer (of inner states of one's own and other beings) and beneficiary of the outcomes of the action performed by an observed agent to an observed object or person (cf. below for further discussion).

Only through acquiring the capacity to dissociate from one semantic role and identifying with other ones in successive strategic oscillations can one get to the level of self-consciousness as we experience it privately and communicate about it with others by the means of language. Correspondingly, role-playing and strategic positioning of oneself in an interpersonally structured mental space become possible on a regular basis only with the advent of language. It is due to this specificity of language that the researchers studying Shared Attention Mechanism (SAM) and Theory of Mind Mechanism (ToMM) cannot avoid representing the structure of the resultant mental representations in a 'propositional' format, i.e., in a moderately disguised format of language (cf. e.g., Baron-Cohen & Swettenham 1996).

## 2.4  The MNS object compared to the linguistic Patient

A further proof that MNS is not fit for communicative purposes and interpersonal understanding, i.e., does not possess the interactional structure (Self + other Self), comes from the analysis of the nature of the *object of action* in MNS. This object (which on a par with the Agent should be named Patient) is not interpreted as an animate entity (individual of the same species). The object is just the thing to be achieved. The thing tends to be of a size capable of being freely manipulated, e.g., by a single hand. If this is the case, there is no possibility to interpret the object as the second interlocutor in interpersonal interaction. For some situation to be taken as self-other interaction would require a re-interpretation of the structure in question that becomes possible only with language (cf. above).

The MNS object is also problematic for a different reason (as viewed from the perspective of language). This is the case because the object may or may not be present. The structural and functional characteristics of the two classes of action to which MNS is primarily sensitive look at first sight incommensurable. This becomes evident if we describe these actions in sentences, e.g., *I pick up a banana (with a hand)* vs. *I talk (with my mouth)*. Only the first class of manual actions seems to be a transitive one, i.e., a one having a direct object. The second one is intransitive if the action by the mouth is supposed to serve a communicative function of emitting vocal gestures. The two classes become on a par, however, in the case we consider the action by mouth to be done for the sake of manipulating a small object, i.e., for the sake of eating a piece of food. This is possible to verify with experiments checking does the MNS activate when a monkey observes somebody eating in front of it. If this turns out to be the case, the structure of the action by the mouth has isomorphic structure – *I eat a banana (with my mouth)*. If this structural parallelism is confirmed, however, the association of MNS with speech becomes more indirect.

On the other hand, if we cut the direct object in the transitive construction as a symbolic correlate of exchanging communicative for behavioral action, both hand and mouth will be interpreted as instruments for communication, e.g., with the means of mouth or manual gesture, as would be the case in *I talk (with my mouth)* and *I talk (with my hands)* (in the language of the deaf). This exchange of an object by a body member to be used as an instrument seems essential for the metamorphosis of the behavioral into a communicative action.

## 2.5  The nature of action implemented by the MNS

In the case of MNS, the structure of the enacted or echoed action is quite rigid. The MN pattern of manipulating a small object by a hand has syntactic and semantic pattern that are incapable of being distinguished from each other. It is impossible to extend and/or modify this structure, e.g., to add another component (argument) to it. Performance of the same manual manipulation not by hand but by an instrument would block the activation of MNS. This is quite unlike the syntactic sentence-format of language which provides ample opportunities for different types of recursive extension and transformation. It is also impossible semantically to re-interpret the formal structure of MNS, as is the case for example with the transitive construction in language (cf. below).

Having in mind the peculiarity of the types of action to which MNS is attuned, a question of high priority is to find a functional motivation for its origin and development. Recently, Gallese (2000a, 2000b) offered a new, considerably broader account of MNS, its place and functions. The basic idea remains that MNS supports the interpersonal 'understanding' by modelling a behavior as an action with the help of a motor equivalence between what another individual does and what the observer does. The question that still remains to be answered is how such a peculiar system like MNS could have evolved at all in the first place, i.e., before being put to use in interpersonal understanding. Here Gallese offers the following proposal which could help to explain the rationale for emergence of such a system without maintaining that the outcome can be put *directly* to 'interpersonal' use (one of the main points of this article is that MNS cannot be used directly for this purpose).

Gallese (2000a) suggests that the primary function of the evolving MNS was to achieve on-line control of the execution of behavioral actions. The latter actions consist as a rule of more than a single motoric movement and for this reason require planning. This can be achieved by the following 'distribution of work' in implementing MNS: in a particular sector of the premotor cortex, the area F of a monkey, there are two distinct classes of neurons that code goal-related hand movements, and which differ in their visual responsiveness – mirror neurons respond to action observation, while canonical neurons to object observation. Thus we have two distinct populations of grasping-related premotor neurons. Once the features of the object to be grasped are specified and translated by canonical neurons into the most suitable motor program enabling a successful action to be produced, a copy of this signal is fed to mirror neurons. This signal would act as a sort of 'simulator' of the programmed action. The function of this action simulation is that of predicting its consequences, thus enabling the achievement of a better control strategy. On the one hand, it serves the purpose of prognosis of the outcome of the action; on the other hand, the simulation binds the action with its goal-directed agent (while previously it was represented in relation to the features of the object only).

This account seems to provide the basis for the much needed functional explanation how such a system like MNS could have ever developed. What still remains to be explained is why the feedback from observation is fed to control structures before the differentiation is made who is doing the action – 'me' or 'my monkey'. It seems quite uneconomical to program the system to activate the control structures responsive for planning each time one sees anybody doing something that looks like a token of a certain class of actions. The point that this may help afterwards to imitation, social learning, etc. would be not a functional but a teleological explanation, a type of explanation which is not favored in cognitive sciences.

As I pointed out in the Section 2.1. above, another possibility is to represent this circuit as the one containing a 'primal sketch' how to deal with a certain class of objects plus a copy of this sketch sent for serialization to MNS. Thus, the circuit can perform autonomously, i.e., it does not need to relate to a separate functional entity serving the role of an 'agent'. If this is the case however, an autonomous brain circuit like MNS may start to look a less likely candidate for a language-trigger mechanism because the MNS becomes in charge of just the serial realization in time of a motor pattern and cannot serve as a structural prototype for development of language structure. But, may be, what was generalized from MNS to both phonology and syntax of natural language was just the hierarchical sequencing of the parts (movements) of action, i.e., the processing (the hierarchical sequencing of movements) but not the representational aspect of MNS. This possibility remains to be studied.

## 2.6 MNS and the embodied self

As the latest research shows (cf. Gallese 2000b and the literature discussed there) it seems that the hand and mouth actions are not the only classes of actions capable of evoking the mirror-matching effect. The whole frontal agranular cortex (constituted by the primary motor cortex, the supplementary motor area and the premotor cortex), as well the parietal cortices are constituted by a mosaic of areas, endowed with peculiar anatomo-functional properties, which interact by means of reciprocal connections within distinct cortico-cortical networks. Each of these networks integrates sensory and motor informations relative to a given body part in order to represent and control the particular body part within distinct spatial reference frame. In other words, we have a case of a multiplicity of cortical representations forming a neural correlate of the body schema. The brain seems to have developed a set of specialized 'mirror matching mechanisms' for representing the body in action in the world. This is a distributed (i.e. an emphatically non-integrated into a whole body schema) brain network, which keeps the body in 'resonance' with the similar bodies in the world here-and-now. Gallese (2000b) names this schema 'shared manifold'.

What must be added (if the analysis of the nature of the 'agent' in MNS turns out to be on the right track; cf. Section 2.1 above) is that the body schema is an 'anonymous' one, in the sense that it is a schema (distributed representation) of the body without the accompanying awareness of a self-attribution. The perceiving-cum-acting brain appears to treat the body as 'the current body acting in the world here-and-now'. This system supports the individual embodiment as a type of embodiment in the world but not as a token in a possession of an 'unique' body awareness and an *ego*. Furthermore (and in a quite non-platonic fashion), this virtual body, unlike the physical one, is not united into a single gestalt – it is a 'mosaic' (modular) one and one that is not 'owned' by a self, but scandalously shared on an online-service basis.

The representation of the embodied agent-self in language, on the other hand, appears to function on a rather different basis. It is remarkable to notice that it is part of the language-specific semantic representation to talk about the whole body doing different things in the world, but not about alienated (modular) or shared body parts acting as autonomous agents on its own account. For example, it is quite normal to say (1) and anomalous (at least in English and all other Indo-European languages I am aware of) to utter (2), while it would be strictly anomalous (in any possible human language) to represent something as done with a shared body member (as in 3) or two body members doing an action in parallel (in resonance) with each other (as in 4):

(1)   I am eating the sandwich.

(2)   *My mouth is eating the sandwich.

(3)   *Our mouth is eating the sandwich.

(4)   *Your mouth and my mouth are eating the sandwiches.

(5a)  *A mouth is eating the sandwich.

(5b)  ?The mouth is eating the sandwich.

(5c)  ?The hand is picking up the banana.

The belief in the unity of the body and the primacy of this unity compared to any component of it is a definitive part of the semantic structure of language. This self is represented in language as the 'owner' of the body with its members. The latter are represented as 'servants' or 'instruments' to be used by the 'embodied (behavioral) self' in order to achieve different aims in different situations in the world. A body member cannot *per se* be represented as the agent of some behavioral action (cf. 5a), unless some quite specific context is generated and the member is used in a metonymic way for the embodied self as a whole (cf. 5b).

If we compare the representational specificity of language structure with MNS, the point is that what looks quite unnatural expressed in natural language IS the way the MNS represents the classes of actions verbalized in (5b) and (5c). This is what their structure looks like, if verbalized with the means of natural language. The most veridical structure of MNS action would look like (5d). In this case, there is no representation or association to a unified agent-self or observer-self of the type we are accustomed to via the use of language.

(5d)**Picking up the hand-banana.

It would be not superfluous at this point to add that the way of talking about the embodied self as a unified and unique entity is specific to language, and not necessarily to the way one consciously can imagine oneself acting in the world. For example, if I am asked to imagine the 'meaning' of (6), I will tend to imagine my own fingers as the profiled member of my body doing the action, while as a base of the profiled body member I may see my own arm (I use here 'profile' and 'base' as concepts from the descriptive apparatus of the cognitive grammar of Langacker 1987):

(6)   I am plucking the rose.

Please note that I will not imagine myself as a whole human agent in the way I can actually see or imagine a third-person agent doing this same action. Thus, the language-specific representation of the self does not necessarily coincide with the structure of the self capable of being visualized (imagined). In other words, the linguistic self-consciousness does not necessarily coincide, i.e., it not identical in its structure, with the spatial-cognition-based self-consciousness. The linguistically based self-consciousness appears to be the most unified one, it is the best candidate for functioning as 'the center of autobiographical gravity' (the other – the imaginable self – looks as less unified mental representations of the self as an agent in the world). In this way, 'one and the same', presumably, psychological function (self-consciousness as observer or agent) is implemented in a at least partially different way in language and explicit spatial cognition.

In the case of MNS, as a matter of fact, there is no agent whatsoever and, therefore, there is no ground for a self-like psychological structure and intersubjectivity either. Correspondingly, the identification in MNS of a cognitive pattern containing an agent or even the analogy both in structure and function may be partially or completely misleading (the point how misleading the psychological analogy can be is made for animal cognition in general by Povinelli, Bering, & Giambrone 2000).

My hypothesis about the functional motivation for the fragmentation of body schema into a set or mirror-neuron-like mechanisms would be that one doesn't need a representation of the body as a whole for any single concrete occasion in the control of one's motor behavior. The same holds for the functional differentiation of the 'self' as a purportedly individuated and unified executive central processor in the mind. The latter also appears to be a luxury from the point of view of carrying out perceptions cum actions to a very high level of development and differentiation of life forms.

### 3.   What makes MNS challenging from linguistic point of view: MNS between speech and language

I already mentioned above some features that make MNS commensurable to language faculty in two main directions. Apparently, this is the case because in MNS two quite different sets of behavioral actions are lumped together – the manual grasping and manipulation of a small object and the action with one's own mouth. The former prefigures a cognitive representation of a behavioral action; the latter establishes potentially a link to a behavioral as well as to a communicative action. The specificity of this dual orientation could be summarized briefly as follows:

1.   there is an apparent analogy in the semantic structure of the behavioral action having an agent and a patient and the linguistic transitive construction with subject and direct object. The MNS as a system responsible for matching manual action looks closer in its *structure* to the representational semantic-syntactic level of language (i.e., language capacity) than to the vocal or gestural level of behavior. The analogy, however, is at least partially misleading (as I tried to show above). The misleading about it is that MNS has representational structure comparable to the conscious mental representations humans have in thought patterns supported and implemented in the sentences of natural language;

2.   at the level of enactive motor performance the MNS seems closer to the vocal and gestural performance because of its explicitness as a behavioral pattern and its deictic relation to an action executed here-and-now. Speech is a behavior that is observable (unlike language with its structures and meanings), as well as explicitly located in the here-and-now of a current situation in the world;

3.   the MNS seems to function faster than thought (with its propositional format), but apparently is not as fast in its speed of processing as speech (cf. Studdert-Kennedy, this volume; for further discussion cf. Liberman & Whalen 2000). It remains to be explored in detail how fast are MNs in tracking actions and action completion. The speed of the MNS functioning does not figure prominently as an object of discussion in the available literature.

It is these peculiar features of MNS that can make it fascinating to researchers in language sciences. It would be fair to point out, however, that the links in question are currently based on a set of rather loose associations between not just speech, MNS and language; these links also include (implicitly or explicitly) the systems for gesture and imitation, the Shared Attention Mechanism (SAM) (cf. Ferrari et al. 2000) and the Theory of Mind Mechanism (ToMM) (cf. Leslie 1987; Baron-Cohen & Swettenham 1996).

There are many questions that remain to be explored in trying to find out a place for MNS in the scenarios of the origin and development of the language faculty, e.g.:

a.   is the Broca's area the only playground of a MN-like system?
b.   if there are other MNSs in other brain areas representing the action potential of other body members, do they tend to conflate different classes of action (from structural and functional point of view), as is the case with the first one to be discovered?
c.   are there other MN-like systems that are Janus-like, i.e., pointing to more than one cognitive mechanism (language and speech in our case)?
d.   can the types of action to which the other MN-like systems seem compatible be of different level of cognitive abstraction (as is the case with speech and language in the case of MNS)?


## 4.   The action structure as represented in language

Below for the benefit of nonlinguist readers I will provide a short presentation what makes the human language faculty unique on the syntactic and semantic level taking as example the cognitive pattern of an agent manipulating an object – the one that looks best comparable to MNS.

First of all, it is important to become aware that language-specific representations of behavioral events have at least three different sources of systematic variation in the contributing cognitive patterns. All of them, unlike MNS, are highly abstract in their cognitive format. Only the mapping of these three tiers of cognitive pattern formation can lead to what we experience as a single holistic act of meaningful experience supported by language structure.

The three tiers in question are implemented by different central systems in WM (e.g., by spatial cognition, abstract language of thought, and language-specific syntax). Each of them requires access to LTM, as well as access to the motor routines involved in the execution and control of inner and outer speech. Language is not supported by a closed modular single brain circuit, as is the case with MNS, but by a vast array of neural networks. One can get a preliminary orientation about the scope of the system in question by identifying some basic parameters of variation in the structure and meaning of just one pattern – the transitive construction.


### 4.1   The event structure as represented by verb's meaning

The first of these sources of variation is associated with the conceptual complexity of cognitive events coded as lexical meanings in the corresponding verbs. Here the following four types of verb meaning (*Actionsart*) are usually distinguished (according to a scale of inner complexity as established by logical analysis):

a.   a STATE is a type of predicate (the correlate of the verb in logic) that denotes properties or non-dynamic circumstances, for example *be sad*, *be afraid* (if we speak about human states);

b.   an ACTIVITY is a type of predicate that indicates a dynamic event in which there is no change of state such as in *sing*, *see*, or *run*;

c.   an ACHIEVEMENT is a type of predicate that indicates a change in a state or dynamic circumstances such as *break*, or *die*;

d.   an ACCOMPLISHMENT is a type of predicate that is characterized by a causative semantics, e.g., *kill* "cause to die" (cf. Vendler 1967).

The verb-specific meaning provides us with the possibilities to name and conceptualize different broad classes of static and dynamic situations in the external and internal world. The language also provides us with the means to discuss verbs' representational potential and develop classifications of their structure with layered nesting of actors (arguments) from one to four. The prototypical construction among them is that of an agent acting in the world of concrete objects as expressed in transitive verbs.

## 4.2   The semantic (theta) structure of the verbally coded event and the roles of the self

The second source of variation provides the implicit executive actor-self with a set of opportunities to explicate itself in different semantic (theta) roles, e.g. as:

| | |
|---|---|
| *Agent* | the participant in an event that causes things to happen (the instigator of action), e.g., *I am talking to my dog*, where some *I* initiates and enacts the act of talking; |
| *Patient* | the participant that undergoes the outcome of some action (seen from third-person perspective). This is the object or person who undergoes the result of the action named by the verb, as in *Peter is beating John* or *Peter is rolling the rock*. The speaker can represent her/himself in this role as, e.g., in *Jill is pushing me out of the room*; |
| *Possessor* | a person possessing an object or characteristics, e.g., *a pencil of mine*; |
| *Experiencer* | a person who consciously, i.e., first-person, undergoes some mental process or state, e.g., *I tremble*; |
| *Beneficiary* | the person who will benefit from the action denoted by the verb, e.g., *Mary bought a rose for me*. |

The important point to realize in analysing the list of semantic roles given above in our context is that the self can be projected *strategically* in any of the semantic roles listed above. This means that the self is not always associated (interpreted) as being the agent of action, but can be ascribed different roles in different classes of actions on different occasions in a flexible way. This also applies for the way one conceptualizes online the position of the other self with whom one communicates. The flexibility in the way of conceptualization and explication of the self seems to be a major achievement due to the semantic structure of language.

If we compare the way the linguistic actor-self acquires meaning with MNS, we see that the latter performs in an autonomous way (as a modular brain circuit), i.e. it does not require access to some separate memory store where the structure and meaning of action is stored, activated and interpreted in each individual case potentially in a different way, etc.

### 4.3 The syntactic structure of the verbally coded events and their meaning potential

The third source of systematic variation in the structure of language is syntax. The function of syntactic structure compared to the semantic one is that of (a) dissociating structure from content, (b) speeding up its cognitive computation, and (c) hiding the process of construction of cognitive structure from the 'visible' to consciousness semantic representation. In other words, the difference between syntax and explicit meaning is not only that of process vs. representation, but also relates to the way of construction and to the accessibility vs. inaccessibility in principle to consciousness.

The level of abstractness of syntactic representation could be once again best illustrated by the structure and function of the transitive construction. The latter is the structural matrix of the prototypical sentence in world's languages. It consists of a subject, verb, and direct object. It is quite remarkable to acknowledge that the single 'transitive relation' is, as a matter of fact, an interface among (or a 'common denominator' of) at least three different prototypical meaning schemata in cognition:

a. *agent* and *patient* form a transitive relationship due to a pattern of action which is initiated by the agent and affecting the patient (animate or inanimate); the patient undergoes some changes due to the realization (complete or incomplete) of the action. This schema is the *schema of self-initiated action*, e.g., *I hit John (with fist)* or *I pick up the scissors*. This type of action can be conceptualized as one between a first-person agent and a third person object;

b. the schema of *dialogical communicative interaction* with *speaker* and *hearer* (or *agent* and *counter-agent* or *agonist* and *antagonist*) as basic roles, e.g., *I will per-*

*suade you* or *You understand me*, etc. This type of interaction is quite unlike the physical one (cf. (a) or (c)) in its 'force dynamics'. The interpersonal basis of motivation and intentionality is a separate and specifically socially determined aspect of the formation of thought, language structure and consciousness. Its dynamics is based on first-person to second-person dynamic (reversible) relationship. The third-person formulated sentences like *Fred criticizes John* can be understood only through 'stepping into the shoes' of either *Fred* or *John*;

c. *figure*$_1$ and *figure*$_2$ (or trajector and landmark; source and goal) determine a relationship of movement, location and change of location of primarily physical entities which is possible to perceive (or visualize), categorize and afterwards express by means of natural language, as in *The ball hits the rock*. This schema is the structural *schema of motion*. This is a representation of two third-person objects (for further discussion and references cf. Stamenov 1997).

The syntactically defined transitive construction is a cognitive pattern supporting at least three different semantic schemata of high level of generalization. The prototypes (a)–(c) are universal because they express the specificity of human motor actions (a), the capacity for social cognition in communicative interaction (b), and for spatial cognition (c). The important thing to realize is the possibility not only to associate but also to dissociate the concepts of agent, actor, figure$_1$ and syntactic subject, on the one hand, and of patient, counter-agent, figure$_2$ and syntactic object, on the other, in representing different actions in the mind. Thus language structure, quite unlike MNS, has the capacity flexibly to match the structure and meaning of different cognitive patterns representing different types of events on a very abstract level.

The basic sentence structure of the natural language shows that the realization of the possibility for multiple pattern-to-pattern mapping is a structural-functional prerequisite of language-specific cognitive representation. The patterns in question must be, on the one hand, compatible to each other and, on the other hand, originate from autonomous central systems of mind each of which can independently compute aspects of the future 'holistic' sentence pattern. This potential for multiple mapping requires for its enactment massive feedback and feedforward computations in WM that require a cognitive architecture of a radically different type compared with MNS.

5. **Conclusion: The problem area the discovery of MNS sets for the study of natural language as a species-specific innate faculty**

The 'ancient' mechanism of MNS was originally not devised for action understanding (because the latter necessarily involves conscious intentionality). It rather aims at the attunement of a body member of a monkey or a primate to acting in the world here-and-now with the shortest possible route in the local neural circuitry. This attunement is numb and dumb, unconscious, selfless and agentless. Who could have envisaged only 10 years ago that such a system waits for its discoverers buried somewhere in the monkey's, primate's and potentially man's brain? In monkeys and primates MNS cannot support understanding, imitation of intentional action, language and role playing in their human sense. The actual structure of MNS seems to consist minimally of two components (how the object looks and how it looks when handled) which are anatomically localized in two or three different brain regions (cf. Fogassi & Gallese, this volume). The first component is a sort of 'representation' of features of subject and object of action in the form of a sort of a 'primal sketch' what one can do with an appropriate body member to a class of objects. The second component is a 'representation' of the action as a serial order of its execution, i.e., in time series. There are other closely related possibilities to interpret the brain circuitry of MNS, e.g. when one of the two components serves functionally as 'representation' the other helps to serialize the required exploratory or motor movements and *vice versa* in a mirror-like fashion within the MNS itself. Still another possibility would be that to the three brain areas identified correspond two 'representations' (in perceiving the object and handling the object) while the third area helps serialize either of them or coordinates their joint serialization. The main point is that in these 'representations' and serializations of movements during action execution there is no functional part or component corresponding to 'Observer' or 'Agent' as a functional unity. If there is a matching or mapping, it is between how the object looks, when explored by perceptual activity, and how it looks when handled by the corresponding body member, e.g., by the hand.

The most challenging aspect of my proposal is associated with the claim that MNS does not perform the same way in monkeys and humans (if we assume a causal role of MNS for language origin). In the latter species it can apparently function not only as a part of a local brain circuit, but also in an unencapsulated way as a component of the central system supporting the processing of speech and language. If this indeed turns out to be the case after further experimental verification – that the MNS in humans is a double-action system – this would entail both good news and bad news.

The bad news would be that one and the same class of neurons functions in different way in two biological species. This means that from studying monkey brains we cannot infer for sure how the human brains perform even on the 'low' level of the way classes of neurons function. This is definitely not a good news, as the majority of the neurological studies of monkeys and primates are made with an eye that the human brain performs the same way.

The good news would be rather more hypothetical in nature and consequences. It involves the construction of a controversial scenario involving the unencapsulation of the serial component of MNS on an evolutionary scale and the generalization of its application to the nascent mechanisms of speech and language. This scenario is a very challenging one as it would change the way we envisage the relationships between body, brain, and mind.

## References

Baron-Cohen, S., & Swettenham, J. (1996). The relationship between SAM and ToMM: Two hypotheses. In P. Carruthers & P. K. Smith (Eds.), *Theories of Theories of Mind* (pp. 158–168). Cambridge: Cambridge University Press.

Chomsky, N. (1993). *Language and Thought*. Wakefield, RI: Moyer Bell.

Chomsky, N. (1994). Naturalism and dualism in the study of language and mind. *International Journal of Philosophical Studies, 2*(2), 181–209.

Chomsky, N. (1995). Language and nature. *Mind, 104*, 1–61.

Fadiga, L., & Gallese, V. (1997). Action representation and language in the brain. *Theoretical Linguistics, 23*(3), 267–280.

Ferrari, P. F., Kohler, E., Fogassi, L., & Gallese, V. (2000). The ability to follow eye gaze and its emergence during development in macaque monkeys. *Proceedings of the National Academy of Sciences of USA, 97*, 13997–14002.

Fodor, J. (1983). *The Modularity of Mind*. Cambridge, MA: MIT Press.

Gallese, V. (2000a). The acting subject: Towards the neural basis of social cognition. In T. Metzinger (Ed.), *Neural Correlates of Consciousness: Empirical and conceptual questions* (pp. 325–333). Cambridge, MA: MIT Press.

Gallese, V. (2000b). The inner sense of action agency and motor representations. *Journal of Consciousness Studies, 7*(10), 23–40.

Gallese, V., & Goldman, A. (1998). Mirror neurons and the simulation theory of mind-reading. *Trends in Cognitive Sciences, 2*, 493–501.

Hauser, M. (1996). *The Evolution of Communication*. Cambridge, MA: MIT Press.

Hurford, J., Studdert-Kennedy, M., & Knight, C. (Eds.). (1998). *Approaches to the Evolution of Language*. Cambridge: Cambridge University Press.

Langacker, R. (1987). *Foundations of Cognitive Grammar*, Vol. 1: *Theoretical prerequisites*. Stanford: Stanford University Press.

Leslie, A. (1987). Pretence and representation: The origins of 'theory of mind'. *Psychological Review, 94*, 412–426.

Liberman, A. M., & Whalen, D. H. (2000). On the relation of speech to language. *Trends in Cognitive Sciences, 4*(5), 187–196.

Lyons, J. (1991). The origin of language, speech and languages. In *Natural Language and Universal Grammar*, Vol. 1: *Essays in Linguistic Theory* (pp. 73–95). Cambridge: Cambridge University Press.

Povinelli, D. J., Bering, J. M., & Giambrone, S. (2000). Toward a science of other minds: Escaping the argument by analogy. *Cognitive Science, 24*, 509–541.

Ramachandran, V. (2000). Mirror neurons and imitation learning as the driving force behind "the great leap forward" in human evolution. http://www.edge.org/documents/archive/edge69.html

Rizzolatti, G., & Arbib, M. A. (1998). Language within our grasp. *Trends in Neuroscience, 21*, 188–194.

Rizzolatti, G., Fadiga, L., Gallese, V., & Fogassi, L. (1996). Premotor cortex and the recognition of motor action. *Cognitive Brain Research, 3*, 131–141.

Rizzolatti, G., Fogassi, L., & Gallese, V. (2001). Neurophysiological mechanisms underlying the understanding and imitation of action. *Nature Reviews Neuroscience, 2*, 661–670.

Stamenov, M. I. (1997). Grammar, meaning, and consciousness: What can the sentence structure tell us about the structure of consciousness? In M. I. Stamenov (Ed.), *Language Structure, Discourse, and the Access to Consciousness* (pp. 277–343). Amsterdam & Philadelphia: John Benjamins.

Stamenov, M. I. (forthcoming). Language and self-consciousness: Modes of self-presentation in language structure. In T. Kircher & A. David (Eds.), *The Self in Neuroscience and Psychiatry*. Cambridge: Cambridge University Press.

Trabant, J., & Ward, S. (Eds.). (2001). *New Essays on the Origin of Language*. Berlin: Mouton de Gruyter.

Vendler, Z. (1967). Verbs and times. In *Linguistics in Philosophy* (pp. 97–121). Ithaca, NY: Cornell University Press.

Weigand, E. (2000). The dialogic action game. In M. Coulthard, J. Cotterill, & F. Rock (Eds.), *Dialogue Analysis VII. Working with dialogue* (pp. 1–18). Tübingen: Niemeyer.

# Altercentric perception by infants and adults in dialogue

## Ego's virtual participation in Alter's complementary act

Stein Bråten

University of Oslo, Norway

## 1. Introduction: The questions

Recent infancy research reveals inborn capacity for attunement to others' gestures and for reciprocal face-to-face interplay. Human infants can engage in protoconversation with their caregivers in the first weeks of life, and can reciprocate caregiving before their first year's birthday. Such preverbal capacity appears to be used also by the verbal mind in conversation. For example, teenagers in face-to-face conversation often reflect one another's gestures in much the same way that we observe in early infant-adult interaction (Figure 1), and sometimes complete one another's speech acts akin to patterns manifested also in preverbal object-oriented manual interplay. Instances to be presented, drawn *inter alia* from my own records of verbal and preverbal interactions, invite these two questions:

(Q1) How may we account for the way in which conversation partners frequently may be heard to complete one another's utterances?

(Q2) What enables 11-month-olds to learn from face-to-face situations to reciprocate their caregivers' spoon-feeding?

Replies offered in terms of virtual participation in the complementary act executed by the other (Bråten 1974, 1998a) may now be supported by the role that 'mirror neurons' and premotor processes have been found to play in the perception of action and speech (cf. Rizzolatti & Arbib 1998, including their reference to Liberman 1993). I shall first present a cybernetic model of conversation partners simulating one another's complementary processes, implying a reply to the first question in these terms: Ego's speaking is monitored by predictive simulation of Alter's listening, and Ego's listening by postdictive simulation of Alter's speaking (Bråten 1973, 1974).

A likely support of such simulation-of-mind ability is the preverbal capacity uncovered in infants for what I term 'altercentric' perception, entailing virtual participation in what Alter is doing as if being jointly done from Alter's stance (Bråten 1996, 1998a,c). This permits a reply, then, to the second question in terms of infant learning by other-centred participant observation. Trevarthen (1998: 16) finds that the efficiency of interpersonal sympathetic engagements in early infancy signals the capacity to " 'mirror' the motivations and purposes of companions". It requires such a mechanism for virtual (other) participation, intersecting with those that subserve execution of own acts:

> The brain mechanisms that represent the human body of the single subject in all its intelligent and emotional activities and states are at the same time very extensive, of ancient lineage, and greatly elaborated. The mirror system that enables the expressions of other individuals' bodies to play a part in regulating emotion and rational activity and learning intersubjectively, the 'virtual other' mechanism (Bråten 1988, [... 1998a]), must be similarly extensive.
>
> (Trevarthen 1998: 46)

I shall present behavioral instances of how virtual (other) participation is manifested in interpersonal interaction. In conjunction with the neurophysiological evidence of a mirror system in the human brain (Rizzolatti, Craighero & Fadiga, this volume; Rizzolatti & Arbib 1998) they invite this question:

(Q3)   In which evolutionary conditions would such a mirror system adapted to subserve learning by altercentric participation have afforded the most critical selective advantage?

Towards the end I venture a speculative reply.[1] Some putative evolution stages towards speech culture (partly in line with Donald 1991) will be compared to precursory and supportive steps towards speech in infant and child development (cf. Bråten (Ed.) 1998). In dialogue, for example, the "attunement to the attunement of the other" (Rommetveit 1998: 360) appears to be prepared for by the mutual, dance-like interplay which we can observe already in the first weeks after birth (Figure 1).

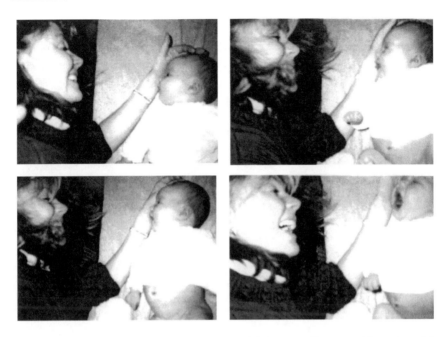

**Figure 1.** Infant girl (11 days) on the nursing table in mutually attuned interplay with her mother (Bråten 1998c: 29).

## 2. Dialogue partners simulating one another's verbal production and understanding

### 2.1 A conversation model

Drawing *inter alia* upon Rommetveit's (1972) emphasis of the complementary nature of speech acts and upon my laboratory and computer simulation studies of dialogues in the early 1970s, I submitted then a model (succinctly expressed in Figure 2) of how conversation partners monitor and adjust their own coding activity without resort to external feedback: while preparing an utterance, the speaker monitors and adjusts own encoding by simulating the listener's decoding, and while processing the utterance, the listener checks and adjusts own decoding by simulating the speaker's encoding process (Bråten 1973, 1974). Anticipating the simulation version of theory-of-mind approaches (cf. Gallese & Goldmann 1998; Harris 1991; Humphrey 1984), this model predicts conversation partners to simulate the reverse processes in one another's minds. Independently of external feedback about possible mismatch between intention (C) and comprehension (Cco), Ego's speaking is monitored by virtual participation in Alter's listening, and Ego's listening by virtual participation in Alter's speaking. The former is partly consistent with Mead's (1934) theory of anticipatory response by perspective-taking, and the latter with Liberman's (1957) original motor theory of speech perception, to which I referred when presenting the model at a meeting on cybernetics (Bråten 1974).

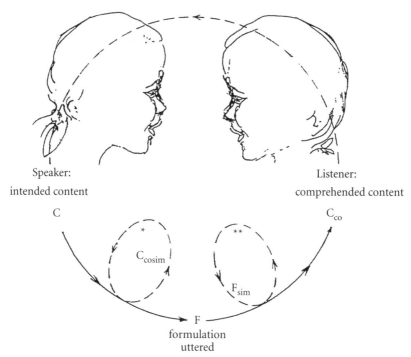

Speaker:
intended content

C

Listener:
comprehended content

$C_{co}$

$C_{cosim}$

$F_{sim}$

F
formulation
uttered

*Legend broken lines circuits:*

*(Left circuit\*)*: the speaker's virtual participation in the listener process by predictory simulation ~ with the utterance candidate F' as input and simulated content $C_{cosim}$ as output compared to the intended content C:
if $C_{cosim}$ = C then F:= F'
else modify F' for another trial

*(Right circuit\*\*)*: the listener's virtual participation in the speaker production by postdictory simulation ~ with the comprehend candidate $C_{co}'$ as input and F' as output compared to the utterance F as heard:
if F' = F then $C_{co}$:= $C_{co}'$
else modify $C_{co}'$ for another trial

**Figure 2.** A conversation model of speaker's and listener's virtual participation in one another's activity by their simulating the complementary process in the other's mind. While producing an utterance, the speaker regulates own encoding by simulating the listener's decoding, and while processing the utterance, the listener checks own decoding by simulating the speaker's encoding (Bråten 1973, 1974). (In the legend specification of such mental simulation loops, the logical become-operator ":=" means that the value of the variable to the right of the operator is assigned to the variable to the left).

Thereby, a reply is implied to the first question (Q1): sentence completion is an overt manifestation of such virtual participation in the other's production. As I would phrase it today in terms of altercentric participation, when you find yourself more or less unwittingly completing what your conversation partner is about to say, you overtly manifest your virtual participation in the other's speech act as if you were a virtual co-author, enabled by a virtual-other mechanism, i.e. by an other-centred mirror system adapted in phylogeny to subserve preverbal and verbal conversational efficiency. Thus, the internal, and more or less unwitting, simulation loops marked in Figure 2 may sometimes be overtly manifested in behaviour.

## 2.2  Illustrations from a moral dilemma processing dialogue

Such completion of what the other is about to say, by conversation partners absorbed in one another and in their topic, may be frequently heard if paid attention to. Below is an example recorded in a laboratory experiment. Paired students were asked to be a moral jury on the moral dilemma involved in *euthanasia*. Presented with a similar case for two physicians at two different hospitals, one physician complying and the other refusing to comply with the plea of their respective patients, suffering from an incurable and unbearable fatal disease, the students' task was to reach by dialogue a moral judgement. Here are two extracts from the beginning of the dialogue between the students *a* and *b* (in which two instances of completion of one another's statements are marked in italics):

a:  *yes...*

b:  *what then to do?* That I am not sure of. It appears terribly difficult to decide at all what judgment we are to make on that

a:  yes, it does make it so, yes

b:  and the one who killed him; I would be quite certain someone would characterize as direct murder

.  —

.  —

b:  and even if a man then says, "I, my personal opinion, now", then even if a man is judged by the physicians to be incurable, there are possibilities. And I have heard about several instances where people have been cured through others sitting down and praying to God.

a:  mm, and is it a fact, actually, from which one cannot escape, that *people have been cured*

b:  *by prayer*

a:  yes, that's right

The above instances of sentence completion may serve to illustrate how virtual participation in the dialogue partner's productions sometimes is overtly manifested.

### 3.   The preverbal mind: Early infant imitation and learning by altercentric participation

Even though higher-order semantic mechanisms were at play in the above exemplified dialogue fragments, such other-centred capacity to virtually participate in the other's complementary activity is also exhibited by the preverbal mind. I shall now turn to this and offer examples of behavioral manifestation.

#### 3.1   When 11-month-olds reciprocate spoon-feeding: Re-enactment from virtual co-enactment

As I have recorded and accounted for in terms of learning by altercentric perception, infants in face-to-face spoon-feeding situations, when allowed to take the spoon with food in their own hand, can reciprocate the spoon-feeding before their first years' birthday. Figure 3 pictures an episodic example: An infant girl (11 1/2 months) is being fed by her mother and – when allowed to take the spoon with food in her own hand – she reciprocates by feeding her mother.

In order for infants to be able reciprocate the spoon-feeding they must have been able to virtually partake in their caregivers' previous spoon-feeding activity as if they were co-authors of the feeding, even though their caregivers have been the actual authors. The virtual-alter mechanism, complementary to the bodily ego, enables the infant to feel a virtual moving with alter's feeding movements from alter's stance, leaving the infant with a procedural memory of having been a virtual co-author of the feeding. Such virtual co-enactment gives rise to what Stern (1999) terms shared temporal *vitality (affects) contours*, reflecting the manner in which the enactment is felt to be co-enacted and the feeling that directs the virtual co-enactment. This virtual co-enactment guides the infant's re-enactment from a procedural memory which I term '*e-motional*' (from the root sense 'out-

Figure 3. Reciprocating her mother's spoon-feeding, an infant (11 1/2 months) demonstrates learning by altercentric perception: her virtual participation in the caregivers' previous spoon-feeding acts has left her with a procedural memory of having been a virtual co-author of their enactments (as if they had been hand-guiding her) and which guides her reciprocal re-enactment (from Bråten 1998c: 15; cf. also Figure 4 (bottom)).

of-motion' and the folk sense 'being moved by'). When virtually moving with the facing other as if facing the same direction and being hand-guided, altercentric co-ordinates are entailed by one's moving from the bodily center of the other, not one's own, and which then have to be reversed to allow for executed re-enactment in a body-centred frame.

The above gives the defining characteristics of learning by altercentric participation, permitting this explanatory reply to the second question (Q2): infants learn to re-enact from altercentric perception of their caregivers' enactment as if they had been hand-guided from the caregivers' stance. When this altercentric capacity is biologically impaired, the mirror reversal required by face-to-face situations will present learning problems, for example, to subjects with autism. Their predicted and revealed difficulties in gestural imitation in face-to-face situations have been explained by their attributed inability to transcend own body-centred viewpoint (Bråten 1993, 1994).[2] When asked to "Do as I do", they have difficulties in reverting and matching the model's gesture, such as grasp thumb, or peek-a-boo, or even simply raising arms (cf. Bråten 1998a:115–118; Ohta 1987; Whiten & Brown 1998:267–271).

## 3.2 Feeder's mouth movements reveal their altercentric perception of the recipient's act

While children with learning problems sometimes require the instructor to sit side by side or use a mirror, face-to-face situations present no problems for gestural imitation and imitative learning in ordinary children. In such settings even nine-month-olds exhibit deferred imitation of object manipulation (Meltzoff 1988; Heimann & Meltzoff 1996) and, as I have shown, 11-month-olds learn to reciprocate spoon-feeding (such as illustrated in Figures 3 and 4 (bottom)). And when they reciprocate by re-enacting the spoon-feeding, their own mouth movements sometimes 'mirror' the mouth movements of the recipient to whom they offer food, thereby manifesting their virtual participation in the caregiver's food-intake from the caregiver's stance. For example, when reciprocating his sister's spoon feeding the Oslo-boy (11 3/4 months) (Figure 4 (bottom, right)) opens his own mouth as his sister opens her mouth to receive the spoon he is offering. The fact that these infants are able to reciprocate indicates their having learnt to re-enact from altercentric perception of their caregivers' previous spoon-feeding, and such altercentric perception is sometimes again manifested in a reverse manner when they become the feeder, showing by their accompanying mouth movements their virtual participation in the recipient's preparing-to-eat movements.

This way of virtually participating in the other's activity, as if co-enacting that activity, is not unique for infants in a spoon-feeding culture. For example,

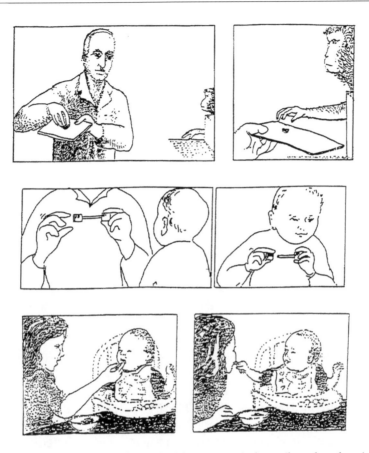

**Figure 4.** Exposure to manual acts inviting, respectively, unilateral and reciprocal matching responses.

*(Top)* When the macaque monkey observes the grasping of a piece of food and when grasping the food by itself, there is a grasp-specific premotor neuron discharge (drawing adapted from Di Pellegrino et al. 1992).

*(Middle)* A toddler (18 months) watches the experimenter's failed effort to pull the dumbbell apart and then, when handed the object, pulls it apart (Meltzoff 1995; cf. also accounts in Bråten (Ed.) 1998: 47–62, 112–115).

*(Bottom)* Infant (11 3/4 months), reciprocating his sister's spoon-feeding is demonstrating by his mouth movements his altercentric perception: he is opening his own mouth as he puts the spoon into her mouth, revealing his virtual participation in her preparing-to-eat act (drawing after video record by Bråten 1996: 2).

in an Amazona tribe Eibl-Eibesfeldt (1979: 15) pictures a Yanomami baby (11 to 12 months) who, while held by the big sister, extends a morsel to the sister's mouth, opening own mouth in the process and tightening own lips as the sister's mouth closes on the morsel. When feeding a baby or a sick person, adult caregivers in Western cultures exhibit the same tendency to be opening their mouth as they put the spoon in the recipient's mouth, thereby unwittingly manifesting that their own executed spoon-feeding act is accompanied by virtual participation in the recipient's complementary act of receiving the food by the mouth. Of course, motivation makes a difference here; no such overt manifestation can be expected of the unmotivated, disengaged observer.

### 3.3   Parallels to conversational efficiency

When mouth movements of the feeder – infant or adult – reflect the corresponding mouth movements of the one being fed we may see a parallel here also to the virtual participation exhibited by partners in verbal conversation (Figure 2), co-enacting one another's complementary acts and sometimes completing one another's utterances.

Another preverbal parallel has been afforded by toddlers (18 months) who successfully complete a target act failed by the experimenter in Meltzoff's (1995) behavioral re-enactment design. The toddler watches the experimenter (fake his) failing to pull a dumbbell apart and, when handed the dumbbell, the toddler pulls it apart – sometimes with a triumphant smile (Figure 4 (middle)). Again, this may be accounted for in the above terms: Virtual participation in the experimenter's failed effort evokes in the toddler a simulated completion of the unrealized target act guiding the executed realization (Bråten 1998a: 112–115). Thus, the internal mirror system response, matching the experimenter's aborted act, enables the toddler to realize by mental simulation the target completion and which in turn invites and guides the toddler's successful target act execution.

## 4.   Discussion

### 4.1   Object display or mechanical demonstration do not evoke response, while directed acts do

Now, with reference to the situations portrayed in Figure 4, it may be objected from a Gibsonian view of perception that the adult model merely demonstrates the affordance of the objects in questions: a piece of food invites to be eaten, a spoon invites to put in a mouth, and a dumbbell invites to pulled apart; the watcher needs no resonant mirror match and learners need no virtual participation in the model's activity in order to respond or to re-enact. Such an interpretation in terms of object affordance is ruled out, however, by the results of control experiments relating to those pictured in Figure 4.

In a control of the dumbbell experiment portrayed in Figure 4 (middle) the experimenter remains passive while a machine with the dumbbell in its claws 'tries' to pull it apart without success. That did virtually not elicit any re-enactment by the watching toddlers. According to Meltzoff and Moore (1998:52) they were denied the opportunity to read any intention and, as I see it, the inanimate object demonstration prevented their virtual participation in any target effort and, hence, there could be no matching response evoking any simulated completion guiding any executed re-enactment.

Also, in the case of the macaque monkey (Figure 4 (top)), no matching mirror response is evoked by mere object-display from a distance: What is required is the sight of act execution. While there is mirror neuron discharge when the monkey watches the experimenter grasp a piece of food in sight and when grasping the food by itself, there is little or no response when the experimenter uses a plier to grasp the food, and no mirror response at all when the food is merely pushed towards the monkey, while out of reach (cf. Rizzolatti & Arbib 1998).

## 4.2 Virtual (other) participation during own act execution in interpersonal interaction

Unlike humans, the monkey is unlikely to have in its motor repertoire a 'give act' which would complement its 'grasp act'. Hence during the monkey's execution of its own grasping the food handed over by the experimenter (Figure 4 (top)) we should expect no parallel activation in the monkey of any virtual give-specific mirror response matching the experimenter's giving.

In humans on the other hand, engaged in reciprocal give-and-take situations, complementary frames of reference for virtual participation and action execution are expected to be concurrently at play. For example, in the children portrayed in Figure 4 (bottom), the spoon-feeding act execution is accompanied by the virtual participation in the recipient's mouth movements. This entails the activation in each subject of two parallel frames of references for coding and re-coding: the body-centred frame of co-ordinates pertaining to own act execution (such as spoonfeeding), and the complementary altercentric frame of reference for Ego's virtual participation in Alter's execution (such as altercentric participation in the Alter's complementary food-intake).[3]

Such reciprocal and complementary features of interpersonal interaction makes for more elaborate systems dynamics than what is entailed by the monkey's unilateral grasp situation (Figure 4, top). While there may have been no selective pressure on non-human primates towards a mirror system adapted to subserve learning by virtual (other) participation in face-to-face interaction, I shall now indicate why such an adaptation may have been critical to human evolution, taking us to the question (Q3).

### 5. In which evolutionary conditions would altercentric perception have afforded the most critical selective advantage?

Based *inter alia* on comparative studies of infant-adult relations in humans and chimpanzees, my speculative reply is this: Compensating for the lost protective and nurturing advantage of offsprings clinging to their mother's body which may be attributed to Miocene apes, such an adapted mirror system would have made a critical difference to Pliocene hominids, and to early Homo erectus before the invention of baby-carriers (Figure 5 (right)) which would have restored some of the lost advantage continued to be enjoyed by modern apes (Figure 5 (left)).

Deprived of this protective and learning body-clinging advantage, with children having to be left hold of (in a tree or on the ground) when bipedal parents needed to use both arms, and given increased infant helplessness and prolonged childhood, those Hominidae/Homo species parents and children would have had a critical selective advantage, I submit, whose resonant mirror system had been adapted to subserve altercentric perception and virtual participation in others' action execution in reciprocal settings. That is, to subserve face-to-face pedagogy and social learning in which own body-centred (egocentric) stance is transcended by teachers and learners by the kind of perspective-taking that Tomasello et al. (1993) attribute to cultural learning in human ontogeny, and which may be accounted for by the virtual-other mechanism enabling altercentric perception.

### 5.1 Comparing adult-infant relations and infant-carrying modalities in humans and chimpanzees

When clinging to the mother's back, offsprings of great apes learn to orient themselves in the world in which they operate from the carrying mother's stance. Moving with her movements, they may even be afforded the opportunity to learn by copying her movements (perhaps in the way that Byrne (1998) terms "program-level imitation") without having to transcend own (egocentric) body-centred perspective. During my eight years of periodically studying chimpanzee-offspring relations in a Zoo and Wild-life Park in Southern Norway, I observe how the infants, when old enough to cling to the mother's back, not only bodily move with the mother's movements but often adjust their head to her movement direction, appearing to be gazing in the same direction as she does. When a mother holds the infant in front of her for grooming (which adults more often do from behind one

**Figure 5.** Clinging to the mother's body, offsprings of great apes, like this infant (5 1/2 months) of a chimpanzee mother in the Kristiansand Zoo and Wildlife Park (who also let the infant cling to her brother (6 years)), are afforded opportunities to learn and orient themselves without transcending own body-centred stance. While probably enjoyed also by ancestral Miocene apes, this protective body-clinging learning advantage would have had to be compensated in Pliocene hominids and early Homo erectus before the attributed invention of a baby-carrying device (of animal skin or plant material, pictured to the right).

another), a sort of face-to-face situation is established, but not for the kind of reciprocal interplay entailing mutual gazing and gesticulation which we observe in human infant-adult pairs. When chimpanzee infants, however, are nursed by human caretakers and sensitized to face-to-face interaction with humans, they appear able, as Bard (1998) has shown, to take after her emotional facial gestures. I have a video record of a chimpanzee infant (39 days) engaging in a sort of turn-taking vocal interplay with his foster parent, but I have never observed this in infant-adult chimpanzee interaction.

In his attachment theory, Bowlby (1984: 184) makes this functional point: "At birth or soon after, all primate infants, bar the humans, cling to their mothers." I submit, then, that this lack of natural means for bodily attachment contributes to the relatively high face-to-face frequency in human infant-adult relations and, hence, affects the nature of cultural learning.

While infants of great apes enjoy the advantage of clinging to their parents for safety and in silence by a natural mode of bodily attachment, hominid infants have had to resort to a sort of early "*mental* clinging" and to learn from engagements in face-to-face pedagogy. This would entail an evolved adaptive mental architecture for interpersonal connectivity that becomes operative even in face-to-face relations at some distance, and which would have been most critical before the invention of baby-carrying bags attributed by Richard Leakey (1995:97) to Homo erectus. If and when they migrated from Africa, they could not have done so without such an invention.

In some contemporary African cultures, unlike many Western cultures, such baby-bags are still in use, for example, in the Gusii culture, where often a sibling of the infant is assigned the task of carrying the infant on the back (LeVine & LeVine 1988), and where face-to-face interplay with infants by looking or talking is relatively rare:

> Virtually all of [the Gusii mothers'] interaction with the babies includes hold-ing or carrying, and they often respond to infant vocal or visual signals with physical contact rather than reciprocal talking or looking. By contrast, the Boston mothers, who hold their babies in less than a third of their interac-tions from 6 months onward, seek to engage their infants in visual and verbal communication at a distance.                    (LeVine et al. 1996:198)

Taking after the facial expressions, vocalizations and gestures of adults is a way to ensure connectivity and learning even at a distance.

## 5.2   Putative stages towards speech-mediated teaching and learning in evolution

In phylogeny, as indicated in Figure 6 (below the line), distinguishing hominid cultures from patterns in other primates, an enhanced mirror system subserving face-to-face pedagogy and learning in situations inviting transition of own body-centred stance, may have facilitated the transition to what Donald (1991) terms 'mimetic culture'. This entails a tool-designing pedagogy and mimetic learning by hand-guidance and, I would add, learning by altercentric participation (virtual hand-guidance).

Donald also argues for the transitional role of a mimetic Homo erectus culture towards a narrative (mythic) culture. An evolved capacity for altercentric percep-tion in face-to-face learning and warning situations may have been precursory and supportive of later conversational speech adaptation in archaic humans. To infants endowed with the capacity for virtual (other) participation in gesticulating and

Symbolic communication (tertiary intersubjectivity) by conversational speech and self-reflective narratives ~ from about 24 months ~ and second-order understanding of others (theory or simulation of mind) from 3–6 years

Object-oriented interpersonal communion (secondary intersubjectivity) about jointly attended objects and states ~ from 9 months or earlier ~ imitative learning by altercentric participation affording virtual hand-guidance in reciprocal face-to-face settings

Subject-subject reciprocal communion (primary intersubjectivity) in which participants attend and attune to one another's facial, manual and vocal gestures inviting semblant re-enactment and affect attunement ~ beginning with neonatal imitation and protoconversation in the first weeks of life and continuing throughout life to support higher-order object-oriented and speech-mediated communication and understanding

---

Face-to-face pedagogy ~ mental 'clinging' replacing bodily 'clinging' in Pliocene hominids, compensating for the loss of the back-carrying advantage enjoyed by ancestral Miocene apes and affording selective advantage to mothers and children capable of mutually attuned teaching and learning transcending own (egocentric) stance

Mimetic tool-designing pedagogy in late Homo erectus (or earlier ?) in a tool-oriented culture reproducing itself by mimesis, with instruction and imitative learning of skills for using and shaping tools by actual hand-guidance and by altercentric participation (virtual hand-guidance)

Speech-mediated pedagogy in Homo sapiens (or earlier?) by verbal instruction and understanding supported by preverbal skills and gestures in an emerging local and unique speech community producing itself by narratives

Figure 6. *(Above the line)* Ontogenetic layers or steps of intersubjective attunement in early childhood with each lower-order layer continuing throughout life to support higher-order layers (cf. Bråten & Trevarthen 1994; Bråten (Ed.) 1998; Stern 1985).

*(Below)* Putative phylogenetic stages of cultural learning and teaching (the mimetic culture distinction is consistent with Donald 1991). See the conclusion about the halting parallel between ontogeny and phylogeny in affording steps towards speech language.

articulating others, an emerging ambient speech language, perhaps accompanied by a speech-mediated pedagogy, would have afforded opportunities for their beginning to attune themselves by altercentric speech perception to the prosody and rhythm of the emerging language culture into which they were born.

## 6.   Steps of intersubjective attunement towards speech in early childhood

In early ontogeny, we have distinguished precursory and supportive steps towards speech and dialogue in terms of these different layers or domains of intersubjective attunement and understanding (cf. Bråten (Ed.) 1998), succinctly indicated in Figure 6 (above the line):

1. *Primary intersubjective attunement* in a dyadic reciprocal format of protoconversation (Trevarthen 1974, 1992, 1998), preparing for and supporting higherorder abilities later in life. For example, Kuguimutzakis (1998) have videorecords of 45-minutes-olds attempting to imitate his uttering /a/, and from her studies of speech perception Kuhl (1998) finds that by 6 months infants have 'pruned' sounds from their perceptual space that make no sense in the ambient language.

2. *Secondary intersubjective attunement* when infants join others in shared attention about objects (Trevarthen & Hubley 1978), learning by imitation to manipulate objects (Meltzoff 1988), and to reciprocate caregivers' acts (Bråten 1996). Such object-oriented cultural learning in a triangular format opens for semantic learning (Akhtar & Tomasello 1998; Hobson 1998).

3. *Tertiary intersubjective understanding* in conversational and narrative speech, entailing predication (Akhtar & Tomasello 1998) and a sense of verbal or narrative self and other (Stern 1985). Understanding of others' minds and emotion (Dunn 1998) opens for perspective-taking and emotional absorption, even in fictional others (Harris 1998), and for simulation of conversation partners' minds (cf. Figure 2).

Thus, the preverbal capacities and opportunities for cultural learning in early infancy, including the capacity for altercentric (speech) perception and perspective-taking, nurture and support the higher-order level abilities for conversational and narrative speech. For example, in an Oxford study, Rall and Harris (2000) find that when 3- and 4-year-olds are asked to retell fairytales, say about Cinderella, they manage best when the verbs in the stories listened to are consistent with the stance of the protagonist with whom they identify, inviting their altercentric participation in 'Cinderella's slippers', as it were. The children have trouble when the verbs in the stories told are used from the reverse perspective, at odds with their perspective-taking.

We cannot yet explain the qualitative leap to children's simulation or theory of mind, correlating with their verbal and conversational ability and entailing second-order understanding of others' thoughts and emotions. But it seems reasonable to assume that a mirror system for matching or simulating others' acts may afford a precursory and nurturing path to simulation of other minds (cf. Bråten 1998a, b; Gallese & Goldman 1998), and that such preverbal capacity for virtual participation in what others are doing are likely to support the kind of inner feedback loops defined by the conversation model (Figure 2).


## 7.  Conclusion

The model implies that conversation partners simulate one another's complementary processes, affording a reply to the first question (Q1) in terms of virtual participation in the partner's executed act. A preverbal parallel pertains to the second question (Q2): infants re-enact from altercentric participation in their caregivers' enactment as if they had been hand-guided from the caregivers' stance. When the listener completes the talker's speech act and when the feeder's mouth movements match the recipient's mouth movements, their virtual participation is even overtly manifested.

The discovery of 'mirror neurons' in experiments of the kind depicted in Figure 4 (top) and the electrophysiological experimental evidences of a mirror system in the human brain (Rizzolatti, Craighero & Fadiga, this volume) lend support to the above explanatory replies. Finding in human subjects that cortical and spinal levels (the latter to a lesser degree) resonate during action viewing, Rizzolatti, Craighero and Fadiga (this volume) suggest that the human viewer's motor system "simulates" what the other is doing as if being (virtually) done by the viewer.

As for the third question (Q3), I have offered a speculation as to why such an adaptation in support of learning by altercentric participation may have afforded a selective advantage in hominid and human evolution, and indicated how the efficient speech perception and learning we find in early ontogeny may reflect putative stages of face-to-face pedagogy and learning in the evolution of speech cultures. While phylogeny, however, entails the generation of new cultural lifeworlds, including the co-evolution of linguistic environments never before in existence, Homo sapiens sapiens infants are born into an already existing local speech community, a linguistic 'sea' in which they rapidly and creatively learn to swim, as it were. Thus, invited by Figure 6, the comparison of phylogenetic stages and onto-genetic steps towards speech is halting. It blurs the distinction between a speech community evolving as a co-created novelty and as environmentally given in ontogeny to those born into the speech community, nurturing their impressive speech learning capacity. This would not have been as efficient, we may now claim, were it not supported by an innate, preverbal virtual-other mechanism enabling altercentric speech perception, and subserved by a phylogenetically afforded and adapted resonant mirror system.

Before I had learnt of the mirror neurons discovery I predicted that "neural systems, perhaps even neurons, sensitized to alter-centric perception will be uncovered in experiments designed to test this prediction" (Bråten 1997).[4] By now, then, this prediction has been partly confirmed. Given the evidence of a resonant mirror system in humans, referred to *inter alia* in the keynote article by Rizzolatti and Arbib (1998), we are beginning to spell out some of the rich and radical implications. No longer can be upheld Cartesian and Leibnizian conceptions of monadic subjects and disembodied minds without windows to one another except as mediated by symbolic or constructed representations. A neurosocial bridgehead has been found that is likely to support the intersubjective arch of virtual (other) participation, and to subserve in adapted form efficient conversation and learning by altercentric speech perception.

## Notes

1. Presented by the author to The Norwegian Academy of Science and Letters, 10 February 2000.

2. On this explanatory ground I further predict significant difference in resonant mirror response when electrophysiological equipment will be improved to allow for measurement in persons interacting face-to-face: there will be little or no matching response in subjects with autism.

3. Human reciprocal give-and-take-food situations have been described to be likely to evoke in parallel or near-parallel *complementary* mirror responses: in the giver, Ego's executed act of giving is accompanied by the virtual participation in Alter's act of receiving, and in the receiver, Ego act of receiving the food is accompanied by virtual participation in Alter's giving. Such parallel activation may specified in these terms: Let C denote the caregiver and B the baby boy or girl. Let E.p denote the executed manual act of feeding and E.q the executed mouth act of taking in the food, while *A.*p and *A.*q mark the virtual altercentric acts of respectively feeding and food-intake, involving the virtual-Alter mechanism *A. Then the situations portrayed in Figure 3 and Figure 4 (bottom) could be described in this way:

C(E.p; *A.*q) & B(E.q;*A.*p) → B (E.p;*A.*q) & C(E.q;*A.*p)

That is, the caregiver C's execution of feeding (while participating in the baby B's food-intake) evokes in the infant B concurrent intake of food and virtual participation in the caregiver's feeding, which in turn invites and enables the baby – when allowed to take the spoon in own hand – to feed the caregiver in a semblant manner (accompanied by the baby's virtual participation in the caregiver's food-intake).

If this is tenable, and there is matching mirror system support, we should expect, for example, that in humans *give*-mirror neurons be activated during own giving and while watching the other give, and that *grasp*-mirror neurons be activated during own grasping and while watching the other grasp. (Presented by the author at a seminar with the Institute of Human Physiology faculty, University of Parma, 4 May 2000).

**4.** This prediction was stimulated *inter alia* by the discovery of allocentric 'place' neurons in animals (cf. O'Keefe 1992) and by my uncovering altercentric perception in infants.

## References

Akhtar, N., & Tomasello, M. (1998). Intersubjectivity in early language learning and use. In S. Bråten (Ed.), *Intersubjective Communication and Emotion in Early Ontogeny* (pp. 316–335). Cambridge: Cambridge University Press.

Bard, K. A. (1998). Social-experiential contributions to imitation and emotion in chimpanzees. In S. Bråten (Ed.), *Intersubjective Communication and Emotion in Early Ontogeny* (pp. 208–227). Cambridge: Cambridge University Press.

Bowlby, J. (1969/1984). *Attachment and Loss* (Vol.1, 2 ed.). Reading: Pelican Books.

Bråten, S. (1973). *Tegnbehandling og meningsutveksling*. Oslo: Scandinavian University Books/ Universitetsforlaget.

Bråten, S. (1974). Coding simulation circuits during symbolic interaction. In *Proceedings on the 7th International Congress on Cybernetics 1973* (pp. 327–336). Namur: Association Internationale de Cybernetique.

Bråten, S. (1986/1988a). Between dialogical mind and monological reason. Postulating the virtual other. In M. Campanella (Ed.), *Between Rationality and Cognition* (pp. 205–236). Turin & Geneve: Albert Meynier (Revised paper presented at the Gordon Research Conference on Cybernetics of Cognition, Wolfeboro, June 1986).

Bråten, S. (1988). Dialogic mind: The infant and the adult in protoconversation. In M. Carvallo (Ed.), *Nature, Cognition and System*, I (pp. 187–205). Dordrecht: Kluwer Academic.

Bråten, S. (1992). The virtual other in infants' minds and social feelings. In A. H. Wold (Ed.), *The Dialogical Alternative* (pp. 77–97). Oslo: Scandinavian University Press/ Oxford University Press.

Bråten, S. (1993). Social-emotional and auto-operational roots of cultural (peer) learning. Commentary. *Behavioral and Brain Sciences, 16*, 515.

Bråten, S. (1994). The companion space theorem. Paper presented at the ISIS Pre-conference on imitation, Paris 1 June 1994. Published revised in S. Bråten, *Modellmakt og altersentriske spedbarn: Essays on Dialogue in Infant & Adult* (pp. 219–230). Bergen: Sigma (2000).

Bråten, S. (1996). Infants demonstrate that care-giving is reciprocal. *Centre for Advanced Study Newsletter*. Oslo (November 1996) No. 2, 2.

Bråten, S. (1997). What enables infants to give care? Prosociality and learning by alter-centric participation. Centre for Advanced Study lecture in The Norwegian Academy of Science and Letters, Oslo 4 March 1997. Published in S. Bråten, *Modellmakt og altersentriske spedbarn: Essays on Dialogue in Infant & Adult* (pp. 231–242). Bergen: Sigma (2000).

Bråten, S. (Ed.). (1998). *Intersubjective Communication and Emotion in Early Ontogeny*. Cambridge: Cambridge University Press.

Bråten, S. (1998a). Infant learning by altercentric participation: The reverse of egocentric observation in autism. In S. Bråten (Ed.), *Intersubjective Communication and Emotion in Early Ontogeny* (pp. 105–124). Cambridge: Cambridge University Press.

Bråten, S. (1998b). Intersubjective communion and understanding: Development and perturbation. In S. Bråten (Ed.), *Intersubjective Communication and Emotion in Early Ontogeny* (pp. 372–382). Cambridge: Cambridge University Press.

Bråten, S. (1998c). *Kommunikasjon og samspill – fra fuudsel til alderdom*. Oslo: Tano-Achehoug.

Bråten, S., & Trevarthen, C. (1994/2000). Beginings of cultural learning. Paper presented at the ZiF symposium on the formative process of society, Bielefeld, 17–19 November 1994. Published in S. Bråten, *Modellmakt og altersentriske spedbarn: Essays on Dialogue in Infant & Adult* (pp. 213–218). Bergen: Sigma.

Byrne, R. W. (1998). Imitation: The contributions of priming and program-level copying. In S. Bråten (Ed.), *Intersubjective Communication and Emotion in Early Ontogeny* (pp. 228–244). Cambridge: Cambridge University Press.

Di Pellegrino, G., Fadiga, L., Fogassi, L., Gallese, V., & Rizolatti, G. (1992). Understanding motor evens: A neurophysiological study. *Experimental Brain Research, 91*, 176–180.

Donald, M. (1991). *Origins of the Modern Mind*. Cambridge MA: Harvard University Press.

Dunn, J. (1998). Siblings, emotion and the development of understanding. In S. Bråten (Ed.), *Intersubjective Communication and Emotion in Early Ontogeny* (pp. 158–168). Cambridge: Cambridge University Press.

Eibl-Eibesfeldt, I. (1979). Human ethology: concepts and implications for the sciences of man. *Behavioral and Brain Sciences, 2*, 1–57.

Gallese, V., & Goldman, A. (1998). Mirror neurons and the simulation theory of mind-reading. *Trends in Cognitive Sciences, 2*(12), 293–450.

Harris, P. L. (1991). The work of imagination. In A. Whiten (Ed.), *Natural Theories of Mind* (pp. 283–304). Oxford: Basil Blackwell.

Harris, P. L. (1998). Fictional absorption: emotional responses to make-believe. In S. Bråten (Ed.), *Intersubjective Communication and Emotion in Early Ontogeny* (pp. 336–353). Cambridge: Cambridge University Press.

Heimann, M., & Meltzoff, A. (1996). Deferred imitation in 9- and 14-month-old infants. *British Journal of Developmental Psychology, 14*, 55–64.

Hobson, R. P. (1998). The intersubjective foundations of thought. In S. Bråten (Ed.), *Intersubjective Communication and Emotion in Early Ontogeny* (pp. 283–296). Cambridge: Cambridge University Press.

Humphrey, N. (1984). *Consciousness Regained*. Oxford: Oxford University Press.

Kugiumutzakis, G. (1998). Neonatal imitation in the intersubjective companion space. In S. Bråten (Ed.), *Intersubjective Communication and Emotion in Early Ontogeny* (pp. 63–88). Cambridge: Cambridge University Press.

Kuhl, P. (1998). Language, culture and intersubjectivity: The creation of shared perception. In S. Bråten (Ed.), *Intersubjective Communication and Emotion in Early Ontogeny* (pp. 307–315). Cambridge: Cambridge University Press.

Leakey, R. (1995). *The Origin of Humankind*. London: Phoenix.

LeVine, R., & LeVine, S. (1988). Mother-child interaction in diverse cultures. Guest lecture, University of Oslo (22 August 1998).

LeVine, R., Dixon, S., LeVine, S., Richman, A., Leiderman, P. H., Keefer, C., & Brazelton, T. B. (1996). *Child Care and Culture. Lessons from Africa*. Cambridge: Cambridge University Press.

Liberman, A. (1957). Some results of research on speech perception. *Journal of Acoustical Society of America, 29*, 117–123.

Liberman, A. (1993). *Haskins Laboratories Status Report on Speech Research, 113*, 1–32.

Mead, G. H. (1934). *Mind, Self, and Society*. Chicago: Chicago University Press.

Meltzoff, A. N. (1988). Infant imitation and memory in immediate and deferred tests. *Child Development, 59*, 217–225.

Meltzoff, A. N. (1995). Understanding the intention of others: Re-enactment of intended acts by 18-month-old children. *Developmental Psychology, 31*, 838–850.

Meltzoff, A., & Moore, M. K. (1998). Infant intersubjectivity: broading the dialogue to include imitation, identity and intention. In S. Bråten (Ed.), *Intersubjective Communication and Emotion in Early Ontogeny* (pp. 47–62). Cambridge: Cambridge University Press.

Ohta, M. (1987). Cognitive disorders of infantile autism: A study emplying the WISC, spatial relationship conceptualization, and gesture imitation. *Journal of Autism and Developmental Disorders, 17*, 45–62.

O'Keefe, J. (1992). Self consciousness and allocentric maps. Paper presented at the King's College Research Centre Workshop on Perception of Subjects and Objects, University of Cambridge, September 17–22, 1992.

Rall, J., & Harris, P. (2000). In Cinderella's slippers? Story comprehension from the protagonist's point of view. *Developmental Psychology, 36*, 202–208.

Rizzolatti, G., & Arbib, M. (1998). Language within our grasp. *Trends in Neurosciences, 21*(5), 188–193.

Rommetveit, R. (1972). *Språk, tanke og kommunikasjon*. Oslo: Universitetsforlaget.

Rommetveit, R. (1974). *On Message Structure*. New York: Wiley.

Rommetveit, R. (1998). Intersubjective attunement and linguistically mediated meaning in discourse. In S. Bråten (Ed.), *Intersubjective Communication and Emotion in Early Ontogeny* (pp. 354–371). Cambridge: Cambridge University Press.

Stern, D. N. (1985). *The Interpersonal World of the Infant*. New York: Basic Books.

Stern, D. N. (1999). Vitality contours: The temporal contour of feelings as a basic unit for constructing the infant's social experience. In P. Rochat (Ed.), *Early Social Cognition* (pp. 67–80). Mahwah, NJ & London: Lawrence Erlbaum.

Tomasello, M., Kruger, A. C., & Ratner, H. H. (1993). Cultural learning. *Behavioral and Brain Sciences, 16*, 495–525.

Trevarthen, C. (1974). Conversations with a two-month-old. *New Scientist, 2*, 230–235.

Trevarthen, C. (1992). An infant's motives for speaking and thinking in the culture. In A. H. Wold (Ed.), *The Dialogical Alternative* (pp. 99–137). Oslo: Scandinavian University Press/Oxford University Press.

Trevarthen, C. (1998). The concept and foundations of infant intersubjectivity. In S. Bråten (Ed.), *Intersubjective Communication and Emotion in Early Ontogeny* (pp. 15–46). Cambridge: Cambridge University Press.

Trevarthen, C., & Hubley, P. (1978). Secondary intersubjectivity: Confidence, confiding, and acts of meaning in the first year. In A. Lock (Ed.), *Action, Gesture, and Symbol* (pp. 183–229). London: Academic Press.

Whiten, A., & Brown, J. D. (1998). Imitation and the reading of other minds. In S. Bråten (Ed.), *Intersubjective Communication and Emotion in Early Ontogeny* (pp. 260–280). Cambridge: Cambridge University Press.

# Visual attention and self-grooming behaviors among four-month-old infants

## Indirect evidence pointing to a developmental role for mirror neurons

Samuel W. Anderson, Marina Koulomzin, Beatrice Beebe, and Joseph Jaffe
New York State Psychiatric Institute and Department of Psychiatry, Columbia University, New York, USA

## 1. Introduction

Oral and manual self-grooming gestures observed in four-month-old infants during the course of mother-infant face-to-face play are found to coincide significantly with prolongation of visual attention to mother's face. It has long been believed that it is the visual presentation of animated facial displays alone that attracts and maintains what is regarded as an essentially passive interest on the part of the infant (Tronick 1989). But although the behavioral repertoire of four-month-olds is small, there is now reason to ask whether these visual displays might elicit from the infant a "motor vocabulary" of prehensile "grasping possibilities" (Gallese et al. 1999), and perhaps other forms of integrated pragmatic activities such as oral exploration of objects and mimicry of the mother's facial displays and manual activities themselves. For example, imitation of tongue protrusion, and perhaps of other facial gestures, is reported among newborns by Meltzoff (1999), consistent with the view, supported by work with both monkeys and human adults, that this kind of binding of visual information with motor action is performed by an innate fronto-parietal circuit that feeds visually derived information to mirror neurons to area F5 of the ventral premotor cortex (Gallese et al. 1996).

Our evidence suggests that episodes of mutual gaze fixation, often occurring while the mother is engaged in kissing, stroking or otherwise manipulating the infant during feeding, caretaking and play, recruit the fronto-parietal system to activate corresponding behavior in the infant. While the exact functional role of mirror neurons in area F5 is not yet clear, we propose that concentration of visual attention to the spatial region of the mother's face is initiated and maintained by premotor intermodally driven neuron ensembles that both recognize the maternal facial displays and grooming activity and enable the infant to enact an approximately equivalent repertoire of mimetic oral and manual "resonance behaviors" that resemble the maternal forms (Rizzolatti et al. 1999; Meltzoff 1999).

## 2.   Method

Mothers were recruited post-partum from a large metropolitan hospital maternity ward in New York City. The infants were firstborns, following medically uneventful pregnancies and delivered without complication within three weeks pre-term and two weeks post-term.

Video recordings of eight mother-infant pairs collected at age four months were chosen for this study by retrospective selection at one year from among a group of 41 dyads taking part in a larger investigation (Koulomzin 1993). The selection criterion was classification as an A-infant by the Ainsworth test for maternal attachment (Ainsworth et al. 1978).

## 3.   Procedure

The mother was instructed to engage her infant in face-to-face "play, as you normally might do at home." Each session lasted for a period of about 10 minutes, during which mother and baby were videotaped by two cameras, one directly facing each partner from opposite sides of the room. The baby was placed in an infant seat about 90 cm from the floor at a horizontal distance of approximately 60 cm from the mother's chair, allowing her to maintain that distance or to lean forward and touch the baby as she might desire. A split-screen image generator combined the signals from the two cameras into a single view, yielding simultaneous frontal displays of mother and infant. For purposes of coding, a second-by-second digital time display was superimposed on the stream of video frames.

The first two minutes of uninterrupted play (during which the face and hands of both mother and infant remained in full view) were coded for each of the 120 seconds of the record generated by the infant according to criteria combined from the Infant Engagement Scale (Beebe & Gerstman 1980) and the Infant Regulatory Scoring System (Tronick & Weinberg 1990). For purposes of this study, only Gaze Direction, Head Orientation and Tactile/Mouthing codes were analyzed.

## 4. Coding of infant behavior

Gaze Direction and Head Orientation codes were conceptualized in two dimensional rotational space, categorized according to three categories of angular rotation in each dimension, expressed as departures from the frontal alignment of mother and infant established by their seating arrangement (see Figure 1). Vertical codes are *Level* (zero degrees), *Up* (+10 degrees) and *Down* (−10 degrees); horizon-

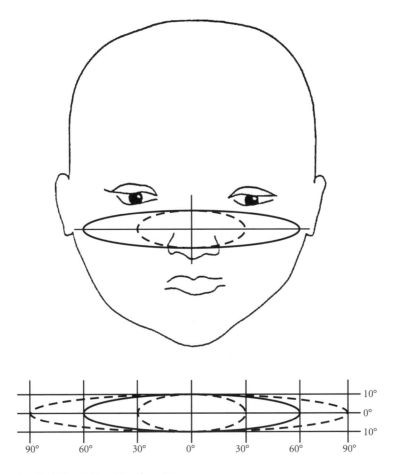

**Figure 1.** Coded rotational head positions.
    Vertical Range:      plus or minus 10 deg.
    Horizontal Range:   plus or minus 90 deg.
    (Solid ellipse identifies range of head positions consistent with sustained frontal attention).

tal codes are *En face* (zero to 30 degrees), *Minor Avert* (right or left between 30 and 60 degrees), and *Major Avert* (right or left between 60 and 90 degrees).

As several of the nine possible combinations of raw gaze and head codes yielded cell frequencies too small for chi-square analysis, they were combined as follows: Gaze *Level En face* was recoded as *Gaze On* mother's face, all other Gaze directions were recoded as *Gaze Off* mother's face; similarly, Head Orientation codes for major and minor avert were combined and analyzed as a single *Avert*, or *Head Off*, category. These recodes yielded a set of 12 possible cells – two for gaze, and three each for horizontal and vertical head orientation (see Figure 1).

*Tactile/Mouthing* behavior was coded as *None*, *Self-directed* (either manual contact with skin or fingering of clothing fabric) or *Other-directed* (either oral or manual contact with mother). In discussing how to interpret these actions, we refer to them without prejudice as "grooming behaviors" to invite interspecies comparisons (see Braten 1998).

## 5.   Data processing methods

It is well known that shifts of visual attention begin with saccadic eye movement, but are completed only after a compensatory head movement follows on, approximately one second later, in accordance with the parameters of an eye/head coordination system specified by Listing's Law (Tweed 1998). Our method makes use of a corollary of Listing's Law: Once a gaze target is achieved and held for two seconds or longer, the system will have established and maintained an *En face* head orientation with respect to that target for as long as it continues to be a focus of interest.

To investigate the consequences of Listing's Law for eye and head movement coordinations observed at one second intervals, the data were modeled as follows:

1.  The time series of accumulating combinations of the 12 gaze/head codes (described above) are cast into the 144 cells of a first-order Markovian transition frequency matrix, which are then normalized to realize a conditional probability matrix for analysis under the assumption that head and eye movements at time t are constrained by their positions just one second earlier (t–1). See Table 1 for an example;

2.  The probability matrix is partitioned into four 6-state submatrices: two "Gaze Stable" matrices (upper left and lower right quadrants) accumulating second-by-second head position data just when gaze remains either on mother or off, respectively, during the 2 second scope identified by each submatrix transition, and two "Gaze Labile" matrices (upper right and lower left quadrants) which track head positioning when eyegaze is shifting either on or off. It should be

noted that each of the partitions preserves a realtime pattern of pairwise sequential events within it just as do the one-step constraints that characterize the transition matrix as a whole.

To test for Listing's Law, corresponding rows of the stable submatrices are compared by the Neyman-Pearson likelihood ratio test, predicting that once the system achieves attentional compliance, continued gazing directly at mother will constrain head movement to no more than minor variations around the *En face* position – a constraint expected to be absent when the infant's attention is directed elsewhere;

3.  The matrix analysis described in 1 and 2 was performed separately for eye/head coordination data collected during active self-grooming (shown in Table 1) and during periods when no such oral or manual activity occurred, repeating the Neyman-Pearson test. Summaries of these data are shown in the form of simplified 4-state frequency tables, in which nonzero cell frequencies predicted by Listing are identified, as seen in Tables 2 and 3.

**Table 1.** Normalized 12-state transition probability model

| | GAZE ON | | | | | | GAZE OFF | | | | | |
|---|---|---|---|---|---|---|---|---|---|---|---|---|
| | EFL | EFU | EFD | AVL | AVU | AVD | EFL | EFU | EFD | AVL | AVU | AVD |
| EFL | .940 | .000 | .003 | .000 | .000 | .000 | .041 | .000 | .003 | .013 | .000 | .000 |
| EFU | .000 | .000 | .000 | .000 | .000 | .000 | .000 | .000 | .000 | .000 | .000 | .000 |
| EFD | .000 | .000 | .778 | .000 | .000 | .000 | .000 | .000 | .222 | .000 | .000 | .000 |
| AVL | .000 | .000 | .000 | .000 | .000 | .000 | .000 | .000 | .000 | .000 | .000 | .000 |
| AVU | .000 | .000 | .000 | .000 | .000 | .000 | .000 | .000 | .000 | .000 | .000 | .000 |
| AVD | .000 | .000 | .000 | .000 | .000 | .000 | .000 | .000 | .000 | .000 | .000 | .000 |
| EFL | .122 | .000 | .000 | .000 | .000 | .000 | .816 | .000 | .041 | .020 | .000 | .000 |
| EFU | .000 | .000 | .000 | .000 | .000 | .000 | .000 | .000 | .000 | .000 | .000 | .000 |
| EFD | .044 | .000 | .011 | .000 | .000 | .000 | .033 | .000 | .912 | .000 | .000 | .000 |
| AVL | .133 | .000 | .000 | .000 | .000 | .000 | .100 | .000 | .067 | .700 | .000 | .000 |
| AVU | .000 | .000 | .000 | .000 | .000 | .000 | .000 | .000 | .000 | .000 | .000 | .000 |
| AVD | .000 | .000 | .000 | .000 | .000 | .000 | .000 | .000 | .000 | .000 | .000 | .000 |

Head Code Abbreviations
EFL = En Face-Level   EFU = En Face-Up   EFD = En Face-Down
AVL = Avert-Level     AVU = Avert-Up     AVD = Avert-Down

**Table 2.** Cumulative transition frequency matrix, self stimulation*

|  | GAZE ON | | GAZE OFF | |
|---|---|---|---|---|
| | H On/On | H On/Off | H On/On | H On/Off |
| **GAZE ON** | <u>299</u> | 1 | <u>13</u> | 5 |
| | H Off/On | H Off/Off | H Off/On | H Off/Off |
| | <u>0</u> | 7 | 0 | 2 |
| | H On/On | H On/Off | H On/On | H On/Off |
| **GAZE OFF** | 12 | 0 | 80 | <u>6</u> |
| | H Off/On | H Off/Off | H Off/On | H Off/Off |
| | 8 | <u>1</u> | 6 | <u>106</u> |

Listing Ratio:   425/546  = 0.778
* Underlined cells predicted to accumulate all entries according to Listing's Law.

## 6.   Results

A comparison of Rows 1 and 7 in the 1st and 4th quadrants of Table 1 yields a significant Neyman-Pearson chi square: = 13.823 (p < .025) . No results were significant apart from the six cell comparisons on just these rows (DF = 5), indicating that gaze-dependent head stability occurs only during head *Level* and in the *En face* orientation, as predicted by the model. This result shows up only during active oral or manual self-stimulation, however. The corresponding chi square computed for eye-head coordination in the absence of oral or manual activity is 9.918, which does not reach significance at the 5% level. Tables 2 and 3 summarize these contrasting effects in eye and head behaviors under the two conditions.

The Listing Ratio – the relative frequency of predicted cell entries – is 0.778 during self grooming activity, while during the absence of oral and manual activity it is 0.640.

**Table 3.** Cumulative transition frequency matrix, no self stimulation*

| | GAZE ON | | GAZE OFF | |
|---|---|---|---|---|
| | H On/On | H On/Off | H On/On | H On/Off |
| **GAZE ON** | <u>72</u> | 0 | <u>12</u> | 4 |
| | H Off/On | H Off/Off | H Off/On | H Off/Off |
| | <u>0</u> | 0 | 0 | 0 |
| | H On/On | H On/Off | H On/On | H On/Off |
| **GAZE OFF** | 12 | 0 | 65 | <u>7</u> |
| | H Off/On | H Off/Off | H Off/On | H Off/Off |
| | 3 | <u>0</u> | 7 | <u>71</u> |

Listing Ratio: 162/253 = 0.640
* Underlined cells predicted to accumulate all entries according to Listing's Law.

## 7. Discussion

Ladavas et al. (1998) have proposed that egocentric visual space is functionally divided into different regions for different behavioral activities in both monkeys and humans. There is evidence that area F5 contains circuits for representing visual space near the face, which she has termed "peripersonal space."

Cross-modal studies in patients with unilateral lesions indicate that some of the F5 neurons have bimodal receptive fields for both visual and tactile inputs such that stimulation in either modality can facilitate or inhibit perceptual extinction, but only provided the stimuli are delivered near the face. The geometric boundaries of peripersonal space have not yet been determined for infants, but it surely includes the region of the baby's hands, at least when they are engaged in grasping objects and moving them for placement in the mouth.

Importantly, the infant's peripersonal space will include the mother's face provided she is positioned within its range. We suggest that our face-to-face placement of mother and infant, at a distance that ranges up to but not beyond 60 cm, places her within that range during much if not all of the recorded play session.

But it should be noted that a distance of 60 cm permits the mother to reach out and touch the infant, but not the converse, a fact that may explain why our grooming effect did not hold for "other directed" manual contact, which requires the mother to lean forward or make other moves not under the infant's control. Such moves make her relatively unavailable to the infant, whereas the infant's own peripersonal space is always available.

Another possibility is that this unidirectional effect, which we have found only among A-infants (Koulomzin 1993), might be due to a tendency among "avoidant" A-infant mothers habitually to spend less time in close tactile proximity, i.e., more time outside of peripersonal space, than do the mothers of "securely attached" B-infants.

## 8.  Conclusion

We have found a significant tendency for infants to maintain focal attention on the mother's active face while performing oral and manual tactile gestures during face-to-face play. This apparent indication of divided attention, from the viewpoint of classical theories of passive visual gaze targeting and active eye-hand coordination, is rejected. Rather, evidence from studies with monkeys and human adults supports an explanation of the finding based on a joint pragmatic and spatial organization among somatomotor and visuomotor neurons within the brain's parieto-frontal circuitry (Fadiga et al. 2000). In this view, such "mirror" neurons respond equally to both to perception and production of pragmatic functions, whether carried out by one person or another, provided the perceptual and productive events occur within an egocentric "peripersonal space".

The fact that eye/head coordination is maximized when a potentially competing task is present rather than when absent is taken as support for the latter view.

## 9.  Summary

We report that self-grooming manual activity among four-month-olds correlates with increased duration of attentive gaze fixation on the mother's face, consistent with Listing's Law (Koulomzin et al. 2002). Eight four-month-old infants were selected from a larger study (by a clinical criterion that is of no interest at present) and were coded second-by-second for infant gaze, head orientation and self-touch/mouthing behavior during face-to-face play with mother.

A comparison of eye-targeting and head motion vis-a-vis mother showed that focused visual attention on mother constrained lateral head movement to within

60 degrees from the line of sight, preserving a stable en face head orientation for 2 seconds or more, consistent with Listing's Law which predicts that steady attention will produce just such a tendency for head orientation to line up with gaze direction. Among these infants it was found that attentive head/gaze coordination was shown to be contingent upon self-touch/mouthing behavior.

It is known that mothers and their very young infants spend much time gazing into each other's faces, during which time oral-to-skin contact and manual grooming of the infant often occur. Episodes of mutual gaze are especially prominent up to the age of 4 months, before which infants exhibit an obligatory, automatic tendency to remain totally gaze-locked on the maternal face (Posner et al. 1998). Regular repetition of these mutual gaze episodes offer ample opportunity for the infant to coordinate the image of the face with her tactile signals of concern and comfort.

If mirror neurons are involved, it is predicted that functional motor structures controlling the mother's manual activities may also be prefigured among neurons in the infant's ventral premotor cortex (Gallese et al. 1996). Indirect evidence for this would be the appearance of maternal grooming patterns – mouthing, stroking, rubbing, etc. – executed by the infants themselves, and later evolving into autonomous resonant motor patterns. Such patterns appear to be what we have observed.

## References

Ainsworth, M., Bleher, M., Waters, E., & Wall, S. (1978). *Patterns of Attachment*. Hillsdale, NJ: Lawrence Erlbaum.

Beebe, B., & Gerstman, L. (1980). The packaging of maternal stimulation. *Merrill-Palmer Quarterly, 26*(4), 321–339.

Bråten, S. (Ed.). (1998). *Intersubjective Communication and Emotion in Early Ontogeny. Studies in emotion and social interaction*, 2nd Series. New York: Cambridge University Press.

Fadiga, L., Fogassi, L., Gallese, V., & Rizzolatti, G. (2000). Visuomotor neurons: ambiguity of the discharge or 'motor' perception? *International Journal of Psychophysiology, 35*(2–3), 165–177.

Fogassi, L., Gallese, V., Fadiga, L., Luppino, G., Matelli, M., & Rizzolatti, G. (1996). Coding of peripersonal space in inferior premotor cortex (Area F4). *Journal of Neurophysiology, 76*, 141–157.

Gallese, V., Fadiga, L., Fogassi, L., & Rizzolatti, G. (1996). Action recognition in the premotor cortex. *Brain, 119*(2), 593–609.

Gallese, V., Craighero, L., Fatiga, L., & Fogassi, L. (1999). Perception through action. *Psyche, 5*(21).

Koulomzin, M. (1993). Attention, affect, self-comfort and subsequent attachment in four-month-old infants. Doctoral Dissertation, Yeshiva University, New York City.

Koulomzin, M., Beebe, B., Anderson, S., Jaffe, J., Feldstein, S. Crown, C. (2002). Infant gaze, head, face and self-touch at four months differentiates secure vs. avoidant attachment at one year. *Attachment and Human Development* (in press).

Ladavas, E., Zeloni, G., & Farne, A. (1998). Visual peripersonal space centered on the face in humans. *Brain, 121*(12), 2317–2326.

Meltzoff, A. (1999). Origins of theory of mind, cognition and communication. *Journal of Communication Disorders, 32*(4), 251–269.

Posner, M. I., & Rothbart, M. K. (1998). Attention, self regulation and consciousness. *Philosophical Transactions of the Royal Society of London – Series B: Biological Sciences, 353*(1377), 1915–1927.

Rizzolatti, G., Fadiga, L., Fogassi, L., & Fadiga, L. (1999). Resonance behaviors and mirror neurons. *Archives Italiennes de Biologie, 137*(2–3), 85–100.

Rizzolatti, G., Fogassi, L., & Gallese, V. (1997). Parietal cortex: From sight to action. *Current Opinion in Neurobiology, 7*(4), 562–567.

Tronick, E. (1989). Emotions and emotional communication in infants. *American Psychologist, 44*(2), 112–119.

Tronick, E., & Weinberg, K. (1990). *Infant Regulatory Scoring System (IRSS).* Boston, MA USA: The Child Development Unit, Children's Hospital.

Tweed, D., Haslwanter, T., & Fetter, M. (1998). Optimizing gaze control in three dimensions. *Science, 281*(5381), 1363–1365.

# The role of mirror neurons in the ontogeny of speech

Marilyn May Vihman
University of Wales, U.K.

## 1. Introduction

It has been suggested that children's own vocal patterns play a key role in the development of segmental representations of adult words (Vihman 1991, 1993; Vihman & DePaolis 2000). The discovery of mirror neurons provides a neurophysiological mechanism for such an 'articulatory filter'. Assuming that only *within repertoire* behaviors can elicit mirror responses, child production of adult-like syllables would be a prerequisite for this kind of matching or filtering. This paper will outline the developmental shift in perception from prosodic to segmental processing over the first year of life and relate that shift to the first major maturational landmark in vocal production, the emergence of canonical syllables. We speculate that it is the activation of the relevant mirror neurons, consequent upon that maturational change, that makes possible the uniquely human shift to segmentally based responses to speech and then to first word production.

## 2. Advances in speech perception: From prosodic to segmental patterns

Over the past decade or so experimental work in infant speech perception has increasingly turned from the early focus on infant capacity for discrimination between speech sounds, whether native or non-native, to attempts to probe advances in familiarity with the ambient language which would imply some kind of longer-term representation for speech. Table 1 provides a summary of those studies. From the division of the table into studies providing evidence of infant knowledge of the *prosody* of speech as opposed to *segmental patterning* it is possible to see a clear developmental trend: For the first six months it is primarily prosodic patterns that

Table 1. Advances in perception and representation of the native language in the first year

| | Child attends more to... | |
| --- | --- | --- |
| | Prosodic patterns | Segmental patterns |
| At birth | ... native language (vs. prosodically dissimilar other language) (Moon et al. 1993) | |
| By 1 mo | ... infant-directed (ID) prosody (vs. adult-directed prosody) (Cooper & Aslin 1990) | |
| By 2 mos | ... native-language narrative passages and short utterances (vs. prosodically dissimilar other language) (Mehler et al. 1989) | |
| By 4 mos | ... 'coincident' clauses (vs. non-coincident clauses (Jusczyk & Kemler Nelson 1996) – but not phrases or words (Jusczyk et al. 1992; Myers et al. 1996) | ... own name (vs. other name: Mandel, Jusczyk, & Pisoni 1995) |
| By 6 mos | ... word list in native language (vs. prosodically dissimilar language, even when low-pass filtered) (Jusczyk et al. 1993) | ... family words for *mama* and *papa*, with matching video (Tincoff & Jusczyk 1999) |
| Between 6 and 8 months | Emergence of canonical (CV) syllables in production, with adult-like timing (Oller 1980; Stark 1980; Lindblom & Zetterström 1986) | |
| By 8 mos | | ... monosyllabic word forms previously trained through narrative passages or word lists (Jusczyk & Aslin 1995) |
| By 9 mos | ... native language stress pattern (even when low-pass filtered) (Jusczyk et al. 1993)<br>... uninterrrupted phrases [even when low-pass filtered] (Jusczyk et al. 1992), but not words (Myers et al. 1996) | ... native language phonotactics (Friederici & Wessels 1993) |
| By 10 mos | | [Fail to discriminate non-native consonant contrasts (Werker & Tees 1984)] |

**Table 1.** (*continued*)

| | Child attends more to… | |
|---|---|---|
| | **Prosodic patterns** | **Segmental patterns** |
| By 11 mos | | … word forms familiar from everyday experience (Hallé & Boysson-Bardies 1994) |
| | | … uninterrupted words [but not when low-pass filtered] (Myers et al. 1996) |
| | | … familiar words [even with a *reversal* of the accentual pattern – but *not* with change to the onset C of the accented syllable] (Vihman et al. 2000) |

underlie a familiarity response to speech. Exceptional evidence of very early (holistic) response to segmental patterns involves stimuli that can be assumed to be imbued with strong affect or 'personal relevance' for infants (e.g., the infant's own name: Mandel, Jusczyk, & Pisoni 1995, and family terms for 'mama' and 'papa': Tincoff & Jusczyk 1999; for elaboration of the notion of 'personal relevance', see Van Lancker 1991).

Infants' apparently greater early memory for prosodic patterns follows naturally from the fact that the foetus gains linguistic experience already in the womb, through hearing the sound of the mother's voice both 'internally' and from the outside, as filtered through the amniotic fluid (DeCasper & Fifer 1980; Querleu, Renard, & Versyp 1981; Hepper, Scott, & Shahidullah 1993). The prosodic information present in the lower frequency bands of the signal can reach foetal ears, once the auditory system is completely formed (by the final trimester of pregnancy), while segmental information, much of which is carried by higher frequencies, cannot. In a series of studies Fernald has shown that prosody must indeed provide the initial entry into language, due not only to its preestablishment as an acoustic signal before birth but also to the intrinsic affective links between particular prosodic patterns and communicative meanings (Fernald 1989, 1992).

With regard to segmental patterning we first see attention to the child's own name (as early as 4 months) and association of parent terms with the appropriate parent (by 6 months), as noted above. Aside from these exceptional word forms, preferential attention to a (trained) segmental pattern is not reported until 7.5 months of age (Jusczyk & Aslin 1995). Thereafter, steady gains in attention to segmental patterning can be seen (e.g., preference for native language phonotactics: Friederici & Wessels 1993, and a narrowing of attention to consonantal contrasts

from broadly 'universal' to native-language only: Werker & Tees 1984; Best 1994).
To complete this picture, Myers, Jusczyk, Kemler Nelson, Charles-Luce, Woodward and Hirsh-Pacek (1996) showed that at 11 months, but not at 4.5 or even 9
months, infants looked longer to 'coincident' passages, which had brief pauses inserted only *between* words, than to 'non-coincident' passages, which included two-
or three-syllable words *interrupted* by such pauses. Unlike earlier studies which
demonstrated greater attention to coherent vs. interrupted units (whether clauses
or phrases) even when the stimuli were low-pass filtered to remove segmental information (Kemler Nelson, Hirsh-Pasek, Jusczyk, & Wright Cassidy 1989; Jusczyk
& Kemler Nelson 1996; Jusczyk et al. 1992), the coherent-word effect was not
obtained when only prosodic information was available.

## 3.   First perceptual representations of speech forms

Despite intensive experimental work on infant responses to speech for over twenty
years, Hallé and Boysson-Bardies (1994) was the first study to examine infant responses to *untrained* speech forms. These authors tested 11-month-old French infants on word patterns expected to be familiar from everyday exposure. The words
were chosen from those produced early in the second year by French infants in an
earlier study and were matched with phonetically similar words of low frequency
in the adult lexicon. Exposed to test lists of 12 words of each kind in the head-turn
preference technique, infants were found to attend longer to the 'familiar' than to
the 'rare' words. In a follow up study, Hallé and Boysson-Bardies (1996) explored
the phonetic basis for the familiarity effect by removing the initial consonant (the
familiarity effect was eliminated), changing voicing or manner of the initial consonant (the familiarity effect was observed), or changing manner of the second
consonant (the effect just failed to reach significance).

Since the accentual pattern of French (iambic, or weak-strong, based primarily on lengthening of word- or phrase-final syllables) is the opposite of the dominant stress pattern found in English content words (trochaic, or strong-weak), we
sought to replicate the French results with infants exposed to British English (Vihman, Nakai, & DePaolis 2000). In a base-line experiment we used lists including
seven trochaic words and five iambic words or phrases (e.g., 'familiar' *apple, baby, a
ball, fall down*, vs. 'rare' *bridle, maiden, a bine, taboo*), the familiar words taken from
previous studies of early word production in children acquiring English. While 11-
month-olds were found to attend longer to the familiar words ($p < .05$), 9-month
olds failed to show a significant difference between listening times to the two lists.

In a second experiment we sought to establish the role of prosody in the familiarity effect for English-learning children by contrasting lists with the unaltered

familiar words versus the same familiar words under altered stress (e.g., **baby** > baby). No significant difference was found, suggesting that the stress pattern did not constitute an essential part of the infants' lexical representations. We validated this finding in an additional experiment in which we contrasted the list of altered familiar words used in the previous experiment with a list of altered *unfamiliar* words; under these conditions the former received significantly longer looks (p < .001), again demonstrating that the infants could 'listen through' the stress pattern to recognize familiar words.

A final pair of experiments was designed to test whether accented syllables are more fully specified in infant word representations than unaccented syllables. We predicted that there would be a difference between infants' representation of early French words, which are iambic, and early English words, which are mainly trochaic. We therefore presented English-learning infants with (trochaic) lists of rare words in contrast with familiar words with, first, a change in manner of articulation affecting all medial consonants (e.g., *bubbles* > *bummles*, *piggy* > *pingy*) or, in a second experiment, all initial consonants (*mubbles*, *figgy*). As anticipated, the infants attended longer to the familiar words despite the change in the second consonant (p < .01) but failed to show significantly longer looking times to the familiar words with changed initial consonants, suggesting that the familiar words were no longer recognizable in that condition. We concluded that the accentual pattern of the adult language may influence which details are noted in early word representations, changes to the initial syllable blocking word recognition in English but not in French. In summary, these experiments suggest that the first representations for words and phrases are influenced by prosodic patterning (hence the ambient-language differences in infant responses to changes in initial vs. second syllable onset-consonant) yet by 11 months prosody itself constitutes a less essential property of these representations than segmental patterning (as found in our stress-change experiment).

## 4.  The articulatory filter and mirror neurons

A plausible source for the shift from a largely prosodic to a primarily segmental basis for attention to speech patterns can be found in the developmental milestones for production in the first year. In hearing infants canonical babbling, or the rhythmic production of consonant-vowel (CV) sequences with adult-like timing, is reliably reported by parents and confirmed by laboratory recordings to occur at about 6–8 months of age (Oller 1980; Stark 1980; Lindblom & Zetterstrom 1986). The coincidence in timing of the production milestone with the shift to first perceptual responses suggestive of segmental representation is striking. Unfortunately, however, infant speech perception studies to date have included neither canonical babbling status for infant participants nor individual perceptual responses, so we lack so much as a correlational study showing the onset of CV production in relation to changes in attention to segmental speech patterns in perception.

The emergence of easily recognized babbled syllables with adult-like timing in the middle of the first year appears to be maturationally based and fits into a broader framework of rhythmic motoric advances that occur around that age (Thelen 1981). One interpretation of the developmental match between the shift to attention to segments and the onset of CV production is the articulatory filter hypothesis (Vihman 1993). On this account, the experience of frequently producing CV syllables sensitizes infants to similar patterns in the input speech stream (note that deaf infants fail to persevere in CV syllable production at the typical age: Oller & Eilers 1988). As in the 'cocktail party' effect produced in adults when their own name occurs in an unattended conversation (Wood & Cowan 1995), particular segmental patterns would now begin to 'pop out' of input speech which previously might have constituted only 'background music' for the infant listener. Infants could be expected to differ in their sensitivity to the putative 'match' of own production patterns to adult input; presumably, the process would not be instantaneous but cumulative, leading eventually to the best-represented adult patterns – those closely resembling the child's own most typical production patterns – forming the basis for first words in expressive infants. Most children could be expected to show some influence of their own incipient adultlike syllable production on their attention to speech, a proposition currently being tested in our lab.

The mirror neuron findings provide unanticipated neurophysiological support for this speculative idea. In the course of making single-cell recordings of the premotor cortex in monkeys di Pellegrino, Fadiga, Fogassi, Gallese and Rizzolatti (1992) discovered that "when the monkey observes a motor action that is *present in its natural movement repertoire,* this action is automatically covertly retrieved" (Fadiga, Fogassi, Pavesi, & Rizzolatti 1995: 2608; my emphasis). Fadiga et al. (1995) provide indirect neurological evidence that "in humans [too] there is a neural system matching action observation and execution... The observation of an action automatically recruits neurons that would normally be active when the subject executes that action" (p. 2609). Practice in performing a particular motor routine (e.g., producing CV syllables) lays the groundwork for the activation of the same motor neurons when similar routines (e.g., adult word forms similar to the infant's babbling patterns) are *produced by others.*

This account provides a natural mechanism for imitation of within-repertoire motor behaviors. Thus, some (but not all) 4–5 month-old infants in an auditory/visual matching experiment involving isolated vowels spontaneously imitated the vowels (Kuhl & Meltzoff 1988). Kuhl and Meltzoff proposed that "infants make an intramodal auditory-auditory match; and second, they develop a set of auditory-articulatory mapping rules" (p. 254). The mirror neurons provide a more direct neuromotor mechanism for effecting imitation as a by-product of attending to vowels at 4–5 months. The later emergence of speech-like CV patterns in an infant's vocal repertoire would then provide the basis for the more sophisticated capacity to pay privileged attention to and represent or remember the particular speech forms of the ambient language. The critical point is the requirement that a movement be present in the individual's natural repertoire before the mirror system can effect a match of observed to potential action patterns. By this account, it is only after the individual infant's neurological system has been prepared by the onset of rhythmic babbling that the mirror system relating heard patterns to potential production patterns can begin to function to highlight a subset of the patterns embedded in the fast-changing input speech signal.

## 5.    First word production as the product of an articulatory filter

It has long been recognized that the first words tend to be relatively accurate, arguably due to selection on phonological grounds (Ferguson & Farwell 1975; Schwartz 1988), and that they resemble babbling patterns, both generally and for individual children (Vihman, Macken, Miller, Simmons, & Miller 1985). Analyses of later word forms indicate that a given child's first well-practiced, consistent supraglottal production patterns ('vocal motor schemes': McCune & Vihman 1987) provide the basis for the later development of 'word templates' (Vihman & Velleman 2000). These word templates abstract from and extend the piecemeal learning evidenced by the selection patterns of first words (Table 2).

The apparent paradox of such early word selection – how do children know which words not to attempt, or which sounds they cannot yet produce? (Stemberger & Bernhardt 1999) – does not arise if we assume that the first words result from infant matching of own vocal patterns to the input speech signal. The evidence from neurophysiology that a mirror system may mediate perception-action links, imitation, and learning, although still speculative, places the notion of an articulatory filter in the first year of life on firmer ground. More conclusive evidence will have to come from ongoing direct empirical research with infants.

**Table 2.** Relationships between adult and child word forms (Alice)

| 9–10 Months | |
| --- | --- |
| Adult Target | Selected child forms |
| *baby* /beɪbi/ | [pɛpɛː] |
| *daddy* /dædi/ | [dæ] |
| *hi* /haɪ/ | [haːi] |
| *mommy* /mami/ | [mːanːə] |
| *no* /noʊ/ | [njæ] |

| 14 Months | | |
| --- | --- | --- |
| Template schema <CVCi> | | |
| Adult Target | Selected child forms | Adapted child forms |
| *baby* /beɪbi/ | [bebi] | |
| *bottle* /baɾəl/ | | [baɹi] |
| *daddy* /dædi/ | [tæɹi] | |
| *hiya* /haɪ/ | | [haːji] |
| *lady* /leɪɾi/ | [jɛiji] | |
| *mommy* /mami/ | [maːɲi] | |

Alice's first spontaneous words, recorded at 9–10 months, are listed here in full. The adult targets for these first words suggest 'selection on phonological grounds': Note that three of the five are disyllables including a single repeated stop or nasal and ending in the vowel /i/. In contrast, some of the words that Alice produced at 14 months fit the template schema she has now evolved (these are the '[pre-]selected' word forms) while others are adapted to fit the schema.

## References

Best, C. T. (1994). The emergence of language-specific phonemic influences in infant speech perception. In J. C. Goodman, & H. C. Nusbaum (Eds.), *The Development of Speech Perception* (pp. 167–234). Cambridge, MA: MIT Press.

Cooper, R. P., & Aslin, R. N. (1990). Preference for infant-directed speech in the first month after birth. *Child Development, 61,* 1584–1595.

DeCasper, A. J., & Fifer, W. P. (1980). Of human bonding. *Science, 208,* 1174–1176.

di Pellegrino, G., Fadiga, L., Fogassi, L., Gallese, V., & Rizzolatti, G. (1992). Understanding motor events. *Experimental Brain Research, 91,* 176–180.

Fadiga, L., Fogassi, L., Pavesi, G., & Rizzolatti, G. (1995). Motor facilitation during action observation. *Journal of Neurophysiology, 73,* 2608–2609.

Ferguson, C. A., & Farwell, C. B. (1975). Words and sounds in early language acquisition. *Language, 51,* 419–439.

Fernald, A. (1989). Intonation and communicative intent in mothers' speech to infants. *Child Development, 60,* 1497–1510.

Fernald, A. (1992). Human maternal vocalizations to infants as biologically relevant signals. In J. H. Barkow, L. Cosmides, & J. Tooby (Eds.), *The Adapted Mind* (pp. 391–428). Oxford: Oxford University Press.

Friederici, A. D., & Wessels, J. M. I. (1993). Phonotactic knowledge of word boundaries and its use in infant speech perception. *Perception & Psychophysics, 54,* 287–295.

Hallé, P., & de Boysson-Bardies, B. (1994). Emergence of an early lexicon. *Infant Behavior and Development, 17,* 119–129.

Hallé, P., & de Boysson-Bardies, B. (1996). The format of representation of recognized words in infants' early receptive lexicon. *Infant Behavior and Development, 19,* 435–451.

Hepper, P. G., Scott, D., & Shahidullah, S. (1993). Newborn and fetal response to maternal voice. *Journal of Reproductive and Infant Psychology, 11,* 147–153.

Jusczyk, P. W., & Aslin, R. N. (1995). Infants' detection of the sound patterns of words in fluent speech. *Cognitive Psychology, 29,* 1–23.

Jusczyk, P. W., Cutler, A., & Redanz, N. J. (1993). Infants' preference for the predominant stress patterns of English words. *Child Development, 64,* 675–687.

Jusczyk, P. W., & Kemler Nelson, D. G. (1996). Syntactic units, prosody, and psychological reality during infancy. In J. L. Morgan, & K. D. Demuth (Eds.), *Signal to Syntax* (pp. 389–408). Hillsdale, NJ: Lawrence Erlbaum Associates.

Jusczyk, P. W., Kemler Nelson, D. G., Hirsh-Pasek, K., Kennedy, L. J., Woodward, A., & Piwoz, J. (1992). Perception of acoustic correlates of major phrasal units by young infants. *Cognitive Psychology, 24,* 252–293.

Kemler N., Deborah G., Kathy Hirsh-Pasek, Peter W. Jusczyk, & Kimberly Wright Cassidy (1989). How the prosodic cues in motherese might assist language learning. *Journal of Child Language, 16,* 55–68.

Kuhl, P. K., & Meltzoff, A. N. (1988). Speech as an intermodal object of perception. In A. Yonas (Ed.), *Perceptual Development in Infancy. The Minnesota Symposia on Child Psychology, 20* (235–266). Hillsdale, NJ: Lawrence Erlbaum.

Lindblom, B., & Zetterström, R. (Eds.). (1986). *Precursors of Early Speech.* Basingstoke, Hampshire: Macmillan Press.

Mandel, D. R., Jusczyk, P. W., & Pisoni, D. B. (1995). Infants' recognition of the sound pattern of their own names. *Psychological Science, 6,* 315–318.

McCune, L., & Vihman, M. M. (1987). Vocal motor schemes. *Papers and Reports on Child Language Development, 26,* 72–79.

Mehler, J., Jusczyk, P. W., Lambertz, G., Halsted, N., Bertoncini, J., & Amiel-Tison, C. (1988). A precursor of language acquisition in young infants. *Cognition, 29,* 143–178.

Moon, C., Panneton-Cooper, R., & Fifer, W. P. (1993). Two-day-olds prefer their native language. *Infant Behavior and Development, 16,* 495–500.

Myers, J., Jusczyk, P. W., Kemler Nelson, D. G., Charles-Luce, J., Woodward, A. L., & Hirsh-Pasek, K. (1996). Infants' sensitivity to word boundaries in fluent speech. *Journal of Child Language, 23,* 1–30.

Oller, D. K. (1980). The emergence of the sounds of speech in infancy. In G. Yenikomshian, J. F. Kavanagh, & C. A. Ferguson (Eds.), *Child Phonology, I: Production* (pp. 93–112). New York: Academic Press.

Oller, D. K., & Eilers, R. E. (1988). The role of audition in infant babbling. *Child Development, 59,* 441–449.

Querleu, D., Renard, X., Versyp, F., Paris-Delrue, L., & Crépin, G. (1988). Fetal hearing. *European Journal of Obstetrics and Reproductive Biology, 29,* 191–212.

Schwartz, R. G. (1988). Phonological factors in early lexical acquisition. In M. D. Smith, & J. L. Locke (Eds.), *The Emergent Lexicon* (pp. 185–222). New York: Academic Press.

Stark, R. E. (1980). Stages of speech development in the first year of life. In G. Yenikomshian, J. F. Kavanagh, & C. A. Ferguson (Eds.), *Child Phonology,* Vol. 1: *Production* (pp. 73–92). New York: Academic Press.

Stemberger, J. P., & Bernhardt, B. H. (1999). The emergence of faithfulness. In B. MacWhinney (Ed.), *The Emergence of Language* (pp. 417–446). Mahwah, NJ: Lawrence Erlbaum Associates.

Thelen, E. (1981). Rhythmic behavior in infancy. *Developmental Psychology, 17,* 237–257.

Tincoff, R., & Jusczyk, P. W. (1999). Some beginnings of word comprehension in 6-month-olds. *Psychological Science, 10,* 172–175.

Van Lancker, D. (1991). Personal relevance and the human right hemisphere. *Brain and Cognition, 17,* 64–92.

Vihman, M. M. (1991). Ontogeny of phonetic gestures. In I. G. Mattingly & M. Studdert-Kennedy (Eds.), *Modularity and the Motor Theory of Speech Perception* (pp. 69–84). Hillsdale, NJ: Lawrence Erlbaum Associates.

Vihman, M. M. (1993). Variable paths to early word production. *Journal of Phonetics, 21,* 61–82.

Vihman, M. M., & DePaolis, R. A. (2000). The role of mimesis in infant language development. In C. Knight, J. Hurford, & M. Studdert Kennedy (Eds.), *The Evolutionary Emergence of Language.* Cambridge: Cambridge University Press.

Vihman, M. M., Macken, M. A., Miller, R., Simmons, H., & Miller, J. (1985). From babbling to speech. *Language, 61,* 395–443.

Vihman, M. M., Nakai, S., & DePaolis, R. A. (2000). The role of accentual pattern in early lexical representation. Poster presented at 12th International conference on Infant Studies, Brighton, UK. July.

Vihman, M. M., & Velleman, S. L. (2000). The construction of a first phonology. *Phonetica, 57,* 255–266.

Werker, J. F., & Tees, R. C. (1984). Cross-language speech perception. *Infant Behavior and Development, 7,* 49–63.

Wood, N., & Cowan, N. (1995). The cocktail party phenomenon revisited. *Journal of Experimental Psychology: Learning, Memory and Cognition, 21,* 255–260.

# Mirror neurons' registration of biological motion

## A resource for evolution of communication and cognitive/linguistic meaning

Loraine McCune

Rutgers University, New Brunswick, NJ, USA

## 1. Mirror neurons' registration of biological motion: A resource for evolution of communication and cognitive/linguistic meaning

Language is a symbolic or representational process. Meanings in a language map the world mentally and are shared among a linguistic community through some external medium. A major developmental and evolutionary puzzle is determination of the processes through which internal meanings integrated with reality emerged historically, develop in children, and are successfully communicated among modern adult humans. A critical constraint on any developmental or evolutionary proposal is the need for demonstrable physiological processes of the brain underlying developmental and evolutionary processes of change. The mirror neuron discoveries may be a missing link contributing to the solution of this puzzle (Rizzolatti & Arbib 1998).

Language is considered here in the context of Searle's (1992) view of consciousness. Comprehending language instantiates a conscious contentful mental state, while production of language is an aspect of a conscious contentful mental state. This view allows for cross-species and developmental analyses of the qualities of such states which are needed to support language (McCune 1999). In the evolution of language pre-human hominids and early humans needed to achieve the capacity for representational consciousness, such that meanings might be symbolized and expressed in some external medium. Human infants face these same challenges with a difference: the prior existence of an ambient language. In humans the capacity for conscious mental states undergoes a transition from limitation to the here and now (perceptual) to possible consideration of past, future, and counterfactual, (imaginal, representational) states, a distinction described by Sartre (1948) and noted as a developmental transition by Piaget (1946/1962). Apes, unlike monkeys exhibit mental representation in cognitive tasks, and have been capable of rudimentary linguistic tasks, except for language production (McCune 1999; McCune & Agayoff 2002), suggesting the existence of such representational communicative capacities in common ancestors of apes and humans.

## 2.    Action, mental representation and neurological development

Children's development of language occurs in tandem with milestones in representational play that have been shown to progress from pre-symbolic recognition of object meanings around 9 months of age, through simple pretend actions and combinations between 13 and 18 months of age, to hierarchical pretending between 18 and 24 months of age (McCune 1995). Comparisons of brain representations, showing distinctive activation but significant overlap for enacted versus imagined actions suggests a developmental trend whereby overt actions can come to be imagined (Grafton, Arbib & Rizzolatti 1996). As children observe their own manipulations of objects during the first year of life opportunities occur for calibrating neural recognition of both object properties and spatial and temporal conditions. Activation of the mirror system in Broca's area might facilitate recognition of the generality of self and other action, providing the neurological basis for a sensorimotor logic of action (Rizzolatti et al. 1996). The imaginal capacity would then develop in relation to an internal structure based upon established sequences of movement in space and time, forming a fertile ground for the acquisition of linguistic meanings.

During the first three years of life children's play interactions with objects and people follow a timetable that is dependent on and contributes to the developmental trends in brain maturation. Sensory areas develop earliest, with vision and hearing functional even before birth, showing rapid physiological development at 3–4 months, reaching a neuronal density 150% of adult levels between 4 and 12 months, and returning to adult levels at 2 to 4 years. The motor areas, including the oral motor come under control gradually with rhythmicities including babbling and banging toys beginning at 5 to 7 months and fine motor manipulations, allowing manual exploration of objects at 6 to 10 months. Gross motor control increases rapidly from 6 to 12 months and beyond, supporting explorations in larger space. If the mirror system, as shown in monkeys, is reactive to these activities a perceptual and motor basis for later representational meaning would be established. Finally, rapid development of the frontal cortex, supporting mental representational functioning (Bell & Fox 1992) reaches peak development after the first year, and Broca's area shows greater dendritic branching than the area for oral motor control for the first time between 1 and 2 years of age, suggesting the onset of specifically linguistic processing (Johnson 1997). These final twin developments, involving areas implicated in the mirror neuron research and in the brain activation of imagined actions occur during the period of onset for representational play and language. No doubt mutual influence between behavioral and neurodevelopmental processes supports the development of both.

## 3.   From autonomic laryngeal vocalization to learning words

Even in language-trained chimpanzees producing oral language elements has proved elusive. However, communicative use of laryngeal vocalization characterizes all primate species, including humans. Recent discovery of a developmental sequence for grunt vocalizations in human infants highlights the role of the larynx in communicative production and development. McCune, Vihman, Roug-Hellichius, Delery and Gogate (1996) found that children produced grunts of effort in the earliest observation sessions at 9 months, followed by grunts accompanying focussed attention in subsequent sessions, and finally used grunts communicatively in the month prior to referential production or comprehension of language used referentially (13–16 months). The larynx serves a critical role in maintaining oxygen levels in the blood, and grunts are associated with metabolic demand in mammal species from rats to horses. Critical for vocal communication, motor unit activation in the laryngeal muscles is evident immediately before vocalization, suggesting the larynx is responsive to the intention to vocalize (Buchtal & Faaburg-Anderson 1964; Kirchner 1987). Communicative grunts with varied functions are common across primate species: all include movement prediction and conspecific acknowledgement, suggesting a basis in metabolic demand (McCune et al. 1996).

McCune (1999) proposed this laryngeal vocalization as a transitional behavior in the shift from pre-linguistic to linguistic communication in human infants. In the context of mental representational ability, communicative grunts seek to convey the child's experienced prelinguistic conscious state to others. The experience of communicative grunts in meaningful contexts then prompts recognition of the relevance of more specific vocal forms of the ambient language. The child then begins to constellate more differentiated states of meaningful consciousness in relation to adult words. Because of the physiological basis of this process it can recur across generations without specific cultural transmission and could have played a prominent role in the evolution of language. A conspecific might recognize the significance of the vocalization via interpretation through its own physiology (Darwin 1872/1965). Laryngeal representation in the mirror system could provide a mechanism for such recognition.

An analogous developmental process has been observed in vervet monkeys and chimpanzees. The former produce grunts of effort and learn appropriate use of the grunt form predicting physical movement prior to appropriate use of the four species-typical alarm calls. Infant chimpanzees in the wild exhibit communicative grunts prior to development of begging gestures and tickling requests. In both species the significance of specific grunt forms are learned over time. Chimpanzees exposed to nonvocal linguistic systems in laboratories produce communicative grunts within these linguistic settings (McCune 1999).

The centrality of Broca's area to both language and the mirror system suggests that vocal interpretation/production of elements of a language benefit from the mirror system (Vihman, in this volume). Iacoboni, Woods, Brass, Bekkering, Mazziotta and Rizolatti (1999) found that the "left frontal operculum (area 44) and the right anterior parietal cortex (PE/PC) have an imitation mechanism" (p. 2527), which should facilitate immediate production of observed actions. Once the relevance of words in the ambient language is recognized, this system may facilitate acquisition. For single words at the transition to language McCune-Nicolich and Raph (1978) found that vocal imitation was progressive: words shifted month by month from occurring only as imitations to free production. Vocal imitation showed a curvilinear trend, with representational development, infrequent at first, and increasing with representational play level until such play became internally mediated, at which point imitation rapidly declined. This result can be understood as a shift from the external imitation reaction demonstrated by Iacoboni et al. (1999) to the capacity to "image" words heard without external practice supported by mechanisms demonstrated by Grafton et al. (1996).

Development of a representational consciousness allows the emergence of meaning, at first in relation to a simple vocal signal, the grunt, which accompanies attended actions, and subsequently in relation to words of the ambient language. Learning to produce the sounds of the language (Vihman 1996, in this volume), and to acquire unfamiliar words by imitation would be facilitated by the mirror system. In addition, broader aspects of meaning regarding space and motion result from young children's observation of their own and others' manipulation of objects and movements in space. The mirror neuron findings point to a neurological basis for a fundamental mapping of meaningful spatial categories through motion during human infants' second year of life.

## 4.  Relational words

Motion detection and interpretation has long been recognized as critical for survival across species, as well as highly salient for infant learning. Infants have a special interest in manipulating objects in space by rotation, tracing object boundaries with a finger, banging objects repeatedly on surfaces, and later exploring potential spatial relationships between objects as these occur over time (Ruff 1982; Sinclair et al. 1989). Their gross motor activities also contribute to perceptual and cognitive learning (Campos et al. 2000; Kermoian & Campos 1988). The new work demonstrating initial functions attributable to mirror neurons and "canonical neurons" located in a traditional language region of the brain provides a neurological mechanism by which children's spatial/temporal knowledge could be derived from observed movements, those of both the infants' own hands and those of other humans, and contribute to the emergence of linguistic categories exposed in the ambient language.

Despite historical research emphasis on syntax, children's first words do not include ordinary action verbs of predication. Verbs are rare in the single word period, with the few observed occurring in accompaniment to the child's own action. When verbs are included in early combinations they are likely to be generic transitive verbs (want, get, eat, drink, make) initially restricted to the child-speaker as implied subject (Ninio 1999). Children's initial single words refer to (a) manipulable objects, (b) biological objects such as pets, parents, and other people, and (c) relational events involving movement in space and time. The latter category, relational words, constellate linguistic meanings interpretable on the basis of a sensorimotor logic of space, movement and time derived from the overt activities of the first year of life. It is the use of these words which may bridge the transition from pre-linguistic transitive meanings in movement and action to eventual syntactic expression.

Talmy (1975) proposed that humans' universal common experience of motion events might form the basis of syntactic understanding and he provided semantic/syntactic analyses conceived by parsing sentences into common semantic constituents of motion. Relational word use expresses, with a single word, a single aspect of a motion event as described in adult sentences. A motion event is considered a "situation containing movement of an entity or maintenance of an entity at a static location", where "movement" includes directed or translative motion that results in a change of location, and "location" includes either static location or a contained movement that results in no overall change in location, such as jumping up and down (Talmy 1985: 60; Choi & Bowerman 1991: 85). The semantic components characterizing motion events expressed by verbs and their associated particles in sentences of adult languages include movement, figure (the moved object), ground (with respect to which movement occurs), path (the direction of motion in relation to the ground), deixis (direction with respect to the speaker), manner (e.g., *walked, ran, rolled* vs. *moved*) and cause (e.g., transitive vs. intransitive) (Talmy 1975).

Motion events also provide the reference point for the initial relational meanings reported for young children. They, along with generic transitive verbs, form the core of children's initial syntactic combinations. The set of relational meanings originally proposed by McCune-Nicolich (1981) can be incorporated into three superordinate categories reflecting the semantics of motion events (McCune, in preparation).

Spatial Direction or Path in the vertical plane (e.g., *up, down*) or the "deictic plane", involves movement in relation to the child's body (*here* used in exchange; *there* accompanying placing actions). Gravity is a defining property of all life on earth and infants' interaction with objects is affected both as they calibrate body movements in lifting and lowering both objects and themselves and when objects drop to the floor. Development of self/other relationships and turn-taking games by one-year-olds highlights movement toward and away from the self. Both of these situations of temporal and spatial reversibility are highly salient to children, leading to the need for words to mark them.

Spatial Relations between Entities (Figure/Ground) are also the focus of early relational words, including reversible aspects of containment (*open, closed, in, out*), attachment (*stuck*) and occlusion (*allgone*). These relationships are experienced by children as they are carried about and objects move into and out of their view. Young children play with these changes in opening and closing, dumping and filling containers, and putting together and taking apart puzzles.

Temporal Event Sequences are coded with relational words indicating a mental comparison of the ongoing state of affairs to a prior, expected, or desired reverse alternative: iteration or conjunction (*more, again*) and potential reversibility or negation (*no, uhoh, back*). The use of these words links the critical linguistic notions of conjunction and negation with prior non-linguistic capacities.

The primacy of these relational meanings is confirmed in the findings of a large body of research from diverse languages including English, Estonian, German and Korean, indicating a consistent semantic space influenced by phonetic and semantic characteristics of the ambient language (McCune & Vihman 1999). Against the backdrop of a capacity for mental representation and vocal expression of meaning the development of these linguistic elements, perhaps supported by the mirror neuron system, may prove the link between single words and syntax.

# References

Bell, M. A., & Fox, N. A. (1992). The relations between frontal brain electrical activity and cognitive development during infancy. *Child Development, 63,* 1142–1163.

Buchtal, F., & Faaborg-Anderson, K. L. (1964). Electromyography of laryngeal and respiratory muscles. *Annals of Otology, Rhinology and Laryngology, 73,* 18–121.

Campos, J., Anderson, D. I., Barbu-Roth, M. A., Hubbard, E. M., Hertenstein, M. J., & Witherington, D. (2000). Travel broadens the mind. *Infancy, 1,* 149–220.

Choi, S., & Bowerman, M. (1991). Learning to express motion events in English and Korean: The influence of language-specific lexical patterns. *Cognition, 41,* 83–121.

Darwin, C. (1872/1965). *The Expression of the Emotions in Man and Animals.* Chicago: University of Chicago Press.

Grafton, S. T., Arbib, L., & Rizzolatti, G. (1996). Localization of grasp representations in humans by positron emission tomography 2. Observation compared with imagination. *Experimental Brain Research, 112,* 103–111.

Iacoboni, M., Woods, R. P., Brass, M., Bekkering, H., Mazziotta, J. C., & Rizolatti, G. (1999). Cortical mechanisms of human imitation. *Science, 286,* 2526–2528.

Johnson, M. H. (1997). *Developmental Cognitive Neuroscience.* Cambridge, MA: Blackwell.

Kermoian, R., & Campos, J. J. (1988). Locomotor experience: A facilitator of spatial cognitive development. *Child Development, 59,* 908–917.

Kirchner, J. A. (1987). Laryngeal reflex systems. In T. Baer, C. Sasaki, & K. Harris (Eds.), *Laryngeal Function in Phonation and Respiration* (pp. 65–70). Boston: Little, Brown and Company.

McCune, L. (1995). A normative study of representational play at the transition to language. *Developmental Psychology, 31,* 198–206.

McCune, L. (1999). Children's transition to language: A human model for development of the vocal repertoire in extant and ancestral primate species? In B. J. King (Ed.), *Origins of Language: What can nonhuman primates tell us?* (pp. 269–306). Santa Fe: SAR Press.

McCune, L. (in preparation). *Becoming Linguistic: A dynamic systems view.*

McCune, L., & Agayoff, J. (2002). Pretending as representation: A developmental and comparative view. In R. Mitchell (Ed.), *Pretending in Animals and Children* (pp. 43–55). Cambridge: Cambridge University Press.

McCune, L., & Vihman, M. M. (1999). Relational words + motion events: A universal bootstrap to syntax? Poster presentation at the biennial meeting of the Society for Research in Child Development, Albuquerque, New Mexico.

McCune, L., Vihman, M. M., Roug-Hellichius, L., Delery, D. B., & Gogate, L. (1996). Grunt communication in human infants. *Journal of Comparative Psychology, 110,* 27–37.

McCune-Nicolich, L. (1981). The cognitive basis of relational words. *Journal of Child Language, 8,* 15–36.

McCune-Nicolich, L., & Raph, J. (1978). Imitative language and symbolic maturity. *Journal of Psycholinguistic Research, 7,* 401–417.

Ninio, A. (1999). Pathbreaking verbs in syntactic development and the question of prototypical transitivity. *Journal of Child Language, 26,* 619–653.

Piaget, J. (1946/1962). *Play, Dreams and Imitation.* New York: Norton.

Rizzolatti, G., & Arbib, M. (1998). Language within our grasp. *Trends in Neurosciences, 21*, 188–194.

Rizzolatti, G., Fadiga, L., Matelli, M., Bettinardi, V., Paulesu, E., Perani, D., & Fazio, F. (1996). Localization of grasp representations in humans by PET 1. Observation versus execution. *Experimental Brain Research, 111*, 246–252.

Ruff, H. (1982). Role of manipulation in infants' responses to the invariant properties of objects. *Developmental Psychology, 18*, 682–691.

Sartre, J. (1948, 1962). *The psychology of Imagination*. New York: Philosophical Library.

Searle, J. (1992). *The Rediscovery of the Mind*. Cambridge, MA: MIT Press.

Sinclair, H., Stambak, M., Lezine, I., Rayna, S., & Verba, M. (1989). *Infants and Objects: The creativity of cognitive development*. New York: Academic Press.

Talmy, L. (1975). The semantics and syntax of motion. In J. Kimball (Ed.), *Semantics and Syntax* (pp. 181–238). New York: Academic Press.

Talmy, L. (1985). How language structures space. In H. L. Pick, & L. P. Acredolo (Eds.), *Spatial Orientation: Theory, research, application* (pp. 225–282). New York: Plenum Press.

Vihman, M. M. (1996). *Phonological Development: The origin of language in the child*. Oxford: Blackwell.

# Looking for neural answers
# to linguistic questions

Bernard H. Bichakjian
University of Nijmegen, The Netherlands

## 1. Toward a concerted approach

Long left almost exclusively to linguists, language is attracting more and more attention from the various branches of biology. For the collective effort to bear fruit, there must be a minimum of shared knowledge. On the one hand, linguists will have to learn to put aside their prejudices about evolution, while, on the other hand, biologists and advocates of biological scenarios will need to become better acquainted with the facts of historical linguistics and the proper interpretation of linguistic diversity, so that the hypotheses they will propose with be compatible with the observational data.

## 2. What linguists should know about evolution

The theory of evolution is not entirely unknown outside the biological sciences, but there is a strong methodological divide. Scholars have indeed heard of Darwin, and they find it perfectly acceptable for biologists to work within an evolutionary paradigm, but they immediately draw a line and categorically oppose any attempt to introduce evolutionary reasoning into the humanities. This emotional attitude is in part understandable. Schleicher's precipitous application of his brand of evolution to language was less than successful (1873:6), while Darwinian sociology remains tarnished for the reasons we know all too well. There is, however, another consideration: in the humanities, it is often felt that the mechanics of evolution may indeed be suitable for bones and body parts, but such reasoning is deemed much too crude and common for the proper appreciation of the noble products of the human mind (see also Restak 1994:73). Sadly enough, these feelings are deep-rooted, but if they are to contribute to the understanding of language, linguists must learn to work with the evolutionary paradigm, because if, as Dobzhansky pointed out, "nothing in biology makes sense except in the light of evolution" (quoted from Mayr 1997: 178), no true understanding of language can be achieved "except in the light of evolution."

Saying that the brain of apes is more developed than that of dogs is not some form of racism against canines; arguing that warm-bloodedness confers selective advantages is not a sign of cruelty against crocodiles; pointing out the benefits of the domestication of plants and animals is not an act of discrimination against hunter-gatherers; and recognizing the greater effectiveness of firearms over bows and arrows is not a human rights violation. Likewise, it must be possible for linguists to assess and compare the production cost and the functional yield of linguistic items and conclude that item X is more advantageous than item Y without their being accused of unethical behavior.

Even in biology, where it is fully accepted, *evolution* is a difficult concept to define. The gradual increase in size and complexity of the brain is an unquestionable case of evolution, and so are the changes undergone by seals, whose ancestors' terrestrial features have been modified into alternatives suitable for a semi-aquatic life. In both cases, the changes are adaptive. Evolutionists would also argue that some changes can be totally neutral, but, by and large, evolution is seen as the accumulation of mutations that confer their bearers selective advantages and eventually lead to the formation of a new species (Mayr 1963: 621 & 1988: 253).

In linguistics, if a given feature – that is, a speech sound, a grammatical distinction, or a syntactic strategy – has been consistently replaced with a new alternative, it will be the task of the linguist to track the selective advantages that the incoming feature has over its antecedent. An easy example is word order. The ancestral word order was head-last or SOV, the modern one is head-first or SVO. This pervasive shift needs an explanation, and it is incumbent upon linguists to uncover the selective advantages of the modern word order.

The pattern of evolution may seem confusing. If, for instance, we focus on the evolution of classes from fishes to mammals, the process looks linear and unidirectional – fishes gave rise to amphibians, amphibians to reptilians, and reptilians to mammals, and never was the process reversed. On the other hand, if the focus is on the specialized leg and head anatomy of woodpeckers, the fish-like features of whales and dolphins or the atrophied wings of kiwis the obvious conclusion is that diversity or even course reversal is possible in the case of ecologically-adapted features. Evolution therefore never implies a neatly linear pattern, nor does it mean the total extinction of ancestral species. The survival of today's reptilians does not belie their being ancestral to mammals, nor the fact that mammals have selective advantages over reptilians. Likewise, returning to the word order example, the existence of extant SOV languages cannot be used to dispute SOV being ancestral to SVO, nor the fact that SVO has selective advantages over SOV.

## 3.   What evolutionists should know about language

If it is necessary for linguists to lose their misgivings about evolutionary reasoning, evolutionists, and indeed other advocates of innate models would make contributions better in line with the linguistic facts if they were to become better acquainted with the developments that have taken place in the history of languages. Unlike archaeologists, linguists do not have artefacts going back to the lower paleolithic. The recorded data, extended with the reconstructions achieved through the comparative method, internal reconstruction and extrapolation take us back only to the early days of the Neolithic. But in spite of its relative shortness, this period displays highly significant developments and a definite pattern.

A considerable number of highly pervasive changes have to a greater or lesser extent and in one form or another occurred in all the human languages during the last 10,000 years (for a detailed presentation see Bichakjian 1999 and 2002):

- Complex stops → Plain stops & fricatives;
- Laryngeals + e → Long/short vowels → More vowels (no length);
- Vowel alternation → Suffixes → Particles;
- Aspect & Modality → Tense;
- Verbs of state → Adjectives;
- Agent/patient → subject/object;
- Active/middle → Active/passive;
- Verbal phrases → Embedded sentences;
- Head-last → Head-first.

These continuous developments clearly suggest that the steady state conception posited by advocates of innatist scenarios (cf., e.g., Pinker & Bloom 1990) has absolutely no empirical support. Nor is there empirical support for Bickerton's two-plateau model (1990:124). The observational data strongly suggest instead that language, like industry, is a continuum, which started as a rudimentary implement of thought and communication, and gradually developed into increasingly powerful and efficient systems. Like other developments, the development of language has not come to a stop. Linguistic features are developing and will continue to develop, just as technology will progress and biological evolution will go on. Admittedly, all the observed and reconstructed data put together cover no more than 10,000 years, and before that we have no empirical material, but the clearly observed developmental pattern is like the tip of an iceberg. We do not see the submerged part, but we can surmise its existence and its being identical in nature with the visible part. We have no possibility of tracing the developments that led to the features found at the dawn of the empirical period, but we have no reason to doubt that they were the product of steady developments of which the observed ones are the continuation.

Scientists who are trying on the basis of fossil indications to determine when in the hominid line of evolution modernlike speech became possible, may rightly argue that in the absence of a given indicator – larger canals for the innervation of the tongue or larger ducts for the innervation of the thoracic muscles – oral fluency could not have been at the present level (cf. Kay et al. 1998; DeGusta et al. 1999; MacLarnon & Hewitt 1999; see also Lieberman 1998 on the position of the larynx). But, while these morphological studies are certainly enlightening, they do not provide evidence for a jump from a protolanguage to a language plateau. The plausible scenario is that the increase in nerve size ran parallel with the development of the language areas in the brain, the general development of cognition and the development of linguistic features, and that these developments were probably cross-fertilizing one another.

Another misunderstanding that needs to be cleared away is the pervasive assumption (cf. e.g. MacNeilage & Davis 2000) that the language of incipient speakers was like child language. Nothing can be further from the truth. Language evolution and language acquisition are two different processes, which in the main run in opposite directions. Language acquisition does not recapitulate language evolution. Like our biological evolution, language evolution is a neotenous process (cf. Bichakjian 1992 and Deacon 1997: 137). If an analogy must be drawn between the two, the correct one is that both are continuous processes. Just as language acquisition is a steady process from the speechless infant (Lat. *infans* meant "nonspeaking") to the articulate adult, so is language evolution from our remote ancestor's first word to today's most convoluted sentence a gradually expanding continuum.

## 4.    What linguists and biologists can achieve together

While the empirical data from historical linguistics do indeed argue against the one or two-plateau innatist models and the naive assumption that the primeval linguistic implements can be equated with the features of present-day baby talk, it must be stressed that language is NOT an abstract entity; it is a set of features that link with a concrete biological interface. Such a premise has recently led neuroscientists to discover that, when a rhesus monkey grasps or observes someone grasp an object, the cerebral activation takes place in the macaque's homologue of Broca's area (Gallese et al. 1996; Rizzolatti & Arbib 1998). The phenomenon is indeed intriguing, but before examining its possible significance, I shall discuss two radical changes – one that has played a major role in the evolution of syntax and a similar one in the developmental history of writing.

## 4.1  Why was the original word order head last?

In syntax, the original order was head last, *i.e.*, verbs came after their objects, auxiliaries after participles, prepositions after their objects, adjectives and genitives before their modified nouns, referents before comparatives, etc. In the course of time this order was reversed and the modern word order is head first. There are admittedly extant head last languages, but wherever word order has evolved, the shift has been from head last to head first, or from SOV to SVO. The unidirectionality of this change had long been observed by empiricists (cf. e.g. Givón 1979: 275–276), but it is worth noting that this fact is now acknowledged by such an influential formalist as Newmeyer (1998).

Reasoning in an evolutionary perspective, one may easily point out that since head first structures are less taxing for the speaker's and the listener's working memory, and therefore more information can be encoded and decoded, it is not surprising that natural selection would have guided languages toward a head first word order. The more intriguing question is why was the ancestral word order head last? The following methodological suggestion was recently made in cosmology:

> The laws of physics generally describe how a physical system develops from some initial state. But any theory that explains how the universe began must involve a radically different kind of law, one that explains the initial state itself. If normal laws are road maps telling you how to get from A to B, the new laws must justify why you started at A to begin with.   (Bucher & Spergel 1999: 48)

A similar approach must also be used in linguistics, and this is where neurology comes in. My hypothesis is that when incipient speakers began cobbling linguistic systems, the experience they brought to the new task was the one they had acquired by observing environmental features and events. They had theretofore functioned essentially in the PERCEPTUAL mode – hence, the need to include in their grammars environmental distinctions such as solid *vs.* liquid, vegetal *vs.* mineral, agent *vs.* patient, etc. Gradually, they eliminated the linguistically irrelevant distinctions, such as compact *vs.* cordlike, or active and middle, and replaced the useful ones with truly linguistic alternatives, such as subject and object, and active and passive. There was, I surmise, a shift from the PERCEPTUAL mode, characteristic of prelinguistic individuals, to the CONCEPTUAL mode, characteristic of the linguistically endowed. Since the perceptual mode is holistic, the shift to the conceptual alternative meant a switch from a synthetic to an analytical modus operandi, and since head last structures are holistic units, whereas head first alternatives allow for an as-you-go processing, it is only logical for the original word order to have been head last, and for the modern one to be head first.

This indeed is logical reasoning done on the basis of what is known of the two cerebral hemispheres (Levy 1974: 167; Posner & Raichle 1994: 94 & 162; Gazzaniga 1992: 130; Deacon 1997: 312). More research and especially targeted experiments are needed to confirm this hypothesis. What cannot be doubted, is that the reversal of word order is a portentous and heuristically important phenomenon which cannot be brushed aside with ad hoc expedients and relativistic platitudes. It is incumbent upon linguists, along with neuroscientists, to explain why the original word order was head last and why it was reversed.

### 4.2   Why was the direction of writing reversed?

Since space does not allow even an outline of the development of writing from hieroglyphs to alphabet, let us focus directly on the shift from right-to-left to left-to-right writing. This reversal and its broad extension is no trivial matter, and we must ask why the shift. The ergonomics of hand movement no doubt played a role (van Sommers 1991: 5, 10–11, 20), but there was perhaps another important factor on which neurology can shed light.

If we bear in mind that pictorial objects are generally perceived by the right hemisphere, and that the right hemisphere normally scans leftward, while the linear processing of abstract detailed representations is the work of the left hemisphere, which directs its attention rightward (Posner & Raichle 1994: 159; also cf. Posner, Walker et al. 1987; Posner, Inhoff et al. 1987), we may have a clue to all three of our questions. (1) Why was the direction of hieroglyphic writing from right to left? (2) Why is the direction of alphabetic writing from left to right? and (3) Why did the reversal occur with the shift to alphabetic writing?

The experimental data from neurology suggest that as long as writing and reading were indeed the holistic presentation and perception of pictorial objects it was natural for the direction to be from right to left, since that is how our brain proceeds when performing such tasks. Likewise, when writing and reading became a matter of encoding and decoding linear sequences of purely conventional signs, it was natural for the serializing to proceed rightward, since that is the direction in which our brain prefers to scan over details. And finally, it is logical that the change of direction occurred when it did, because that is when the old pictorial model was forever abandoned in favor of a fully analytical system using purely arbitrary signs.

## 4.3   The significance of mirror neurons for language origin

The firing of neurons in the simian homologue of Broca's area, both when the monkey is grasping an object and when it sees the grasping of the object, has prompted observers to hypothesize that the performance and/or the witnessing of an action could have been the stimulus that led to the emergence of a system of communication. They realize of course that there is a gap between action and speech, and in order to bridge it they argue that, since Broca's area also controls brachiomanual movements, "the first open system [of communication] to evolve *en route* to human speech was a manual gestural system that exploited the observation and execution matching system" (Rizzolatti & Arbib 1998: 192).

Perhaps, but it should be borne in mind that Broca's area is the control center of *articulatory movements* and partially at least of grammatical organization, and that for communication to take place there has to be both a system of transmission and a message to transmit. With its action-gesture-articulation scenario, the Mirror Neurons Theory has a hypothesis for the transmission part, but how about the message itself? How did the sense of meaning and the faculty to compose a message develop?

A recent study using positron emission tomography has come to the conclusion that the cerebral center for verbal intelligence is located "in the lateral frontal

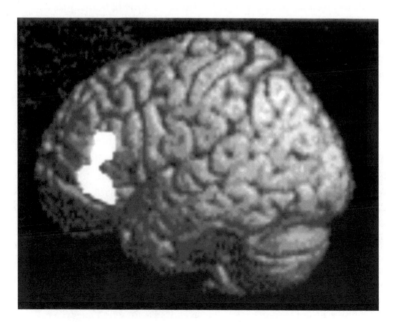

**Figure 1.** Verbal intelligence activation area (Duncan et al. 2000: 459)

cortex of the left hemisphere" with the activation pattern "closely corresponding" that of the spatial intelligence test, which also displays an activation pattern in the right hemisphere (Duncan et al. 2000:457). The near congruence of spatial and verbal intelligence areas in the left hemisphere, and the fact that they fall largely outside Broca's area seem to suggest that verbal intelligence and speech articulation are two different things, and the rise of the latter does not explain the development of the former.

Instead of the hypothesis whereby mirror neurons are made to provide a bridgehead for a gestural and later an oral mode of communication, the more plausible alternative would be to see the brain center for oro-facial and oro-laryngeal movements being pressed into the role of controlling the motor skills needed for speech *as* the language faculty was developing.

## 5.   Toward a concerted effort

The jury may still be out on the significance of mirror neurons for the study of language origins, but there is absolute certainty about the need for linguists and neurologists to work together. For the concerted effort to produce the best results, linguists will have to cast away their prejudices against evolution, and biologists and advocates of innatist scenarios will need to become better acquainted with the historical record. If such a mutual understanding is reached, and if language is seen as having a linguistic and a biological interface that should be studied in reference to each other, then we will be able to understand the true nature of language and explain the most portentous developments that have taken place.

## References

Bichakjian, B. H. (1992). Language evolution: Evidence from historical linguistics. In J. Wind, B. H. Bichakjian, A. Nocentini, & B. Chiarelli (Eds.), *Language Origin: A multidisciplinary approach* (pp. 507–526). Dordrecht: Kluwer.

Bichakjian, B. H. (1999). Language evolution and the complexity criterion. *Psycoloquy*. http://www.cogsci.soton.ac.uk/psyc-bin/newpsy?10.033

Bichakjian, B. H. (2002). *Language in a Darwinian Perspective*. Frankfurt: Peter Lang.

Bucher, M. A., & Spergel, D. N. (1999). Inflation in a low-density universe. *Scientific American, 280*(1), 42–49.

Deacon, T. W. (1997). *The Symbolic Species*. New York: W.W. Norton.

Duncan, J., Seitz, R. J., Kolodny, J., Bor, D., Herzog, H., Ahmed, A., Newell, F. N., & Emslie, H. (2000). A neural basis of general intelligence. *Science, 289*, 457–460.

DeGusta, D., Gilbert, W. H., & Turner, S. P. (1999). Hypoglossal canal size and hominid speech. *Proceedings of the National Academy of Sciences, USA, 96*, 1800–1804.

Gallese, V., Fadiga, L., Fogassi, L., & Rizzolati, G. (1996). Action recognition in the premotor cortex. *Brain, 119*, 593–609.

Gazzaniga, M. S. (1992). *Nature's Mind*. New York: Basic Books.

Givón, T. (1979). *On Understanding Grammar*. New York: Academic Press.

Kay, R. F., Cartmill, M., & Balow, M. (1998). The hypoglossal canal and the origin of human vocal behavior. *Proceedings of the National Academy of Sciences, USA, 95*, 5417–5419.

Levy, J. (1974). Psychobiological implications of bilateral asymmetry. In S. J. Dimond, & J. Graham Beaumont (Eds.), *Hemisphere Function in the Human Brain* (121–183). London: Elek Science.

Lieberman, P. (1998). *Eve Spoke*. New York: W.W. Norton.

MacLarnon, A. M., & Hewitt, G. P. (1999). The evolution of human speech the role of enhanced breathing control. *American Journal of Physical Anthropology, 109*, 341–363.

MacNeilage, P. F., & Davis, B. L. (2000). On the origin of internal structure of word forms. *Science, 288*, 527–531.

Mayr, E. (1963). *Animal Species and Evolution*. Cambridge, MA: Harvard University Press.

Mayr, E. (1988). *Toward a New Philosophy of Biology*. Cambridge, MA: Harvard University Press.

Mayr, E. (1997). *This is Biology*. Cambridge, MA: Harvard University Press.

Newmeyer, F. J. (2000). On the reconstruction of 'proto-world' word order. In C. Knight, M. Studdert-Kennedy, & J. Hurford (Eds.), *The Evolutionary Emergence of Language: Social function and the origins of linguistic form* (pp. 372–390). Cambridge: Cambridge University Press.

Pinker, S., & Bloom, P. (1990). Natural language and natural selection. *Behavioral and Brain Sciences, 13*, 707–784.

Posner, M. I., Inhoff, A. W., Friedrich, F. J., & Cohen, A. (1987). Isolating attentional systems: A cognitive-anatomical analysis. *Psychobiology, 15*, 107–121.

Posner, M. I., & Raichle, M. E. (1994). *Images of Mind*. New York: Scientific American Library.

Posner, M. I., Walker, J. A., Friedrich, F. J., & Rafal, R. D. (1987). How do the parietal lobes direct covert attention? *Neuropsychologia, 25A*, 135–146.

Restak, R. M. (1994). *Receptors*. New York: Bantam Book.

Rizzolatti, G., & Arbib, M. A. (1998). Language within our grasp. *Trends in Neurosciences, 21*, 188–194.

Schleicher, A. (1863/1873). *Die Darwinsche Theorie und die Sprachwissenschaft*. Weimar: Böhlau.

Sommers, P. van (1991). Where writing starts: The analysis of action applied to the historical development of writing. In J. Wann, A. M. Wing, & N. Sõvik (Eds.), *Development of Graphic Skills: Research perspectives and educational implications* (pp. 3–38). London: Academic Press.

# Mirror neurons and cultural transmission

India Morrison

Department of Humanities, University of Skövde, Sweden

Their spirits are so married in conjunction with the participation of society,
that they flock together in consent, like so many wild-geese... It is certain, that
either wise bearing, or ignorant carriage, is caught, as men take diseases, one
of another. Jack Falstaff,
in William Shakespeare's "King Henry IV", Part II, Act V scene i.

The question of what drives cultural change has attracted the interest of researchers
in many fields. Analysis of cultural change from anthropology has been abundant
over the last century or so. However, it is only within the last two decades that
researchers investigating culture have begun to perceive the necessity of regarding cultural transmission not only via anthropological approaches, but psychologically and evolutionarily as well (Sperber 1996; Plotkin 1997). Also, the behavior
of culture itself has begun to be looked upon as a potentially quantifiable dynamic
system, quite interesting to study in its own right (e.g. Cavalli-Sforza & Feldman
1981). What is emerging is a very complex set of demands upon our empirical
characterization of cultural change, and a need to coordinate findings across disciplines ranging from anthropology and psychology to evolutionary biology and
computer science.

A rather new, but I hope promising, approach to cultural dynamics contem-
plates cultural transmission from a neurophysiological vantage. Neurons have recently been discovered in the macaque prefrontal cortex which fire both when a
monkey observes an action and when it performs the action itself (e.g. Rizzolatti
et al. 1996). These neurons – mirror neurons – will probably prove indispensible
to a neurophysiological approach to culture. Indeed, they might even become central to it, as further research on primate mirror systems unfolds. There are many
ways in which mirror neurons could play a role in cultural transmission. Language,
as this volume testifies, is one. Although macaque mirror neuron activity has not
been observed in imitation contexts, mirror neurons are also likely to play a part

in the sharing of mental representations – essential for the transmission of cultural information from one person to the next.

Here I shall focus on two rather more quirky aspects of human cultural behavior, "catchy memes" and "migratory mannerisms." Catchy memes are those apparently meaningless yet very contagious bits of cultural information that can become irrevocably caught in one's ongoing mental experience, such as tunes or catch-phrases. Migratory mannerisms are idiosyncratic movements, ways of enunciating, etc., which become "picked up" and in turn used by individuals who did not originate the mannerism. (If a neurophysiological approach to these specific aspects of cultural transmission is taken up in future studies, it will be necessary to have a formal definition of what it means for a behavioral impulse to be "stuck" or "picked up." I hesitate to attempt that here, and so will rest my description on the intuitive notions.) When we use secondhand mannerisms, we are often surprised to find ourselves doing so, though we are aware that the mannerism came from another individual and can often easily identify the source.

As cultural phenomena, catchy memes and migratory mannerisms are particularly intriguing in several respects. For example, catchy memes seem relatively resistant to the scheduling ordinarily exerted by frontal and supplementary motor areas over forms of socially relevant behavior. Indeed, catchy memes are especially notorious for persisting in one's mind and becoming translated into action despite efforts to refrain; unless one's attention becomes powerfully engaged elsewhere, a catchy meme remains somewhat near the forefront of executive priority. Migratory mannerisms are also most often executed with a lack of conscious deliberation. What is remarkable about both is that behavior is reproduced with impressive fidelity after as little as one exposure to the behavior in question, yet without seeming to employ any of the usual major classes of learning familiar to psychology (Heyes 1994). In this respect the existence of mirror systems in humans hints at promising explanations. However, an oddity of both is the conspicuous lack of goal-directedness, which complicates matters.

Because these phenomena tend to elude adaptive explanations, one theory of cultural change has defined this class of apparently purposeless behavior in terms of the reproductive fitness of the information itself, rather than that of human genes. This hypothesis centers around the "meme", a postulated unit of cultural replication analogous to the gene as a biological replicator (Dawkins 1976). Meme theory brought these enigmatic behaviors into the light of evolutionary theory. I have accordingly borrowed the word "meme" in the name "catchy memes," though here I am using it to refer to a very specific type of behavior, not just culturally-transmitted information in general. The prevailing description of a meme is as an informational entity that exploits neural resources as a means of reproducing itself, in the manner of a parasite or even a virus (Dennett 1995; Blackmore 1999). My use of the term meme should be taken somewhat independently of this sense; "meme" is a useful existing word for certain phenomena that are familiar to everyone.

One of the hopes of a cognitive neuroscience of cultural transmission is to illuminate mechanisms of cultural transmission while also providing an adaptive explanation for catchy memes and migratory mannerisms that does not necessarily stipulate the interests of entities apart from humans. (But this does not preclude the possibility that culture is nevertheless a Darwinian evolutionary system, a very important possibility that can also be explored in light of neurophysiological mechanisms.) A good way of beginning this endeavor would be to ask, "What neurophysiological factors can influence the likelihood that an observed action will be repeated by the observer?" There is a host of mechanisms that can be swept under this broad heading, but three of the most pivotal are those of:

1. Memory;
2. Patterns of excitation and inhibition;
3. Emotion.

The role of mirror neurons, or the human equivalent, would be pervasive among these influential factors. In memory, they would come into play in the perception of behavior that is remembered and later performed by the observer. Their role in patterns of excitation and inhibition of the motor system might determine not only a behavior's *susceptibility* to being reproduced, but also the *types* of behavior that tend to become reproduced. Of the above three factors, the first has most immediately to do with catchy memes, and the other two with migratory mannerisms.

In macaques, mirror neurons fire at an early visuomotor stage of processing in social perception, so memory processing of this information probably occurs downstream from mirror perception. The way memory and mirror perception interact will be important to our understanding of cultural dynamics, especially if mirror systems are broader in the human brain than in the macaque brain. With regard to catchy memes, short-term mechanisms probably have the greatest claim to importance over, for instance, working memory or modifications maintained by gene regulation in long-term memory. The associated hemodynamic activity of Broca's area in mirror phenomena (Iacoboni et al. 1999) suggest that speech (and I venture to add music as well) is processed audiomotorically. A catchy meme, then, is probably represented as a motor image (Jeannerod 1997) even upon perception, whether it becomes overtly expressed as behavior or not.

Catchy memes tend to be brief musical or spoken phrases, with a strong dose of periodicity (rhythm, rhyme, etc.) and a dominant motor flavor that often does result in overt behavior. If a catchy meme is a) learned at least in part via mirror perception, and b) represented as a motor image, this would open new avenues for a neurophysiological analysis of this kind of phenomenon. One such avenue would be determining the part played by areas implicated in human mirror activity, such as superior temporal, inferior frontal, and posterior parietal cortex, in the acquisition and execution of catchy-meme behavior. Another would be the way motor representation in the cortex interacts with memory systems.

Work with melodic pattern recognition has suggested that "initial experiences of the melody form a template in the plane defined by pitch and time. When the tune is repeated, it is compared with this template, and for the appropriate starting time there will be matches between what is received and what is predicted from the template..." (Wong & Barlow 2000:952). It is conceivable that perceptual patterns in the auditory domain are detected partly by reference to memories created via audiomotoric mirror perception, in the form of motor representations. Another contributing factor might be that brief and uncomplex spoken or musical phrases seem to be custom-made to dominate short-term memory and associative processing. For instance, they tend to swallow their own tails. The mental experience of the phrase may be sufficient to cause the template to become re-matched, instigating indefinite repetitions of the same match in short-term memory. The neurophysiology of music and the evolution of rhythmic behavior, especially rhythmic social behavior, will need to be better understood before a neurophysiological account of catchy memes can mature.

Patterns of motor excitation and inhibition within the premotor and motor cortices are also likely to influence cultural transmission patterns to some degree. Both excitatory and inhibitory signalling interact in the production of action. Selective facilitation and inhibition during motor imagery, and probably during action perception as well, begins at the segmental spinal level (Jeannerod 1997). Negatively congruent mirror neurons in the macaque premotor cortex discharge inhibitory signals in response to a certain stimulus, whereas the activity of other neurons are positively congruent for that stimulus (Rizzolatti et al. 1996). If such complex patterning is already under way during the perception of conspecific social behavior, it will affect cultural dynamics. Some types of behavior may more readily excite, facilitate, or disinhibit corresponding circuits in the perceiving brain, whereas others may have difficulty exciting or overcoming inhibition of them. Weak or incomplete inhibition, as sometimes occurs during motor imagery (Jeannerod 1997), can give rise to the whole or partial behavioral execution of a motor representation. For a variety of reasons that affect excitation-inhibition patterns, certain body parts also lend themselves more easily than others to transmissible motor representation. For example, facial and hand gestures are common in culture; expressions of the knee or stomach are less common.

In these respects migratory mannerisms are well-suited to an analysis in terms of mirror systems. However, they are not directed towards an obvious goal. In macaques, mirror neurons respond only to goal-directed actions. Mirror neurons seem primarily to be a perceptual detection mechanism for behaviorally relevant goals in the social environment in the context of *agent-object* interactions. In humans, mirror systems may be more expansive (for example, involving different sensory domains; unlike macaques we also have an understanding of functional equivalence important for tool use). For humans, *agent-agent* interactions are quite prevalent in our social environment and cognitive landscapes. The environment of our ancestors (and of anatomically modern humans as well) probably consisted of small social groups of interacting individuals. A mirror system which would allow individuals to understand agent-agent interactions in a comparable way to agent-object understanding would have become adaptive in a social mileu like the one in which our ancestors probably lived.

However, a goal in an agent-agent interaction is not the same as a goal in an agent-object interaction. Most notably, it is not concretely manifested. There are at least two ways, then, to describe catchy memes and migratory mannerisms with respect to mirror phenomena. Either: a) they are not goal-directed phenomena at all; or b) they involve a different class of goals. Understanding the intentional interactions of conspecifics might involve non-object-oriented goal representation of *social* goals. If human mirror systems are involved in the representation of social goals, observed agent-agent intentional interactions would be translated into body-centered motor coordinates. That means that although social goals have no material object, they can nevertheless be described somatosensorily.

If human agent-agent interactions are perceived in terms of the motor system, then every interaction, exchange, and conversation in which humans engage will inspire a welter of ongoing motor representations. Many of the movements which occur in social interactions are interpreted as meaningful signals, even very swift ones like fleeting changes in facial expression or gaze. Indeed, a good deal of movement in social interaction is perceived on a millisecond scale, unaccompanied by conscious awareness. Film research on social interactions has indicated that movements between interacting people are highly organized, and often synchronized (e.g. Kendon 1970). The degree to which synchrony occurs is strongly correlated with the relative status of the people in the interaction. The closer two people are in status, the more closely timed their movements are (there may also be gender-related effects in conversational synchrony, see Boker & Rotondo, this volume). Synchrony can thus serve as a social signal indicating either social parity – as when movement is highly synchronized – or disparity, as when little or no synchrony occurs. Facial mimicry also occurs (e.g. Dimberg et al. 2000). It is plausible that mirror systems play a role in movement symmetry between two people, via apprehension of intentional states and representation of social goals during interactions.

Movement during conversation is structured around speech (Goodwin 1981). Because paralinguistic movement is attended to as well as speech in interpreting an interlocutor's meaning, intention, and social position during conversation, there is high information value not only in the movements themselves but in the way they are organized in the syntax of conversational movement. Social interactions enlist a suite of cognitive mechanisms. For instance, complex executive processing occurs in the frontal lobe, and emotional processing is distributed throughout the brain. Emotion influences future behavior by coupling perception with disposition to act (e.g. LeDoux 1998). Rather than having an immediate influence on future behavior, mirror perception more likely contributes at a relatively early stage to a cascade of responses which couple perception with action, and action disposition with memory. Brothers (1995) has proposed that complex shades of perceptual-emotional response in social interactions are reflected by correspondingly complex profiles of action disposition in the brain. Expanding Damasio's term, she suggests that emotional action dispositions are thus "somatically marked" (Damasio 1998; Brothers 1995). This means that certain action dispositions become better remembered by virtue of the strength and character of the emotional response that produced them. This may go a long way towards accounting for why less dominant people tend to imitate those more dominant and socially impressive, rather than vice-versa.

Social movement, then, has many important features: it is highly structured, coordinated between interlocutors, susceptible to mirror representation, and sensitive to the influence of emotional response on action dispostition. Therefore there may be "hot spots" during a conversation when a mannerism is more likely to be performed, attended to, or mnemonically "copied." This would increase the chances of future execution by the observer. In humans, mirror representation is probably crucial to learning what is appropriate in the organization of movement during social interactions. This learning is highly specific and capable of indefinite refinement. It may not be limited just to the coarse parameters of social movement coordination, but even encompass idiosyncratically detailed movements such as mannerisms. Objects and artifacts may also be used as markers of social movement organization, so the mannerisms surrounding them are likewise susceptible (the way someone emphatically waves a cigarette, for example).

If mirror systems exist in humans, they probably do play a role in cultural transmission. It might be indirect, yet certainly indispensible. Mirror systems in humans may not have any single discrete function, but contribute to a gamut of social cognitive phenomena. One of their roles could be in understanding conspecific behavior with respect to nonconcrete, as well as object-oriented, goals. Cultural transmission requires *representing* (not necessarily "understanding" in the intellectual sense) the intentions and actions of others. At any rate, action perception-execution is not a monolithic phenomenon, but involves the interaction of many neural populations and signaling systems. Insofar as they are separable, each of these systems possesses a different evolutionary history, confers a different set of adaptive advantages, and performs a different functional role. But as systems interacting in an innumerable variety of complex social circumstances, their interaction may give rise to some behavioral tendencies that have not been directly selected for. The spread of catchy memes and migratory mannerisms may be an instance of this. A cognitive neuroscience of cultural transmission would need to integrate findings from several areas of research, not least in memory and action perception-execution systems. In doing so it would enrich our understanding of cultural dynamics by elucidating how the workings of the brain contribute to the workings of culture. Mirror neurons hold great promise for such an endeavor.

## Acknowledgement

My thanks go to David Chalmers and Henry Plotkin for their helpful comments on an earlier draft.

## References

Blackmore, S. (1999). *The Meme Machine*. Oxford: Oxford University Press.

Brothers, L. (1995). Neurophysiology of the perception of intentions by primates. In M. Gazzaniga (Ed.), *The Cognitive Neurosciences*. Cambridge, MA: MIT Press.

Cavalli-Sforza, L. L., & Feldman, M. W. (1981). *Cultural Transmission and Evolution: A quantitative approach*. Princeton, NJ: Princeton University Press.

Damasio, A. (1998). The somatic marker hypothesis and the possible functions of the prefrontal cortex. In A. C. Roberts et al. (Eds.), *The Prefrontal Cortex*. Oxford: Oxford University Press.

Dawkins, R. (1976/1989). *The Selfish Gene*. Oxford: Oxford University Press.

Dennett, D. C. (1995). *Darwin's Dangerous Idea: Evolution and the meaning of life*. New York: Simon and Schuster.

Dimberg, U. et al. (2000). Unconscious facial reactions to emotional facial expressions. *Psychological Science, 11*(1), 86–89.

Goodwin, C. (1981). *Conversational Organization: Interaction between speakers and hearers*. New York: Academic Press.

Heyes, C. (1994). Social learning in animals: categories and mechanisms. *Biological Review, 69*, 207–231.

Iacoboni, M. et al. (1999). Cortical mechanisms of human imitation. *Science, 286*, 2526–2528.

Jeannerod, M. (1997). *The Cognitive Neuroscience of Action*. Oxford: Blackwell Publishers.

Kendon, A. (1970). Movement coordination in social interaction: some examples described. *Acta Psychologica, 32*, 101–125.

LeDoux, J. (1998). *The Emotional Brain*. London: Weidenfield and Nicolson.

Plotkin, H. (1997). *Evolution in mind*. London: Penguin Books.

Rizzolatti, G. et al. (1996). Premotor cortex and the recognition of motor actions. *Cognitive Brain Research, 3*, 131–141.

Sperber, D. (1996). *Explaining Culture: A naturalistic approach*. Oxford: Blackwell Publishers.

Wong, W., & Barlow, H. (2000). Tunes and templates. *Nature, 404*, 952–953.

PART IV

# Applications

# Mirror neurons and the neural basis for learning by imitation

## Computational modeling

Aude Billard and Michael Arbib
University of Southern California, Los Angeles, USA

## 1. Introduction

> Nothing is so contagious as an example, and we never do such good acts and such bad acts that we do not produce similar ones. We imitate the good actions by emulation, and the bad ones by the malignity of our nature, that shame kept prisoner, and that the example let free.
>
> La Rochefoucauld, *Reflexion Morales* (1678 edition)

This maxim of the 17th century French philosopher François duc de La Rochefoucauld introduces the core idea behind imitation. Learning by imitation is fundamental to social cognition. It is at the basis of the animal's ability to interpret the behavior of others and – though the phrasing is at the risk of anthropomorphism – to attribute intentions, to deceive and to manipulate others' states of mind. Imitative learning takes many forms across species, reaching its fullest complexity in humans.

In order to better understand the leap between the levels of imitation in different animals, there is a need to better describe the neural mechanisms underlying imitation. Our work uses computational neuroscience to analyze the different cognitive processes involved in imitation. We focus, in particular, on the capacity for representation and symbolization which underlies that of imitation and investigate the neural mechanisms by which they are generated. In the long term, this study might shed some light on the question of which species are endowed with symbolic thought, a key question for those who study the origin of symbolic communication and in particular of human language. This paper sketches out the key ideas behind our work and summarize the different models we have implemented.

## 2.    What is imitation?

There is still debate concerning what behaviors the term "imitation" refers to and in which species it is exhibited (Byrne & Whiten 1988; Tomasello 1990). "True" imitation is contrasted to mimicry. Imitation is more than the mere ability to reproduce others' actions; it is the ability to replicate and learn skills which are not part of the animal's usual repertoire simply by the observation of those performed by others. The current agreement is that only humans and possibly apes are provided with the ability for true imitation, although recent data (Myowa-Yamakoshi & Matsugawa 1999) suggest that imitation of manipulatory actions in chimpanzees is quite limited compared to that of humans. Typically, chimpanzees took 12 trials to learn to "imitate" a behavior, and in doing so paid more attention to where the manipulated object was being directed, rather than actual movement of demonstrator.

Simpler forms of imitation and mimicry have, however, been shown in rats (Heyes 1996), parrots (Moore 1996), mynah birds (Nottebohm 1976), dolphins (Hermann 2000), and monkeys (Kawamura 1963; Visalberghi 1990). Imitation is interesting to developmental psychologists because it underlies the human child's growing capacity for representation and symbolization. The development of complex imitation abilities in children accompanies the growth of the child's communicative and linguistic competences (Nadel et al. 1999; Speidel 1989; Trevarthen et al. 1999) and is viewed as a marker of the child's normal cognitive development (Piaget 1969). Studies by Meltzoff and Gopnik (1989) of imitation of facial expression in newborns reopened the debate over whether infants' imitation is innate (Meltzoff's position) or learned (Piaget's position).

While human *neonates* can imitate other's facial gestures (Meltzoff & Moore 1977), this form of imitation disappears or declines at 2 to 3 months of age. It is not until 8–12 months that imitation of facial gestures arises again. Our position is that more complex mechanisms, which allow human learning of complex and novel sequences of actions and involve mapping across multiple modalities, are not expressed by the neonate. We hypothesize that "true" imitation, as found in humans, has an iterative nature and requires the ability for 1) recognizing familiar actions when performed by others, 2) for memorizing (internal representation) the observed movements as variations on, and/or coordinated compositions and sequences, of familiar actions, and 3) for reproducing the actions and tuning the movements such as to perfect the reproduction. We test these hypotheses using computational neuroscience. Next, we briefly summarize our results in developing a computational model of primate ability to learn by imitation. The model addresses points 1 and 2 above. Our current work focuses on point 3.

## 3. A computational model of learning by imitation

The model (see Figure 1) is biologically inspired in its function, as its component modules have functionalities similar to those of specific brain regions, and in its structure, as the modules are composed of artificial neural architectures (see Billard 1999, for a complete description). It is loosely based on neurological findings in primates and incorporates an abstract model of the spinal cord, the primary motor cortex (M1) and premotor cortex (PM) and the temporal cortex (STS). New extension of the model (not described in Billard 1999) emphasizes the role of the supplementary motor area (SMA) in sequence learning, with the cerebellum learning to improve the coordination of movements, decomposing what was modelled as a unitary function of the cerebellum in the earlier model. The model was validated in a dynamic simulation of a 65 Degrees of freedom humanoid avatar.

Each part of the model is implemented at a connectionist level, where the neuron unit is modeled as a first order differential equation (leaky-integrator neuron (Hopfield 1984) or a variant of it (Billard & Hayes 1999). Motor control is directed by the spinal cord module and the primary motor cortex module, both of which have direct connections to motor neurons. Motor neurons activate the avatar's muscles. There are two muscles per degree of freedom per joint. Each muscle is represented as a spring and a damper. Visual recognition of human limb movements is done in the temporal cortex and in the primary motor cortex. Neurons in M1 respond also to corresponding motor commands produced by PC. Learning of new combination of movements is done in PC and the cerebellum modules. These modules were implemented using the Dynamical Recurrent Associative Memory Architecture (DRAMA) (Billard & Hayes 1999) which allows learning of times series and of spatio-temporal invariance in multi-modal inputs.

The model's performance was evaluated for reproducing human arm movements, using video data of human motions. Results showed a high qualitative and quantitative agreement with human data (Billard & Mataric 2000). In particular, it was shown that the imprecision of the model's reproduction is better or comparable to that of humans involved in the same imitation task. Our current work carry out psychophysical experiments to record human imitative performance. The data of these experiments will be used to further develop the model. We will introduce a tuning mechanism to allow the avatar to improve its imitative performance, similarly to what is observed in the human data. The mechanism will shift the focus of attention, during repeated reproduction, so as to pay more attention to those visual and proprioceptive features of the movements which are incorrect.

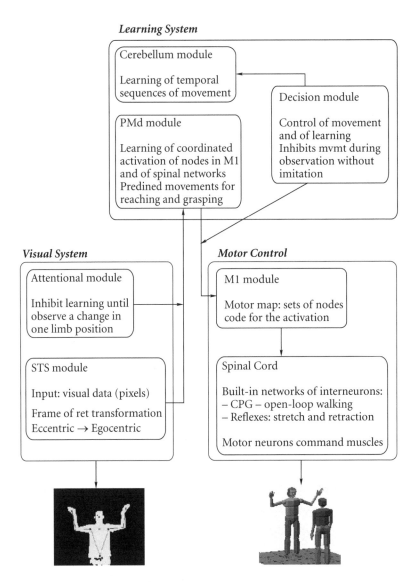

**Figure 1.** The model consists of three parts for visual recognition, motor control and learning and is composed of biologically inspired modules, namely the temporal cortex (STS), the spinal cord, the primary motor cortex (M1), the premotor cortex (PM) and the cerebellum. The bottom left panel indicates the taking of data from visual input (extracting the arm and body movements of a human); the right hand panel shows an avatar imitating the observed movement.

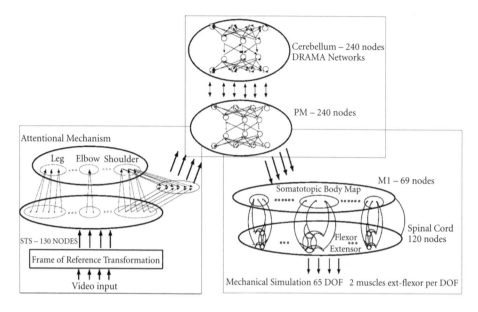

**Figure 1.** *(continued)*

## 4. Neural mechanisms behind imitation

The confusion behind the different definitions of imitation results in part from a lack of information concerning the neural mechanisms at the basis of this ability in animals. Motor skill imitation relies on the ability to recognize conspecifics' actions and to transform visual patterns into motor commands. While there is an important body of literature on visual recognition of movements and on motor learning, little is yet known concerning the brain's ability to match visual and motor representations.

Mirror neurons, as observed in the premotor cortex of monkeys, have been proposed as providing the neural system responsible for matching the neural command for an action with the neural code for the recognition of the same action executed by another primate (Gallese et al. 1996; Rizzolatti et al. 1996a; di Pellegrino et al. 1992). While the above work was done in monkeys, evidence was recently provided that an action observation/execution matching system, similar to that found in monkeys, exists also in humans. These studies are based on transcranial magnetic stimulation (TMS) (Fadiga et al. 1996), positron electron tomography (PET) (Krams et al. 1998) and functional resonance magnetic imaging (fMRI) (Iacoboni et al. 1999).

Erhan Oztop and Michael Arbib develop a detailed model of the monkey mirror neuron system based on monkey neurophysiology and related data (Oztop & Arbib 2002). The model extends the FARS model on monkey grasping (Fagg & Arbib 1998) and reproduces data from Rizzolatti and co-authors on mirror neuron cells firing during specific grasps (precision, power and pinch grasps). Current work of Oztop and Arbib develops a mechanism for self-organization of the observation/execution matching system of the model. This will test hypotheses on how mirror neurons may instruct parietal and inferotemporal neurons to give visual patterns their motor significance.

The discovery of the mirror system in monkeys is very exciting to those who wish to understand the neurological processes behind imitation. It is important to note, however, that research on the mirror system and its relation to imitation is still in its early stages. So far mirror neurons have been observed only for simple actions such as grasping, twisting and tearing by monkeys; while a recent fMRI study (at the regional rather than neuronal level) of imitation in humans has focused on finger tapping. It remains to be shown that mirror neurons exist for other movements than that of the arms and the hands and that they are fundamental to "true" imitation of complex movements. We speculate that once appropriate experiments are done, mirror systems will be found for a wide range of behaviors in a wide range of species (birdsong, for example, looks like an excellent candidate).

With this caveat in mind, our ongoing work on imitation takes a major part of its inspiration on the mirror neuron system to build a computational model of human ability to learn by imitation. In building upon the model of Figure 1, we hypothesized (Arbib et al. 2000) that the human mirror system is located in a network of brain areas which show similar (mirror) activity during both observation, rehearsal and production of the same movements. Note that such a distributed view of the mirror neuron system is more in agreement with imaging data, which have shown mirror activity in several brain areas. These areas are the left Broca's area (area 45) (Krams et al. 1998; Rizzolatti et al. 1996b), the left dorsal premotor area 6, left inferior frontal cortex (opercular region) and the rostral-most region of the right superior parietal lobule (Iacoboni et al. 1999). These areas in the model are the premotor cortex, the supplementary motor area (new module of the network) and the cerebellum. In Arbib et al. (2000), we showed that the activity of these model's areas was comparable to the activity of the corresponding brain areas recorded in fMRI experiments.

## 5. Mirror neurons, imitation and language acquisition

We started this brief article by mentioning that imitation underlies cognitive processes fundamental to social cognition. To conclude this article, we briefly summarize an important goal of our research, namely, to link the ability to imitate to that for language. This relates directly to the conference on "Mirror neurons and the evolution of brain and language" from which the book follows. The conference itself was inspired in part by the paper by Rizzolatti and Arbib (1998). The observation that the area F5 (which contains the mirror neuron system) in monkeys could correspond to Broca's area in the corresponding area 6 of the human motor cortex led the authors to propose that "human language [...] evolved from a basic mechanism [the mirror neuron system] that was not originally related to communication". Arbib (2001) extended this idea by postulating that the ability to imitate provided a crucial bridge from the monkey mirror neuron system to the human ability for language.

It is interesting to relate these ideas to the psycholinguistic literature (Garton 1992; Lock 1978; Meltzoff & Gopnik 1989; Nadel et al. 1999; Trevarthen et al. 1999) which stresses the importance of social cues, such as coordinated behavior (of which imitation is one instance), as the precursor to language development in infants (see Billard 2002, for a review). Imitation has been attributed three different roles in infants' language development: in the motor control of speech, in the infants' internal cognitive development, and in the infants' social interactions. In the last, imitation is a social factor which guide the infants' cognitive development. This developmental step is "a marker for the child's development of more complex form of verbal communication exchanges" (Nadel et al. 1999).

To further accentuate this point, we recall robotic experiments in which we showed that endowing the robot with imitative capabilities would enhance its performance at learning a simple language (Billard 2000; Billard & Hayes 1999). In these experiments, a robot (a doll-shaped or a wheeled-based robot) imitated the motion of the teacher (see Figure 2). The imitation game constrained the robot's attention to the stimuli upon which it could be taught.

Returning to psychology and ethology, there is evidence of precursors to language and imitation in different species. These precursors include processes for (1) recognizing conspecifics and for interpreting their actions in terms of ours (weak version of theory of mind), (2) memorizing others' behaviors by a process of internal representation (a first step towards symbolization); and (3) reproducing (reconstructing) the observed actions and composing new actions (a first step towards the ability for creating novel behaviors by composing observed and learned ones and, in particular, towards linguistic compositionality). This leads us to formulate the hypothesis that similar cognitive processes were present in species, such as

**Figure 2.** The doll-shape robot mirrors the movements of the arm and head of the human demonstrator. While directing the robot, the demonstrator instructs the robot (saying e.g. "You move your right arm"). Through training the robot learns to dissociate the meaning of each word in the demonstrator's sentences and to associate it with its perceptions.

monkeys and chimpanzees, before they could be fully expressed, as it is the case in humans.

Our future work will address these different issues. In previous work (Billard 2000; Billard & Hayes 1999), we followed an engineering approach, exploiting the imitation game to enhance the robustness of the system and exploiting the fact that imitation and learning of a regular language could be programmed using the same artificial neural network. In future work, we will follow a neuroscience approach and study how the same computational model can be used to produce both complex imitative learning and learning of a variety of linguistic properties.

## Acknowledgment

Aude Billard is supported by a personal fellowship from the Swiss National Science Foundation. Michael Arbib's work is supported in part by a grant from the Human Frontier Science Program.

## References

Arbib, M. A. (2002). Mirror neurons, imitation and language: an evolutionary perspective. In C. Nehaniv, & K. Dautenhahn (Eds.), *Imitation in Animals and Artifacts* (pp. 229–280). Cambridge, MA: MIT Press.

Arbib, M. A., Billard, A., Iacoboni, M., & Oztop, E. (2000). Synthetic brain imaging: Grasping, mirror neurons and imitation. *Neural Networks, 13*, 975–997.

Billard, A. (1998). DRAMA, a connectionist model for robot learning: Experiments on grounding communication through imitation in autonomous robots. PhD thesis, Dept. of Artificial Intelligence, University of Edinburgh, U.K.

Billard, A. (2002). Imitation: A means to enhance learning of a synthetic proto-language in an autonomous robot. In C. Nehaniv, & K. Dautenhahn (Eds.), *Imitation in Animals and Artifacts* (pp. 281–310). Cambridge, MA: MIT Press.

Billard, A. (1999). Learning motor skills by imitation: a biologically inspired robotic model. *Cybernetics & Systems Journal,* special issue on *Imitation in Animals and Artifacts, 32*(1–2), 155–193.

Billard, A., & Hayes, G. (1999). Drama, a connectionist architecture for control and learning in autonomous robots. *Adaptive Behavior, 7*(1), 35–64.

Billard, A., & Mataric, M. (2000). Learning human arm movements by imitation: Evaluation of a biologically-inspired connectionist architecture. In *Proceedings, First IEEE-RAS International Conference on Humanoid Robotics* (Humanoids-2000) (on CD-rom). Cambridge, MA, The MIT, September 7–8, 2000.

Byrne, R. W., & Whiten, A. (1988). *Machiavellian Intelligence: Social expertise and the evolution of intellect in monkeys, apes, and humans.* Oxford: Clarendon Press.

di Pellegrino, G., Fadiga, L., Fogassi, L., Gallese, V., & Rizzolatti, G. (1992). Understanding motor events: a neurophysiological study. *Experimental Brain Research, 91*, 176–180.

Fadiga, L., Fogassi, L., Pavesi, G., & Rizzolatti, G. (1995). Motor facilitation during action observation: A magnetic simulation study. *Journal of Neurophysiology, 73*, 2608–2611.

Fagg, A. H., & Arbib, M. A. (1998). Modeling parietal-premotor interactions in primate control of grasping. *Neural Networks, 11*, 1277–1303.

Gallese, V., Rizzolatti, G., Fadiga, L., & Fogassi, L. (1996). Action recognition in the premotor cortex. *Brain, 119*, 593–609.

Garton, A. F. (1992). Social interaction and the development of language and cognition. In *Essays in Developmental Psychology.* Mahwah, NJ: Lawrence Erlbaum Associates.

Heyes, C. M. (1996). *Social Learning in Animals: The roots of culture.* San Diego: Academic Press.

Hopfield, J. J. (1984). Neurons with graded response properties have collective computational properties like those of two-state neurons. *Proceedings of the National Academy of Sciences of USA, 81*, 3088–3092.

Iacoboni, M., Woods, R. P., Brass, M., Bekkering, H., Mazziotta, J. C., & Rizzolatti, G. (1999). Cortical mechanisms of human imitation. *Science, 286*, 2526–2528.

Kawamura, S. (1963). The process of sub-culture propagation among Japanese macaques. *Primates Social Behavior, 2*, 82–90.

Krams, M., Rushworth, M. F. S., Deiber, M.-P., Passingham, R. E., & Frackowiak, R. S. J. (1998). The preparation, execution and suppression of copied movements in the human brain. *Experimental Brain Research, 120*(3), 386–398.

Lock, E. D. (1978). *Action, Gesture and Symbols: the emergence of language.* New York: Academic Press, Harcourt Brace Jovanovich.

Meltzoff, A. N. (1990). The human infant as imitative generalist: a 20-year progress report on infant imitation with implications for comparative psychology. In Heyes C. M., & B. G. Galef (Eds.), *Social Learning in Animals: The roots of culture* (pp. 347–370). New York: Academic Press.

Meltzoff, A. N., & Gopnik, A. (1989). On linking nonverbal imitation, representation and language: Learning in the first two years of life. In G. E. Speidel, & K. E. Nelson (Eds.), *The Many Faces of Imitation in Language Learning* (pp. 23–52). Berlin: Springer Verlag.

Meltzoff, A. N., & Moore, M. K. (1977). Imitation of facial and manual gestures by human neonates. *Science, 198*, 75–78.

Moore, Bruce R. (1996). The evolution of imitative learning. In C. M. Heyes, & B. G. Galef (Eds.), *Social Learning in Animals: The roots of culture* (pp. 245–265). New York: Academic Press.

Myowa-Yamakoshi, M., & Matsuzawa, T. (1999). Factors influencing imitation of manipulatory actions in chimpanzees (*Pan troglodytes*). *Journal of Comparative Psychology, 113*, 128–136.

Nadel, J., Guerini, C., Peze, A., & Rivet, C. (1999). The evolving nature of imitation as a format for communication. In J. Nadel, & G. Butterworth (Eds.), *Imitation in Infancy* (pp. 209–234). Cambridge: Cambridge University Press.

Nottebohm, F. (1976). Vocal tract and brain: a search for evolutionary bottlenecks. *Annals of New York Academy of Sciences, 280*, 643–649.

Oztop, E., & Arbib, M. A. (2002). Schema design and implementation of the grasp-related mirror neuron system. *Biological Cybernetics* (to appear).

Piaget, J. (1962). *Play, Dreams and Imitation in Childhood*. New York: Norton.

Rizzolatti, G., & Arbib, M. (1998). Language within our grasp. *Trends in Neurosciences, 21*, 188–194.

Rizzolatti, G., Fadiga, L., Gallese, V., & Fogassi, L. (1996). Premotor cortex and the recognition of motor actions. *Cognitive Brain Research, 3*, 131–141.

Rizzolatti, G., Fadiga, L., Matelli, M., Bettinardi, V., Perani, D., & Fazio, F. (1996). Localization of grasp representations in humans by positron emission tomography: 1. Observation versus execution. *Experimental Brain Research, 111*, 246–252.

Speidel, G. S. (1989). Imitation: A bootstrap for learning to speak. In G. E. Speidel & K. E. Nelson (Eds.), *The Many Faces of Imitation in Language Learning* (pp. 151–180). Berlin: Springer Verlag.

Tomasello, M. (1990). Cultural transmission in the tool use and communicatory signaling of chimpanzees. In S. T. Parker & K. R. Gibson (Eds.), *Language and Intelligence in Monkeys and Apes: Comparative developmental perspectives* (pp. 274–311). Cambridge: Cambridge University Press.

Trevarthen, C., Kokkinaki, T., & Fiamenghi, Jr. G. A. (1999). What infants' imitations communicate: With mothers, with fathers and with peers. In J. Nadel & G. Butterworth (Eds.), *Imitation in Infancy* (pp. 61–124). Cambridge: Cambridge University Press.

Visalberghi, E., & Fragaszy, D. (1990). Do monkey ape? In S. T. Parker & K. R. Gibson (Eds.), *Language and Intelligence in Monkeys and Apes: Comparative developmental perspectives* (pp. 247–273). Cambridge: Cambridge University Press.

# Mirror neurons and feedback learning

Steve Womble and Stefan Wermter
Centre of Informatics, University of Sunderland, U.K.

## 1.  Introduction

The neuroscience evidence reviewed in (Rizzolatti & Arbib 1998) suggests that mirror neurons are involved in the comparison of '*goal directed actions*' and the perception of them during competent performance by others. Goal directed actions invariably involve the processing of sequences of more primitive actions. The complex manual tasks such as those discussed in (Rizzolatti & Arbib 1998) share some similarities with simple syntax acquisition. In either case the task is to produce or recognise a useful sequence out of primitive elements. Our model of the mirror system is a synergy of this.

Classical interpretations of language acquisition typically lead to connectionist models of syntax acquisition as the passive acquisition of implicit knowledge concerning a syntax (Reber 1989; Cleeremans 1993). It is also explicit in the axioms on which Gold based his formal learning theorem (Gold 1967). It seems highly unlikely to us that psychologically speaking such an interpretation is correct. After all, what is the point in acquiring knowledge if it is not to *use* it? The model mirror system described in the next section makes *active* use of knowledge already acquired during further learning. In the next section we also discuss a suitable test for the system. We then detail our results which are subsequently discussed. Finally we present our conclusions.

## 2.  Overview of our model system

We propose that as knowledge begins to be acquired through passive adaptation to predominantly correct data, this knowledge is actively used by the learner. In our model this utilisation occurs in two ways. It occurs through the attempted production of syntactically correct sentences. Feedback can then be provided, in the form of recognition of correct constructions, and this additional knowledge can be integrated into the production process. Secondly, we can measure how well the system estimates it has already stored the information contained in the example presented. It can then modify the degree to which it adapts itself to optimise for novel information. Note that this measure does *not* require feedback itself to be calculated. It is simply an estimate generated by the learner on its own use of acquired knowledge. It does not measure if that knowledge is naïve or incorrect only the degree to which it is used. However as the learner's acquired knowledge improves this utilisation representation maps into a confidence measure. This can be *quantified* by feedback.

A high level representation of our model is given in Figure 1. The parts in the square boxes represent the '*mirror neuron system*' that we have developed our abstract model of. In our model the learner examines a newly generated representation and deems the produced sequence to be worthy of production only if the utilisation measure is sufficiently high for all parts of the sequence *and* the learner knows that its knowledge is good. This requires the calculation of a threshold for a filter. This is calculated in a computationally efficient non-neural manner. It represents a minimum firing rate necessary for all neurons to be firing at in order to drive a mirror neuron.

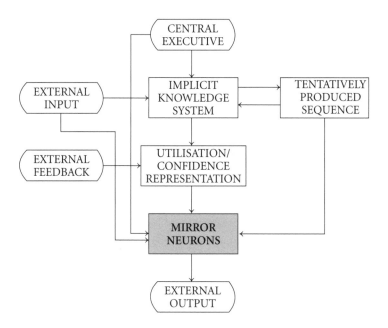

**Figure 1.** High level representation of model mirror system.

Experimental neuroscience results indicate that mirror cells are highly selective, only firing when their specific associated goal directed action occurs (Rizzolatti & Arbib 1998). In order to facilitate this within our model we require that in addition to input from the utilisation system the mirror field receives high level input from both the sequence production mechanism and from the central executive[1] or external input fields. The former selectively stimulates individual mirror neurons at a sub critical level allowing for full activity to occur when additional stimulation is received from the utilisation system. The latter is necessary to form associations between high level goals and sets of individual sequences.

To test our model we used the formal deterministic stochastic finite state grammar (DSFSG) displayed in Figure 2. It was developed by Reber (1989) in 1965. It was designed to be just complicated enough to take a little over an afternoon of exposure to learn by competent humans. It has been used in a series of psychology experiments by Reber and his colleagues over a number of years and was used in a sequence prediction task for a connectionist neural network by Cleeremans (1993). This formal language task sits nicely on the bridge between action sequence production such as has been reported on in the papers of Rizzolatti and others (Rizzolatti & Arbib 1998; Gallese, Fadiga, Fogassi, & Rizzolatti 1996), by macaque monkeys, and language processing classically associated with Broca's area in humans.

Syntactically correct 'Reber strings' are generated by walking through the finite state system shown in Figure 2. The grammar possesses six nodes. For a given subsequence generated from Figure 2 it requires both the current and the preceding term in the sequence to accurately identify the current node during a transversal.

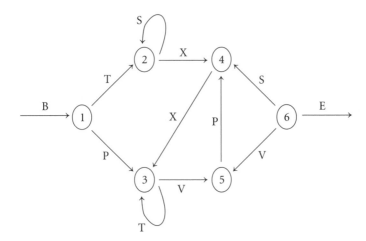

**Figure 2.** The DSFSG devised by Reber.

The learning task that we set our system was to acquire enough information from the environment concerning the grammar to be able to produce strings that conform to the grammar in a manner that is indistinguishable from the production of strings directly from the DSFSG. This was to be achieved by the coupled use of positive examples and feedback concerning the correctness of tentatively produced strings by the learning agent. Externally supplied strings can also be analysed by the system. Hence in our model there is a link between the evaluation of external input and of internally generated sequences.

We used a simple recurrent neural network (SRN) for the sequence prediction task (Elman 1991). Such networks are known to be well suited to learning DSFSGs since, if they are not over complicated, the SRN's associated error surface is efficiently minimised by learning to represent the nodes of the DSFSG as the internal states of the network. We used three hidden nodes, since three bits span eight binary states and there are six nodes in the Reber grammar. We used standard logistic sigmoid functions as the activation function for the hidden nodes and normalised exponential functions for the outputs. This choice of output generates a controllable 'n-of-c' probability distribution estimate with $n \mapsto 1$ as the temperature parameter of the normalised exponentials $T \mapsto 0$. We used a negative log likelihood function with subtracted entropy as the error function to be minimised. We used backpropagation for training. Training sequences were generated stochastically directly from the DSFSG. We summed errors over a sequence before updating the network and used an explicit momentum term, both to help smooth convergence towards a local optimum.

Elsewhere we have investigated the effects of training this system using a competent teacher which can vary its inputs to improve the acquisition of the network in response to behaviourally realistic output from the learning agent (Womble 2000), where further details concerning the system can also be found. In this work we concentrate on the effects of applying feedback to the system.

To summarise, we wish to compare performance of the basic system conventionally trained using randomly generated sets of strings to both the case where the system analyses its own generated strings to reject those that are likely to be wrong and using self reflexive learning to selectively enhance information stored in the network.

In order to fully analyse these paradigms we investigated the following criteria for successful performance:

–   We use a 1-norm measure on the predictive error against the *correct probability distribution* of the next character within a sequence. This measure is summed over a test set of strings and normalised, both with respect to the different lengths of different sequences and over the length of the test set. We generated a test set directly from the DSFSG, selecting the first $m$ most probable strings to be generated.

- There are three related measures concerning the utilisation and effects of the self reflexive analysis of the learning agent's own production performance. These measures are calculated over a set of strings. This set of strings can be generated by the learning agent in which case they correspond to internal reflection on performance, or the strings could be provided externally. The measures are

  - the largest utilisation measure for an incorrectly selected term in a sequence (denoted max{*fail*});
  - the smallest utilisation measure for a correctly selected term in a sequence (denoted min{*pass*});
  - and the relative values of each; in particular we are interested in states for which

$$\min_{S \in T}\{\min\{pass\}\} > \max_{S \in T}\{\max\{fail\}\}, \tag{1}$$

where $S$ denotes a sequence in a test set $T$.

- There are three 'behavioural' measures. These examine only what would be externally available to interacting agents (including human observers) and corresponds to the '*t1*' 'toy' Turing test as defined by Harnad (2001). Ultimately these measures are the most important concerning the apparent performance of the system. These three measures are

  - the raw (non-filtered) string production success rate;
  - the filtered string production success rates;
  - the '*t1*' test itself which is a measure on the difference in the distribution of sets of strings generated by the learning agent and those generated by a competent agent. We weight the contribution of each sequence with respect to its asymptotic frequency in the set of strings generated directly from the DSFSG, so that more frequently occurring strings contribute proportionally more to the measure

$$d_{bias} = \sum_{i=1}^{M} f_{RG}^{\infty}(S_i). \left| f_{RG}^{\infty}(S_i) - f_{MS}^{N}(S_i) \right|, \tag{2}$$

where $f$ denotes a frequency, $RG$ refers to the Reber Grammar, $MS$ refers to the mirror system, $S_i$ is the $i$-th *different* sequence in a test set of $N$ sequences for which there are $M$ *different* sequences, and $|\ldots|$ denotes the 1-norm distance measure.

## 3.  Results

The 1-norm error was carefully calculated using the 755 most frequent strings generated by the DSFSG, with each string weighted according to the asymptotic frequency distribution. For this value a little over 97% of the asymptotic frequency distribution of the infinite set of Reber strings is spanned. This is a deep search, with the probability of the least frequent of these strings occuring naturally being only $2^{-15} \approx 1$ in 32000. We found it quite easy to reduce the 1-norm error per element in a sequence, per string to below 0.1, and with a little more work to around 0.05. However to get below this value a significant amount of searching is required. Our best results using standard backpropagation generated a 1-norm error of 0.0173. To get this we had to perform many searches, and used adaptation of the learning rate $\eta$ to facilitate convergence to the best local minimum we could find. The results for the system in this state along with results for our lowest 1-norm error system trained using the full feedback system, are given in the following table:

| Training Paradigm | 1-norm Error | $d_{bias}$ Mean | $d_{bias}$ SD | Filtered Success |
|---|---|---|---|---|
| BP & Filter | 0.0173 | 22.87e-06 | 1.65e-06 | 100.0% |
| MN System | 0.0085 | 12.32e-06 | 1.04e-06 | 100.0% |
| DSFSG | 0.0000 | 4.904e-06 | 1.35e-06 | 100.0% |
| | Incorrect Rejections | Unfiltered Success | Min Pass | Max Fail |
| BP & Filter | 0.00% | 92.05% | 0.4905 | 0.0795 |
| MN System | 0.00% | 91.63% | 0.8646 | 0.1361 |
| DSFSG | 0.00% | 100.0% | 1.0000 | 0.0000 |

The results quoted are based on 5000 self generated test strings for all measures based on system production, and were repeated 10 times for the calculation of $d_{bias}$ means and standard deviations.

Figure 3 show detailed results comparing difference measures for these systems to that of the DSFSG itself. For the best backpropagation trained system we found that the difference measure remains fairly indistinguishable to external analysis up to a test set size of about 100 generated strings, the mean for the learning system lies at a single standard deviation from the DSFSG at about the 150 string size, and the learning system becomes clearly distinguishable (the ± single standard deviation bands no longer overlap) at about 300 string test sets. While for the lowest 1-norm error feedback trained system, these test set sizes are about 200, 700 and 1000 respectively, and at the 5000 string test set size the full active learning mirror system

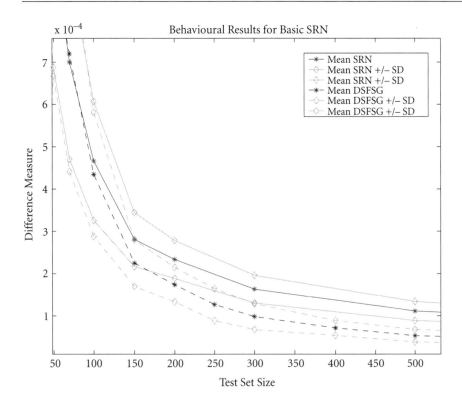

**Figure 3.** Plot of difference measure for minimum 1-norm error found using the basic neural network. The difference measure for the DSFSG is given for reference. The mean and standard deviations (SD) were calculated from 20 trials for each size of test set.

has a difference measure of only about 54% of the best results using just filtering. Finally we note that without the filtering mechanism the system generates illegal strings about 8% of the time, and is thus clearly distinguishable from the DSFSG.

## 4. Discussion

The results from the standard backpropagation training show that it is possible to train the SRN to perform the prediction task quite well. The minimum pass results show that our net trained in this way comfortably satisfies the test criteria discussed by Cleeremans (1993).

A useful way to decompose the contributions to the 1-norm error generated by the SRN is to split it into errors in the probability distribution for the next potentially correct characters, and the error caused by non-zero components of

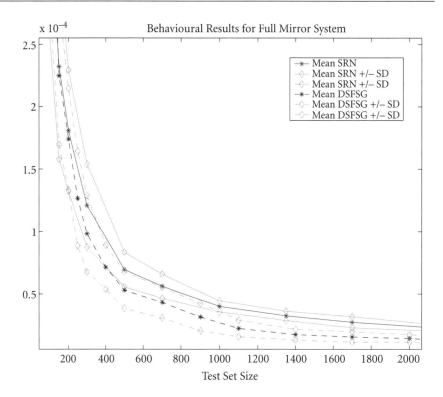

**Figure 4.** Plot of difference measure for minimum 1-norm error found using the abstracted mirror system. The difference measure for the DSFSG is given for reference. The mean and standard deviations (SD) were calculated from 20 trials for each size of test set.

the distribution associated with incorrect next selections. This splits the error into components associated with metrical and topological error respectively. Under this view a *simple* filter mechanism will produce complete success when the probability for any correct next character is greater than the worst topological error. An approximate condition for this is given by (1).

Our results show that for standard backpropagation it is hard to find a solution for which criterium (1) holds. However for our best network it did. An examination of the utilisation measure shows that topological error has been minimised very well for this network, but that there still exists a behaviourally significant metrical error, something that the filter mechanism set up utilising basic feedback cannot improve upon. However our results clearly show that using full feedback learning as described by our model mirror system the metrical information can be significantly improved. This is due to the modification in the batch learning technique we have introduced in the feedback mechanism. Using the utilisation measure, contributions to the direction in weight space through which the network adapts is biased in favour of contributions for which the utilisation was low i.e. for which the network has a relatively poor representation. The effect of this is to improve the performance of the SRN *selectively* around the regions where the metrical information is poor. However this is at the cost of some interference. The reason for the significant improvement in the behavioural error measures lies in the fact that providing the condition (1) holds then any topological error can be filtered out from the perceived external output of the learning agent.

It is noteworthy that a system which learns how to process sequences in this manner will, when producing strings, suffer from errors reminiscent of characteristic errors in classical Broca's aphasics, when the filtering system is disabled. This provides further circumstantial support for the argument that while biological mechanisms may be significantly different at a low level, at least at a modular level our connectionist network contains some of the characteristics of the biological system from which it was inspired.

The results reported here provide empirical evidence supporting the claim that the apparent problems of learning from only positive examples (Gold 1967) can be neatly circumvented using feedback learning, and that this approach is a plausible mechanism for L1 acquisition in humans. While our system is not a detailed model of a biological mirror system at a neural level, we do claim that the *high level* mechanisms used in our model for learning (positive examples, feedback, and optionally intelligent teaching by a competent teaching agent) are plausible mechanisms during infant language learning, and our results show that they have the potential to be successful. As a minimum our results indicate that at least for context free grammars up to the complexity of Reber grammars, positive examples and feedback are *sufficient* for the acquisition process to succeed.

## 5.   Conclusion

The results clearly show that contrary to the claims made by Cleeremans (1993) the use of positive only data for the implicit acquisition of the DSFSG of Reber, when applied to an SRN hand crafted to use the ideal network topology for the acquisition of the grammar, is a difficult task. Given that it is usually thought that the gradient decent backpropagation techniques used during the adaptation of the artificial system are more powerful than those available to the biological system, and that the Reber Grammar is obviously significantly simpler than any natural language we argue that the positive only learning mechanism is likely to be insufficient for language acquisition, in line with Gold's formal analysis (Gold 1967). However our results show that when feedback is available and is used by our abstracted mirror neuron system to both analyse tentative production and to modify the learning process the production performance of a learning agent on the syntax acquisition task presented by the Reber Grammar can become behaviourally indistinguishable from that of a competent agent. Finally we note that since the system developed here is equally applicable to goal directed actions as it is to syntax acquisition the mirror neuron production/perception comparison system on which it is based could quite plausibly provide an explanation for the emergence of modern natural language processing from a mechanism previously adapted for complex goal directed actions synthesized from a *vocabulary* of more basic actions.

## Note

1. For the purposes of our model we mean by central executive only that an instruction to spontaneously generate a particular sequence is initiated externally to the mirror system.

## References

Cleeremans, A. (1993). *Mechanisms of Implicit Learning. Connectionist Models of Sequence Processing*, 35–74. Cambridge, Massachusetts: MIT Press.

Elman, J. L. (1991). Distributed representations, simple recurrent networks, and grammatical structure. *Machine Learning, 7*, 195–224.

Gallese, V., Fadiga, L., Fogassi, L., & Rizzolatti, G. (1996). Action recognition in the premotor cortex. *Brain, 119*, 593–609.

Gold, E. M. (1967). Language identification in the limit. *Information and Control, 16*, 447–474.

Harnad, S. (2001). Minds, machines, and Turing: The indistinguishability of indistinguishables. *Journal of Logic, Language, and Information*, Special Issue on 'Alan Turing and Artificial Intelligence': (in press).

Reber, A. S. (1989). Implicit learning and tacit knowledge. *Journal of Experimental Psychology: General, 118*(3), 219–235.

Rizzolatti, G., & Arbib, M. A. (1998). Language within our grasp. *Trends in the Neurosciences, 21*, 188–194.

Rizzolatti, G., Camarda, R., Fogassi, L.,Gentilucci, M., Luppino, G., & Matelli, M. (1988). Functional organization of inferior area 6 in the macaque monkey: II. Area F5 and the control of distal movements. *Experimental Brain Research, 71*, 491–507.

Rizzolatti, G., Fadiga, L., Fogassi, L., & Gallese, V. (1996). Premotor cortex and the recognition of motor actions. *Cognitive Brain Research, 3*, 131–141.

Womble, S. P. (2000). Connectionist model of a mirror neuron system http://www.his.sunderland.ac.uk/womble/tech2000a. Technical report, University of Sunderland, Hybrid Intelligent Systems Research Group, Informatics Centre, School of Computing and Technology, Sunderland, United Kingdom.

# A connectionist model which unifies the behavioral and the linguistic processes

## Results from robot learning experiments

Yuuya Sugita and Jun Tani

Brain Science Institute, RIKEN / Graduate School of Arts and Sciences, University of Tokyo

## 1. Introduction

How can robots communicate with humans or the other robots using natural language? To achieve this, it is necessary for robots to understand the meaning of sentences. However, what does "the meaning of sentences" mean? We investigate this problem from the viewpoint of embodiment and intersubjectivity. In other words, we consider that not only the symbolic interaction but also the behavioral interaction among humans and robots are indispensable in the processes of co-acquisition of language and behavior.

To understand the processes of co-acquisition of language and behavior, we take a synthetic approach, constructing linguistically communicative behavioral agents in the real world. Our approach is to construe a language which emerges from communication based on embodiment, the way or the tendency to perceive the environment. Concretely, we have proposed and implemented the neural-net architecture to unify behavioral and linguistic processes, and have conducted some experiments using real robot systems.

The development of computer technologies induces the wide-spread thought that our brains are best understood by using computational metaphor. This situation also elicits an application of the synthetic approach for investigation of human intelligence, which comprises the language. In early days, most of such studies were carried out by using artificial intelligence (AI) schemes in which manipulations of symbols were emphasized. At a glance, this approach seems to be affinitive with the language. Moreover, language centrism on human intelligence like "the most significant characteristic of the human intelligence is to use language", has induced the active investigation of language based on the AI approach aiming at understanting the intelligent behavior or constructing and construing intelligence. However a system equivalent to AI which can keep sustainable communication with human-beings has not been achieved yet since the symbols pre-assigned in the systems often cause a frame problem as well as a symbol–grounding problem (Harnad 1990, 1993).

To avoid these problems, there are roughly two strategies: (1) denying the necessity of the internal representation and avoidance of symbols, and (2) autonomously generating the symbols which would be used for internal representation. In the range of the robotics research, it is notable that the description of the environment should be constructed from the robot's own view. That is to say, the robot itself should generate the interface between itself and its environment through the iterative interactions, not that the human beings prepare the world model a priori.

As an example of the former strategy, there are studies of reactive robotics (Brooks 1986, 1991). In these studies, robots are controlled in reflex manner by using the incoming sensor signals. Although some researches have argued that the appropriate combination of these reflex actions could lead to the realization of truly intelligent behaviors, we speculate that such intelligence would inevitably be limited since the system cannot afford to utilize the internal modeling.

On the other hand, some studies of cognitive robotics, conducted by Tani and others (Sugita & Tani 1998; Tani 1996), give an instance of the latter strategy, in which the role of dynamical systems in cognition are emphasized by the ideas of Pollack (1991) and Beer (1995). These studies show that the internal model of the environment can be represented as embedded in the attractor of the internal neural dynamics. Concretely, time series of the robot's sensory-motor information generated through robot's interaction with the environment, is learned as a forward model (Jordan & Rumelhart 1992; Kawato, Furukawa, & Suzuki 1987) by using recurrent neural networks (RNN). Such acquired forward models are utilized for mental manipulations including prediction, planning and rehearsal of the experienced environmental interactions. As shown in Elman's studies of grammar learning (Elman 1990), the significant point of RNN learning is that a symbolic structure of an environment is self-organized in the internal dynamical structure. Especially for robot systems, only the required symbols for learning and interpreting the actual experiences are generated, so that the system does not face the frame problem as well as the symbol – grounding problem. Moreover, the internal dynamics of the robot is entrained by the actual dynamics, therefore temporal abnormal sensory input does not lead to unrecoverable inconsistency. The internal system can be "re-situated" by continuing the interactions among the environment even when the robot loses its current context (Tani 1996).

In this chapter, we propose a novel model which represents the linguistic process and the behavioral process as co-dependent dynamical systems. The essential arguments in our modeling are summarized as follows.

– Behavior denotes the structure of the sensory-motor sequences which arise from iterative interaction between the robot and its environment. Language denotes the structure in the word sequences which arise in the constraint of syntax and semantics.

– Language and behavior correspond with each other in terms of many-to-many mapping. They inevitably become co-dependent systems while such mapping is self-organized.

– Both behavior and language proceed in a context-dependent manner. In the behavioral processes, the current action and sensation can be explained from their accumulated historical information. In the same manner for the linguistic processes, what the current dialog means can be understood from the past sequences of dialog exchanged.

– The recognition of language evokes the corresponding behavior. The recognition of behavior evokes the corresponding language.

– The co-dependent systems of language and behavior could go back and forth between two extremes of "poetic" ungrounded worlds and "realistic" grounded worlds.

There are some related works to be mentioned. Billard and her colleagues have studied multiple robot communication (Billard & Dautenhahn 2000) and human-robot (Billard & Hayes 1998) communication using a neural – net model called DRAMA in the context of imitation learning. Billard has shown that mapping between a sequence of sensory data and its corresponding symbol can be learned. Ogata and Sugano (1999) also studied human – robot communication. They used the Kohonen network for clustering multiple sensory inputs where the self-organized clusters could be regarded as symbols. The drawback of their approach is that only the spatial relations in the sensory patterns are utilized. It is considered that the temporal relations are also dispensable for generating symbolic structures. Steels and his colleagues have studied the evolution of language by conducting naming-game experiments using physical robots (Steels & Kaplan 1999). It was demonstrated that diverse mappings between sensory-states and names of objects can be self-organized through evolutionary processes. Although these studies demonstrate the importance of the embodiment for the acquisition and usage of language, they have not yet been successful in achieving the context-dependent many-to-many mapping between behavioral processes and linguistic processes which this chapter aims for.

The rest of this chapter is organized as follows: In Section 2, we propose our new recurrent neural network architecture which implements the many-to-many mapping between sentences and behaviors, and then Section 3 will depict the experimental configuration and explain the experimental results. Finally in Section 4 we will discuss and summerize the experimental results.

## 2. Proposed RNN architecture

The proposed architecture is explained on the basis of our prototype robot-learning experiments. In our experimental setup, a robot is taught how to behave corresponding to a sequence of command sentences given to the robot. The robot has to learn a many-to-many mapping between the command sentences and the behavioral sequences in the context-dependent manner such that an appropriate action sequence plan can be generated for the inputs of a command sentence as situated to the current behavioral context.

Our proposed RNN architecture consists of two RNN modules, one is for predicting sensory-motor sequences and the other is for predicting words and sentences in the form of inputs/outputs mapping as shown in Figure 1. We employ Jordan's idea of context re-entry which enables the network to represent the internal memory (Jordan & Rumelhart 1992). The current context input is a copy of the previous context output: by this means, the context units remember the previous internal state. This context re-entry mechanism allows the network to self-organize the hidden state representation through learning. It is known that the true state of

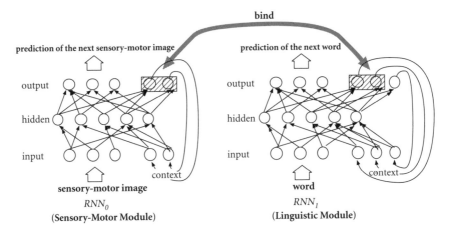

**Figure 1.** Our proposed RNN architecture: Two RNNs are constrained each other through a part of their context units. $RNN_0$ is for behavior, and $RNN_1$ is for language. Each RNN module predicts the input image of the next time step from the current input utilizing the context units activations.

the system can be identified not by external input values but by context activation values.

A scheme called "context-binding" is developed for organizing the many-to-many mapping between two modalities. The main idea behind this scheme is that the context activation of the two RNNs should become the same after each termination of sentence and its corresponding behavior. Consider the following example. The robot is told, "Look at the apple" as the first command sentence, and then told "Take it" as the second sentence. The context activations in the behavioral RNN after the robot actually turns its head toward the apple and in the linguistic RNN after the first sentence is terminated should become the same. This context-binding should be made again after the second sentence "Take it" is terminated. The essential idea behind this scheme is that the right mapping can be achieved when the topologically equivalent structures are self-organized in these two RNN modules.

The learning is proceeded by using back-propagation through time algorithm (Rumelhart, Hinton, & Williams 1986). On each context-binding step, the context activation of two RNN modules are constrained to be the same. For this purpose, an average vector of the two context activation vectors are computed. This average vector is used as the target activation for the context of each RNN module. The error between the target and actual activation of the context is back-propagated through time for each RNN module. The context activation of the two RNN modules become similar when the learning converges.

For the planning of action sequences corresponding to a given command sentence, an idea of the inverse dynamics (Jordan & Rumelhart 1992) is employed. With the input of a sequence of words, the linguistic RNN computes context activation in the forward – dynamics manner. The context activation obtained after the command sentence is completed is utilized as the target context activation for the behavioral RNN. Here, the goal of planning is to search an appropriate sequence of actions which can generate the target context activation in the subsequent step. This search can be conducted by means of the inverse dynamics as well as random search. The current studies employed the inverse dynamics scheme.

## 3.   Experiment

Our proposed model was examined through the experiments using a real mobile robot equipped with a vision system as shown in Figure 2.

**Figure 2.** Our vision-based mobile robot.

## 3.1 Setting

Figure 3 shows the environment where the robot learning experiments were being conducted. There are two objects in the environment – a red postbox and a blue block. After each command sentence is given to the robot, the robot is trained to generate an appropriate sequence of actions in a supervised learning manner.

The robot has the vision sensor which recognizes the color and the size of the objects. The robot can maneuver by changing the rotation speed of left and right wheels. It can attend to the objects by rotating the camera head horizontally and vertically. It can also push the objects forward with its simple arm equipped in the front side. In the environment, the blue block is placed right in front of the wall and therefore the block cannot be moved by the robot's pushing action. On the other hand the postbox is placed at the other end of the table and the postbox falls down to the floor when pushed by the robot's arm. The postbox disappears out of sight of the robot when it falls down.

The robot is implemented with three pre-programmed behavioral modules. Those are (a) turning toward a new object, (b) moving toward an object and (c) push an object. The behavioral RNN has three action input units corresponding

**Figure 3.** The environment for our experiment: A blue block (left) and a red postbox (right) have been placed.

to these action modules. It also has five sensory input units which represent visual information (color and size of the objects stared). The sensory inputs are fed into the behavioral RNN at each completion of the action.

Five words are used to compose command sentences, which are "watch," "reach," "push," "block" and "postbox." The word sequence are fed into the linguistic RNN using the 5 input units allocated.

Command sentences are given in a sequence such that the robot can actually achieve the goals of commands. For instance, when the sentence "watch block" is given while the robot looked at the postbox as shown in Figure 4, it is expected that the robot turns to the direction of the block (arrow (a) in Figure 4), and then, receiving the sentence "reach block", the robot should move to the block and stop just in front of it (arrow (b) in Figure 4).

The behavioral RNN consists of 8 input units and 8 output units, 4 context units and 12 hidden units whereas linguistic RNN consists of 5 input units, 9 context units and 30 hidden units.

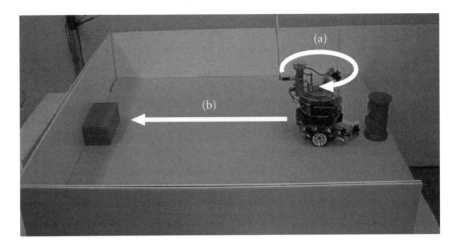

**Figure 4.** An example of a robot's behavior in the environment.

## 3.2  Results

The co-learning experiments were conducted using 60 sequences of sentences and their corresponding behavioral sequences. The learning process took 400,000 iterative steps before the convergence.

Then the phase plot diagram using the context units activations are generated both for the behavioral RNN and the linguistic RNN. The phase plots were generated by iteratively activating the RNNs in the closed-loop mode with inputs comprising 300 steps of possible action sequences. The generated sequences of the context units activation are plotted in the two dimensional phase space as shown in Figure 5.

In this figure, no structural correspondence can be seen between the two phase plots. In order to see the correspondences, the context values only at the binding steps should be plotted. Figure 6 shows the phase plots of the linguistic process using the context values only at the binding steps. It is seen that this phase plot of the linguistic processes and the one of the behavioral processes shown in Figure 5(a) are mostly identical.

Now we examine the possible correspondences between these clusters and the robot situations. The possible robot situations in the adopted task are as follows:

1.   Looking at the block from the side of the block.
2.   Looking at the block from the side of the postbox.
3.   Looking at the postbox from the side of the postbox.
4.   Looking at the postbox from the side of the block.
5.   Not looking at anything (after pushing the postbox).

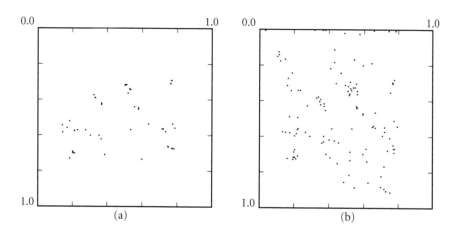

**Figure 5.** The phase plot diagrams for the behavioral RNN on the left side and the linguistic RNN on the right side after 400,000 iterative steps of co-learning.

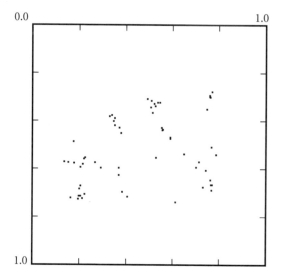

**Figure 6.** The phase plot diagram for the linguistic RNN after 400,000 iterative steps of co-learning. The context values of the binding steps are plotted.

We also studied how the context states transit among these clusters as the robot situations change. Explicit correspondences between the clusters and the robot situations can be seen in Figure 7. In this figure the transitions among the clusters are denoted by arrows. It was found that the transitions among the clusters exactly follow the way the robot situation changes.

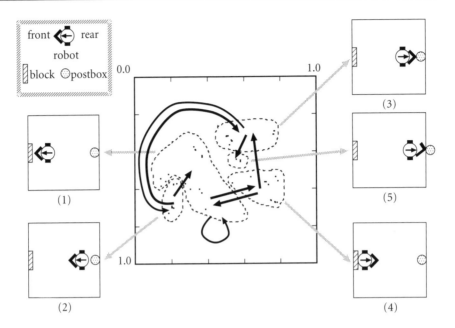

**Figure 7.** The correspondence between the clusters in the phase plots and the situations of the robot. Arrows denote the possible state transitions.

Furthermore, we examined the action planning and generation capability corresponding with given sequence of command sentences. It was found that action sequences are generated appropriately for all rational command sentences. To sum up, it can be said that the context-dependent correspondence between behavioral processes and linguistic processes are generated correctly.

## 4.    Discussion

### 4.1  Structure of an acquired canonical internal representation

In order to examine how the internal dynamical structure of the linguistic RNN are generated as corresponding to the grounded internal representation self-organized in the behavioral RNN, we compared two cases of learning results: (1) co-learning of linguistic sequences and behavioral sequences using our proposed architecture which we have analyzed previously and (2) independent learning of linguistic sequence using a single RNN. Figure 8 shows the context phase plots self-organized in the independent learning cases.

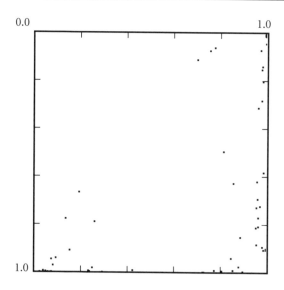

**Figure 8.** The phase plots generated after 200,000 iterative steps in independent learning.

In the case of the independent learning of linguistic sequences, the attractor appears scattered like clouds, and no cluster structures can be observed. This outcome implies that independent linguistic learning results in rote-learning of the training data rather than generalized learning with extracting structures.

These results are due to the fact that the sequences of sentences given to the robot are not merely random, but they are given in a context-dependent manner based on the situated interactions between the robot and its environment. For example, there is a constraint in our experiment that a sentence "watch postbox"cannot be repeated twice. This linguistic constraint is due to the behavioral constraint such that the robot never asked to watch a postbox already being watched by the robot. However, this type of context-dependent structures are hardly learned only by using the linguistic modality, since this sort of learning requires reconstructions of long chains of hidden states in the phase space. In such cases, the sequences experienced are likely to be learned exactly as they are without causing any generalizations.

In the case of the co-learning, each situation becomes more explicit by utilizing both constraints from the linguistic and the behavioral ones, by which a generalized model of the robot-environment interactions can be extracted. In other words, it can be said that the process of co-learning makes the linguistic context structures grounded to the behavioral ones, which were initially ungrounded. It is argued that the behavioral structures heavily affect the interpretation of the linguistic structures where the language can be acquired in an embodied manner.

## 4.2 Floating between the grounded and the ungrounded worlds

Contrary to the previous discussion, there could be an opposite process that the behavioral structures are highly affected by the linguistic structures. Ikegami (Ikegami & Taiji 1998) as well as Tani (1998) have shown the analyses that the arbitrariness in RNN learning could turn out to be the source of generating behavioral diversity. These studies have shown that the RNN learning of a set of complex sequences could result in generating arbitrary different interpretative structures.

A similar situation could take place in the co-learning of linguistic and behavioral structures. When highly expressive linguistic structures are attempted to be learned, the over-articulation in the learning could generate arbitrary fake memory in the linguistic structures. This fake memory in linguistic structures may enforce to generate the fake memory of the behavioral ones through the processes of co-learning. These fake memory structures would cause the robot to generate diverse imaginary behaviors. It can be said that linguistic memory structures tend to generate fake and imaginary patterns because of its highly expressive power while the behavioral structures become more regimental since they are constrained by the grounded interactions. As a result the mental process of the robot is supposed to go back and forth between the ungrounded imaginary world governed by language and the one grounded to reality through behaviors.

## 4.3 Perspectives

We consider that the model proposed in this paper is general enough to support other inter-modality functions such as the models discussed in the mirror neuron studies (Gallese & Goldman 1998; Rizzolatti & Arbib 1998). The context-binding scheme proposed in our model can naturally explain spatio-temporal correspondences between a subject's observation of others and his/her imitating behavior of them.

## 5. Conclusion

The current chapter introduces a novel general connectionist model which can learn to achieve many-to-many correspondences among multiple modalities in a context-dependent manner. The proposed model was examined through an embodied language experiment using real robots. This experiment shows that the robot acquires a simple language set as grounded to its behavioral contexts. Furthermore, our discussion based on the dynamical systems analysis of the experiment has shown the possibility that the arbitrariness of the acquired linguistic structures could generate diverse behavioral interpretations of the world.

Although this study has shown an important first step for understanding the essence of the embodied language using the synthetic approach, it is also quite true that the current experimental status is limited in its scaling. In future studies, we plan to introduce more complexity in behavior and language such that we examine the processes of behavioral structures affected by highly expressive semantics generated in linguistic structures.

## Acknowledgment

The authors wish to thank Ryu Nishimoto (Univ. of Tokyo) and Takanori Komatsu (Univ. of Tokyo) for their comments on the initial manuscript.

## References

Beer, R. D. (1995). A dynamical systems perspective on agent-environment interaction. *Artificial Intelligence, 72*(1), 173–215 .

Billard, A., & Dautenhahn, K. (2000). Experiments in social robotics: grounding and use of communication in autonomous agents. *Adaptive Behavior, 7*(3/4), in press.

Billard, A., & Hayes, G. (1998). Grounding sequences of actions in an autonomous doll robot. *Proc. of Adaptive Behavior Dynamic Recurrent Neural Nets, SAB 98 Workshop*.

Brooks, R. (1986). A robust layered control system for a mobile robot. *IEEE J. Robotics and Automat., RA-2*(1), 14–23.

Brooks, R. (1991). Intelligence without representation. *Artifical Intelligence, 47*, 139–159.

Elman, J. L. (1990). Finding structure in time. *Cognitive Science, 14*, 179–211.

Gallese, V., & Goldman, A. (1998). Mirror neurons and the simulation theory of mind-reading. *Trends in Cognitive Science, 2*(12).

Harnad, S. (1990). The symbol grounding problem. *Physica D, 42*, 335–346.

Harnad, S. (1993). Grounding symbolic capacity in robotic capacity. In L. Steels, & R. Brooks (Eds.), *The "artificial life" route to "artificial intelligence." Building situated embodied agents*. New Haven: Lawrence Erlbaum.

Ikegami, T., & Taiji, M. (1998). Structure of Possible Worlds in a Game of Players with Internal Models. In *Proc. of the Third Int'l Conf. on Emergence* (pp. 601–604). Japan: Iizuka.

Jordan, M. I., & Rumelhart, D. E. (1992). Forward models: supervised learning with a distal teacher. *Cognitive Science, 16*, 307–354.

Kawato, M., Furukawa, K., & Suzuki, R. (1987). A hierarchical neural network model for the control and learning of voluntary movement. *Biological Cybernetics, 57*, 169–185.

Ogata, T., & Sugano, Sh. (1999). Emotional Communication Between Humans and Robots: Consideration of Primitivee Language in Robots. In *Proc. of IEEE/RSJ International Conference on Intelligent Robots and Systems (IROS'99)* (pp. 870–875).

Pollack, J. B. (1991). The induction of dynamical recognizers. *Machine Learning, 7*, 227–252.

Rizzolatti, G., & Arbib, M. A. (1998). Language within our grasp. *Trends in Neurosciences, 21*, 188–194.

Rumelhart, D. E., Hinton, G. E., & Williams, R. J. (1986). Learning internal representations by error propagation. In D. E. Rumelhart, & J. L. Mclelland (Eds.), *Parallel Distributed Processing*. Cambridge, MA: MIT Press.

Steels, L., & Kaplan, F. (1999). Situated grounded word semantics. In T. Dean (Ed.), *Proceedings of the Sixteenth International Joint Conference on Artificial Intelligence IJCAI'99* (pp. 862–867). San Francisco, CA: Morgan Kaufmann Publishers.

Sugita, Y., & Tani, J. (1998). Emergence of Cooperative/Competitve Behavior in Two Robots' Games: Plans or Skills? In *Proc. of Adaptive Behavior Dynamic Recurrent Neural Nets, SAB'98 Workshop*.

Tani, J. (1996). Model-Based Learning for Mobile Robot Navigation from the Dynamical Systems Perspective. *IEEE Trans. on SMC (B), 26*(3), 421–436.

Tani, J. (1998). An interpretation of the "self" from the dynamical systems perspective: a constructivist approach. *Journal of Consciousness Studies, 5*(5–6), 516–542.

# Name index

# Subject index

In the series *Advances in Consciousness Research* the following titles have been published thus far or are scheduled for publication:

34  FETZER, James H. (ed.): Consciousness Evolving. 2002. xx, 253 pp.

33  YASUE, Kunio, Mari JIBU and Tarcisio DELLA SENTA (eds.): No Matter, Never Mind. Proceedings of Toward a Science of Consciousness: Fundamental approaches, Tokyo 1999. 2002. xvi, 391 pp.

32  VITIELLO, Giuseppe: My Double Unveiled. The dissipative quantum model of brain. 2001. xvi, 163 pp.

31  RAKOVER, Sam S. and Baruch CAHLON: Face Recognition. Cognitive and computational processes. 2001. x, 306 pp.

30  BROOK, Andrew and Richard C. DEVIDI (eds.): Self-Reference and Self-Awareness. 2001. viii, 277 pp.

29  VAN LOOCKE, Philip (ed.): The Physical Nature of Consciousness. 2001. viii, 321 pp.

28  ZACHAR, Peter: Psychological Concepts and Biological Psychiatry. A philosophical analysis. 2000. xx, 342 pp.

27  GILLETT, Grant R. and John McMILLAN: Consciousness and Intentionality. 2001. x, 265 pp.

26  Ó NUALLÁIN, Seán (ed.): Spatial Cognition. Foundations and applications. 2000. xvi, 366 pp.

25  BACHMANN, Talis: Microgenetic Approach to the Conscious Mind. 2000. xiv, 300 pp.

24  ROVEE-COLLIER, Carolyn, Harlene HAYNE and Michael COLOMBO: The Development of Implicit and Explicit Memory. 2000. x, 324 pp.

23  ZAHAVI, Dan (ed.): Exploring the Self. Philosophical and psychopathological perspectives on self-experience. 2000. viii, 301 pp.

22  ROSSETTI, Yves and Antti REVONSUO (eds.): Beyond Dissociation. Interaction between dissociated implicit and explicit processing. 2000. x, 372 pp.

21  HUTTO, Daniel D.: Beyond Physicalism. 2000. xvi, 306 pp.

20  KUNZENDORF, Robert G. and Benjamin WALLACE (eds.): Individual Differences in Conscious Experience. 2000. xii, 412 pp.

19  DAUTENHAHN, Kerstin (ed.): Human Cognition and Social Agent Technology. 2000. xxiv, 448 pp.

18  PALMER, Gary B. and Debra J. OCCHI (eds.): Languages of Sentiment. Cultural constructions of emotional substrates. 1999. vi, 272 pp.

17  HUTTO, Daniel D.: The Presence of Mind. 1999. xiv, 252 pp.

16  ELLIS, Ralph D. and Natika NEWTON (eds.): The Caldron of Consciousness. Motivation, affect and self-organization — An anthology. 2000. xxii, 276 pp.

15  CHALLIS, Bradford H. and Boris M. VELICHKOVSKY (eds.): Stratification in Cognition and Consciousness. 1999. viii, 293 pp.

14  SHEETS-JOHNSTONE, Maxine: The Primacy of Movement. 1999. xxxiv, 583 pp.

13  VELMANS, Max (ed.): Investigating Phenomenal Consciousness. New methodologies and maps. 2000. xii, 381 pp.

12  STAMENOV, Maxim I. (ed.): Language Structure, Discourse and the Access to Consciousness. 1997. xii, 364 pp.

11  PYLKKÖ, Pauli: The Aconceptual Mind. Heideggerian themes in holistic naturalism. 1998. xxvi, 297 pp.

10  NEWTON, Natika: Foundations of Understanding. 1996. x, 211 pp.

9  Ó NUALLÁIN, Seán, Paul Mc KEVITT and Eoghan Mac AOGÁIN (eds.): Two Sciences of Mind. Readings in cognitive science and consciousness. 1997. xii, 490 pp.

8  GROSSENBACHER, Peter G. (ed.): Finding Consciousness in the Brain. A neurocognitive approach. 2001. xvi, 326 pp.

7  MAC CORMAC, Earl and Maxim I. STAMENOV (eds.): Fractals of Brain, Fractals of Mind. In search of a symmetry bond. 1996. x, 359 pp.

6  GENNARO, Rocco J.: Consciousness and Self-Consciousness. A defense of the higher-order thought theory of consciousness. 1996. x, 220 pp.

5  STUBENBERG, Leopold: Consciousness and Qualia. 1998. x, 368 pp.

4  HARDCASTLE, Valerie Gray: Locating Consciousness. 1995. xviii, 266 pp.

3  JIBU, Mari and Kunio YASUE: Quantum Brain Dynamics and Consciousness. An introduction. 1995. xvi, 244 pp.

2  ELLIS, Ralph D.: Questioning Consciousness. The interplay of imagery, cognition, and emotion in the human brain. 1995. viii, 262 pp.

1  GLOBUS, Gordon G.: The Postmodern Brain. 1995. xii, 188 pp.